Drywall
Level One

Trainee Guide
First Edition

PEARSON

nccer

New York, New York
Columbus, Ohio

NCCER

President: Don Whyte
Director of Curriculum Revision and Development: Daniele Dixon
Drywall Project Manager: Carla Sly
Production Manager: Tim Davis
Quality Control Coordinator: Debie Ness
Editors: Rob Richardson and Matt Tischler
Desktop Publishing Coordinator: James McKay

NCCER would like to acknowledge the contract service provider for this curriculum:
Topaz Publications, Liverpool, New York.

This information is general in nature and intended for training purposes only. Actual performance of activities described in this manual requires compliance with all applicable operating, service, maintenance, and safety procedures under the direction of qualified personnel. References in this manual to patented or proprietary devices do not constitute a recommendation of their use.

10 9 8 7 6 5 4 3 2 1

PEARSON

ISBN 10   0-13-604512-X
ISBN 13   978-0-13-604512-0

PREFACE

Walk into almost any home, apartment complex, or commercial building and look around. The odds are good that drywall applicators installed the walls and ceilings and placed insulation, sound-proofing, and firestopping materials behind and onto those walls and ceilings. They may also have applied textures and trims to enhance both the interiors and exteriors of the buildings.

There were approximately 150,000 drywall applicators working in the United States in 2004. By 2014, the number is expected to grow by 10,000. Depending on the level of experience and region of the country, annual wages range from $22,000 to $60,000, with most earning $28,000 to $36,000. Careers can progress from installer to specialty finisher to business owner, and related professions include sheetrock applicator and acoustical carpenter.

This is the first installment of a two-level curriculum that meets the requirements of a standard drywall applicator apprenticeship program (a minimum of 288 hours of instruction and 4,000 hours of on-the-job training). In the first level of drywall training, you'll learn the basics, including insulation, installation, and finishing. In the second level, you'll explore specialized topics, such as steel framing, acoustical ceilings, and specialty finishes. In both levels, you'll learn about the materials and tools used in the drywall profession.

This curriculum was developed by a panel of subject matter experts from the drywall industry and produced by NCCER.

By choosing to begin drywall training, you are taking a step toward a satisfying and rewarding career. Continuing your craft education is important, as technology and the materials with which you'll work are changing all the time.

We wish you success as you begin your construction career in the drywall trade, and hope that you'll continue your training outside of this series. By taking advantage of training opportunities as they arise, you'll demonstrate initiative and a desire to learn—qualities that are present in the industry's best professionals.

We also invite you to visit the NCCER website at **www.nccer.org** for the latest releases, training information, the *Cornerstone* magazine, and much more. You can also reference the Pearson product catalog online at **www. crafttraining.com**. Your feedback is welcome. You may email your comments to **curriculum@nccer.org** or send general comments and inquiries to **info@nccer.org**.

NCCER STANDARDIZED CURRICULA

NCCER is a not-for-profit 501(c)(3) education foundation established in 1996 by the world's largest and most progressive construction companies and national construction associations. It was founded to address the severe workforce shortage facing the industry and to develop a standardized training process and curricula. Today, NCCER is supported by hundreds of leading construction and maintenance companies, manufacturers, and national associations. The NCCER Standardized Curricula was developed by NCCER in partnership with Pearson Education, Inc., the world's largest educational publisher.

Some features of NCCER's Standardized Curricula are as follows:

- An industry-proven record of success
- Curricula developed by the industry for the industry
- National standardization providing portability of learned job skills and educational credits
- Compliance with the Office of Apprenticeship requirements for related classroom training (*CFR* 29:29)
- Well-illustrated, up-to-date, and practical information

NCCER also maintains a Registry that provides transcripts, certificates, and wallet cards to individuals who have successfully completed modules of NCCER's Standardized Curricula. *Training programs must be delivered by an NCCER Accredited Training Sponsor in order to receive these credentials.*

Contents

NCCER Standardized Curricula

NCCER's training programs comprise over 80 construction, maintenance, and pipeline areas and include skills assessments, safety training, and management education.

Boilermaking
Cabinetmaking
Carpentry
Concrete Finishing
Construction Craft Laborer
Construction Technology
Core Curriculum:
  Introductory Craft Skills
Drywall
Electrical
Electronic Systems Technician
Heating, Ventilating, and
  Air Conditioning
Heavy Equipment Operations
Highway/Heavy Construction
Hydroblasting
Industrial Coating and Lining
  Application Specialist
Industrial Maintenance Electrical
  and Instrumentation Technician
Industrial Maintenance
  Mechanic
Instrumentation
Insulating
Ironworking
Masonry
Millwright
Mobile Crane Operations
Painting
Painting, Industrial
Pipefitting
Pipelayer
Plumbing
Reinforcing Ironwork
Rigging
Scaffolding
Sheet Metal
Signal Person
Site Layout
Sprinkler Fitting
Tower Crane Operator
Welding

MARITIME

Maritime Industry Fundamentals
Maritime Pipefitting
Maritime Structural Fitter

GREEN/SUSTAINABLE CONSTRUCTION

Building Auditor
Fundamentals of Weatherization
Introduction to Weatherization
Sustainable Construction
  Supervisor
Weatherization Crew Chief
Weatherization Technician
Your Role in the Green
  Environment

ENERGY

Alternative Energy
Introduction to the Power Industry
Introduction to Solar Photovoltaics
Introduction to Wind Energy
Power Industry Fundamentals
Power Generation Maintenance
  Electrician
Power Generation I&C
  Maintenance Technician
Power Generation Maintenance
  Mechanic
Power Line Worker
Power Line Worker: Distribution
Power Line Worker: Substation
Power Line Worker: Transmission
Solar Photovoltaic Systems Installer
Wind Turbine Maintenance
  Technician

PIPELINE

Control Center Operations, Liquid
Corrosion Control
Electrical and Instrumentation
Field Operations, Liquid
Field Operations, Gas
Maintenance
Mechanical

SAFETY

Field Safety
Safety Orientation
Safety Technology

SUPPLEMENTAL TITLES

Applied Construction Math
Tools for Success

MANAGEMENT

Fundamentals of Crew Leadership
Project Management
Project Supervision

SPANISH TITLES

Acabado de concreto: nivel uno
Aislamiento: nivel uno
Albañilería: nivel uno
Andamios
Carpintería:
  Formas para carpintería, nivel tres
Currículo básico: habilidades
  introductorias del oficio
Electricidad: nivel uno
Herrería: nivel uno
Herrería de refuerzo: nivel uno
Instalación de rociadores: nivel uno
Instalación de tuberías: nivel uno
Instrumentación: nivel uno, nivel
  dos, nivel tres, nivel cuatro
Orientación de seguridad
Paneles de yeso: nivel uno
Seguridad de campo

Acknowledgments

This curriculum was revised as a result of the farsightedness
and leadership of the following sponsors:

Associated Builders and Contractors,
Cornhusker Chapter

Associated Builders and Contractors,
New Mexico Chapter

Baker Drywall

C.J. Coakley Co., Inc.

Carolinas
Associated General Contractors

Greenville Technical College

J. Harley Bonds Career Center

Johnson Drywall

Tufly Company

This curriculum would not exist were it not for the dedication
and unselfish energy of those volunteers who served on the Authoring Team.
A sincere thanks is extended to the following:

David Ackerman
Glen Bruning
Bob Consroe
Joe Halcarz

Brad Minden
Perry Moore
Robert Pelletier

NCCER PARTNERS

American Fire Sprinkler Association

Associated Builders and Contractors, Inc.

Associated General Contractors of America

Association for Career and Technical Education

Association for Skilled and Technical Sciences

Construction Industry Institute

Construction Users Roundtable

Construction Workforce Development Center

Design Build Institute of America

GSSC – Gulf States Shipbuilders Consortium

ISN

Manufacturing Institute

Mason Contractors Association of America

Merit Contractors Association of Canada

NACE International

National Association of Minority Contractors

National Association of Women in Construction

National Insulation Association

National Technical Honor Society

National Utility Contractors Association

NAWIC Education Foundation

North American Crane Bureau

North American Technician Excellence

Pearson

Pearson Qualifications International

Prov

SkillsUSA®

Steel Erectors Association of America

U.S. Army Corps of Engineers

University of Florida, M. E. Rinker School of
Building Construction

Women Construction Owners & Executives,
USA

45101-07

Orientation to the Trade

45101-07
Orientation to the Trade

Topics to be presented in this module include:

Overview

The vast majority of homes and businesses in this country have used gypsum drywall panels as the finish for their walls and ceilings. This means that drywall installation and finishing mechanics will be needed on just about every building that will be constructed in the future. Drywall panels are usually applied directly to wood or steel framing members using screws or nails. In order to provide a smooth finished appearance, the joints between the drywall panels are finished with special tapes and finishing compounds. When a job has been finished by professional drywall mechanics, regardless of the number of panels used, the wall or ceiling will appear as though it is made from one single sheet. In order to achieve this level of competence, the drywall mechanic must be thoroughly trained in the use of specialized tools, materials, and techniques.

Objectives

When you have completed this module, you will be able to do the following:

1. Describe the history of the drywall trade.
2. Identify the aptitudes, behaviors, and skills needed to be a successful drywall specialist.
3. Identify the training opportunities within the drywall trade.
4. Identify the career and entrepreneurial opportunities within the drywall trade.
5. Identify the responsibilities of a person working in the construction industry.
6. State the personal characteristics of a professional.
7. Explain the importance of safety in the construction industry.

Trade Terms

Firestopping
Gypsum
Joint
Joists

Lath
Plaster
Studs

Required Trainee Materials

1. Pencil and paper
2. Appropriate personal protective equipment

Prerequisites

Before you begin this module, it is recommended that you successfully complete *Core Curriculum*.

This course map shows all of the modules in the first level of the *Drywall* curriculum. The suggested training order begins at the bottom and proceeds up. Skill levels increase as you advance on the course map. The local Training Program Sponsor may adjust the training order.

DRYWALL

LEVEL ONE
45105-07 Drywall Finishing
45104-07 Drywall Installation
45103-07 Thermal and Moisture Protection
45102-07 Construction Materials and Methods
45101-07 Orientation to the Trade

CORE CURRICULUM: Introductory Craft Skills

101CMAP.EPS

1.0.0 ◆ INTRODUCTION

Gypsum drywall panels are used to finish walls and ceilings on most residential and commercial buildings. Drywall is typically manufactured as 4' by 8' panels. These dimensions are compatible with the spacing of the framing to which the panels are fastened. For example, a common spacing for wall studs and ceiling joists is 16 inches on center (OC). Drywall panels are sold in other sizes and can be obtained in a variety of sizes by special order. Once installed, the joint between the panels must be finished so that the surface has a flat, smooth appearance. A drywall mechanic is a person who has learned the specialized techniques needed to properly install and finish gypsum drywall panels. *Figure 1* shows examples of drywall work.

Opportunity is driven by knowledge and ability, which are in turn driven by education and training. The NCCER *Drywall* curriculum was designed and developed by the construction industry for the construction industry. It is the only nationally accredited, competency-based construction training program in the United States. A competency-based program requires that the trainee demonstrate the ability to safely perform specific job-related tasks in order to receive credit. This approach is unlike other apprentice programs that merely require a trainee to put in the required number of hours in the classroom and on the job.

The primary goal of NCCER is to standardize construction craft training so that both employers and employees will benefit from the training, no matter where they are located. As a trainee in an NCCER program, you will become part of the NCCER Registry. You will receive a certificate for each level of training you complete. If you apply for a job with any participating contractor in the country, a transcript of your training will be available. If your training is incomplete when you make a job transfer, you can pick up where you left off because every participating contractor is using the same training program. In addition, many technical schools and colleges are using the program.

2.0.0 ◆ HISTORY OF DRYWALL

By historic standards, gypsum drywall is a newcomer to the construction industry. Until the middle of the twentieth century, interior walls were generally finished with plaster that was applied over narrow strips of rough wood known as lath. One advantage of plaster is that it retarded the spread of fire. The gypsum drywall of today is known for the same quality. In fact, gypsum dry-

HANGING DRYWALL

FINISHED DRYWALL

COMPLETED PROJECT

101F01.EPS

Figure 1 ◆ Drywall work.

wall has been incorporated into standards that define the construction of fire-rated walls for residential and commercial construction.

Originally, carpenters, painters, and other trades installed and/or finished gypsum drywall. The increased use of the product, combined with the importance of installing and finishing it correctly, led to the emergence of a specialized trade known as drywall mechanics.

3.0.0 ◆ MODERN DRYWALL WORK

Today's drywall panels are manufactured using an inner core of wet gypsum plaster, along with some additives such as fiberglass and fire retardants. The core is sandwiched between sheets of heavy paper or fiberglass mats and allowed to harden. In the manufacturing process, the drywall is a continuous sheet, and is machine-cut to the required length at the end of the process.

The most common type of gypsum drywall in use today is ⅝-inch Type X fire-resistant panels. These panels can be layered, combined with wood or steel studs, and separated by insulation in order to achieve specified fire and sound transmission ratings. Fire ratings are specified in hours. For example, a wall made with a single layer of ⅝ Type X attached to wood studs would have a one-hour rating, while two layers of ⅝ Type X over steel studs would have a two-hour rating. The fire rating defines the time it would take for fire to breach the wall. The rating assumes that any penetrations in the wall, such as those required for piping or wiring runs, have been sealed with a firestopping material.

Over time, standardized methods have evolved for placing and fastening drywall panels. For example, industry experts have determined the correct placement and spacing for fasteners when installing drywall panels. Ceiling panels require more fasteners than wall panels. Fewer fasteners are required if drywall screws are used instead of nails.

A drywall mechanic must learn to use a variety of specialized tools. Drywall is cut to size using a special carbide cutting tool, a utility knife, or a drywall saw (*Figure 2*). Other specialized tools such as the power cutout tool in *Figure 3* are used to cut openings in the panels. For drywall installation, the primary tool is the screwgun (*Figure 4*). The power screwdriver is designed to hold slotted-head screws. A magnetic screwdriver bit is located inside the nose piece. The nose piece is adjustable so that the depth of penetration can be set. The screw gun has a clutch mechanism that disengages the drive when the screw head is below the paper surface of the drywall panel.

Drywall finishing requires a variety of materials and tools. Joints between drywall panels are finished using a paper or fiberglass mesh tape (*Figure 5*). The tape is usually embedded in the joint with joint compound. This compound, called

INSIDE TRACK   **Drywall**

The concept of drywall panels originally came from a plasterer named Augustine Sackett in the latter part of the nineteenth century. Sackett sandwiched wet plaster of Paris between sheets of heavy paper and allowed it to dry. Although this method achieved some success, it wasn't until after World War II that gypsum drywall became popular. Since then, it has become a mainstay of the construction industry.

mud by those in the trade, comes in powder and premix form (*Figure 6*). A joint typically receives three coats of mud, including the tape bedding coat. Each coat is smoothed and sanded.

The mud is applied using a taping knife (*Figure 7*). Successive coats of mud are finished with increasingly broader knives.

CARBIDE CUTTING TOOL

UTILITY KNIFE

DRYWALL SAW

101F02.EPS

Figure 2   Tools used to cut drywall.

Figure 3 ◈ Power cutout tool.

101F03.EPS

101F04.EPS

Figure 4 ◈ Power screwdriver.

Figure 6 ◈ Joint compound.

101F06.EPS

101F05.EPS

Figure 5 ◈ Drywall joint tape.

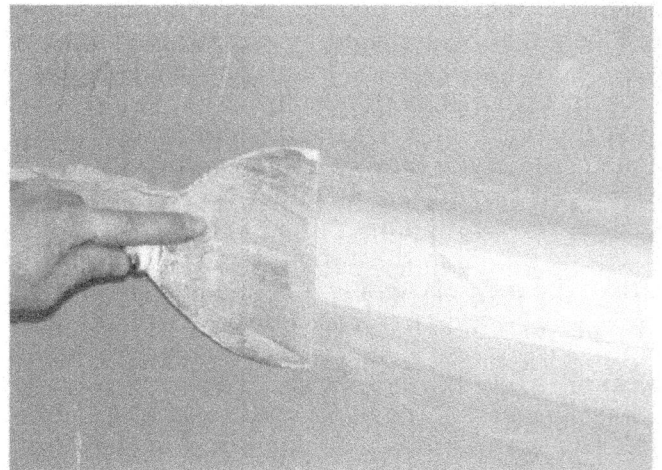

101F07.EPS

Figure 7 ◈ Taping knife.

Production Work

For large jobs, some companies prefer to use automatic taping and finishing tools. The automatic taping tool applies tape and mud at the same time. The flat finisher is filled with mud and automatically applies and levels the mud as it moves along the joint. Other automatic tools are made especially for corner work.

CORNER FINISHER

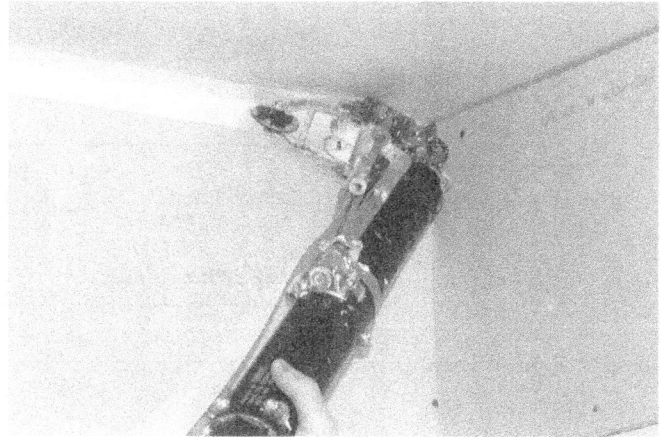

AUTOMATIC TAPING TOOL

FLAT FINISHER

101SA01.EPS

4.0.0 OPPORTUNITIES IN THE CONSTRUCTION INDUSTRY

The construction industry employs more people and contributes more to the nation's economy than any other industry. Our society will always need new homes, roads, airports, hospitals, schools, factories, and office buildings. This means that there will always be a source of well-paying jobs and career opportunities for drywall

mechanics and other construction trade professionals. As shown in *Figure 8*, anyone working in the construction industry has many opportunities for growth if they are willing to commit the time and energy to learn. Once a person has completed their initial training, they have the opportunity to progress from apprentice through several levels:

```
CRAFT
    APPRENTICESHIP
    (2 – 3 YEARS)

    TECHNICAL
    COLLEGE/OJT          TECHNICIAN    TECHNICIAN    TECHNICIAN
    (2 YEARS)            LEVEL 1       LEVEL 2       LEVEL 3

MANAGEMENT
    COLLEGE &
    APPRENTICESHIP
    (2 – 5 YEARS)

                         FOREMAN → SUPERVISOR → PROJECT    → BUSINESS      → CEO/
    UNIVERSITY                                  MANAGER      MANAGEMENT      COMPANY
    (4 YEARS)                                                                OWNER

                         B.S.          MASTER'S DEGREE   PH.D.
                         CONSTRUCTION  CONSTRUCTION      CONSTRUCTION
                         MANAGEMENT    MANAGEMENT        MANAGEMENT
                         ARCHITECTURE  ARCHITECTURE      ARCHITECTURE
                         ENGINEERING   ENGINEERING       ENGINEERING
```

101F08.EPS

Figure 8 ◆ Opportunities in the construction industry.

- *Journeyman* – After successfully completing an apprenticeship, a trainee becomes a journeyman. The term journeyman originally meant to journey away from the master to work alone. A person can remain a journeyman or advance in the trade. Journeymen may have additional duties such as supervisor or estimator. With larger companies and on larger jobs, journeymen often become specialists.
- *Foreman* – This individual is a front-line leader who directs the work of a crew of craft workers and laborers.
- *Supervisor* – Large construction projects require supervisors who oversee the work of crews made up of foremen, apprentices, and journeymen. They are responsible for assigning, directing, and inspecting the work of construction crew members.
- *Safety manager* – An individual responsible for project safety and health-related issues, including development of the safety plan and procedures, safety training for workers, and regulatory compliance.
- *Project manager/administrator* – Business management and administration deal with controlling the scope and direction of the business and dealing with such concerns as payroll, taxes, and employee benefits. Larger contracting firms may have one or several managers/ administrators. This person is responsible for worker output and must determine the best methods to use and the way to apply workers to accomplish the job. A project administrator is responsible for a contractor's support operations, such as accounting, finance, and secretarial work.
- *Estimator* – Estimators work for contractors and building supply companies. They make careful estimates of the materials and labor required for a job. Based on these estimates, the contractor submits bids for jobs. Estimating requires a complete understanding of construction methods as well as the materials and supplies required. Only experienced mechanics who possess good math skills and the patience to prepare detailed, accurate estimates are employed to do this work. This is a highly responsible position since errors in estimates can result in financial losses to the contractor. Depending on the size and type of the business, the job of estimating may be done by the owner, manager, administrator, or an estimating specialist. Today's estimators need solid computer skills because advances in computer software have revolutionized the field of estimating.
- *Architect* – An architect is a person who is licensed to design buildings and oversee their construction. A person normally needs a specialized degree in architecture to qualify as an architect.

Careers

Apprentice training is the first step in a career that has endless possibilities. You can gain knowledge of many different trades and skills while developing your craft. This broad set of skills is a valuable asset in the construction industry and will open the door to a wide variety of exciting career opportunities.

- *General contractor* – A general contractor is an individual or company that manages an entire construction project. The general contractor plans and schedules the project, buys the materials, and usually contracts with carpentry, plumbing, electrical, and other trade contractors to perform the work. The general contractor usually works with architects, engineers, and clients and/or the client's construction manager in planning and implementing a project. The general (prime) contractor is also responsible for safety on site.

- *Construction manager* – The role of the construction manager (CM) is different from that of the general contractor. The CM is usually hired by the building owner to represent the owner's interests on the project. The CM is the individual who works with the general contractor and architect to ensure that the building meets the owner's requirements.

- *Contractor/owner* – Construction contractors/owners are those who have established a contracting business. Generally, they hire apprentices, journeymen, and master carpenters to work for them. Depending upon the size of the business, contractors may work with the crew or they may manage the business full-time. Very small contractors may have only one or two people do everything, including managing the business, preparing estimates, obtaining supplies, and doing the work on the job. This group includes specialty subcontractors who perform specialized tasks such as framing, interior trim work, and cabinet installation.

More than any other construction worker, the drywall mechanic is likely to become knowledgeable about many trades. This makes drywall work interesting and challenging and creates a great variety of career opportunities.

The important thing to learn is that a career is a lifelong learning process. An effective drywall mechanic needs to keep up-to-date with new tools, materials, and methods. If you choose to work your way into management or to someday start your own construction business, you need to learn management and administrative skills on top of keeping your drywall skills honed. Every successful manager and business owner started the same way you are starting, and they all have one thing in common: a desire and willingness to continue learning. The learning process begins with apprentice training.

A person working in the drywall trade may work as a drywall installer or a drywall finisher. In addition, the work performed by drywall mechanics varies from company to company and region to region. For example, many drywall contractors perform other work in addition to drywall work. This includes installation of steel framing, suspended ceilings, and thermal insulation (*Figure 9*). Some drywall contractors also install interior doors. Therefore, it's possible that once the building structure is framed, a single contractor will build and finish all the interior walls. In some locations, drywall mechanics install the drywall, but other trades such as painters do the drywall finishing.

As you develop your skills and gain experience, you will have the opportunity to earn greater pay for your services. There is great financial incentive for learning and growing within the trade. You can't get to the top, however, without learning the basics.

4.1.0 Formal Construction Training

Over the past twenty years, the rate of formal training within the construction industry has been declining. Until the establishment of the NC-CER, the only opportunity for formal construction training was through the U.S. Department of Labor, Bureau of Apprenticeship and Training (BAT). The *National Apprenticeship Act* of 1937, commonly referred to as the *Fitzgerald Act*, officially established BAT. The federal government recently created the Office of Apprenticeship that consolidated both BAT and new employer–labor relations responsibilities.

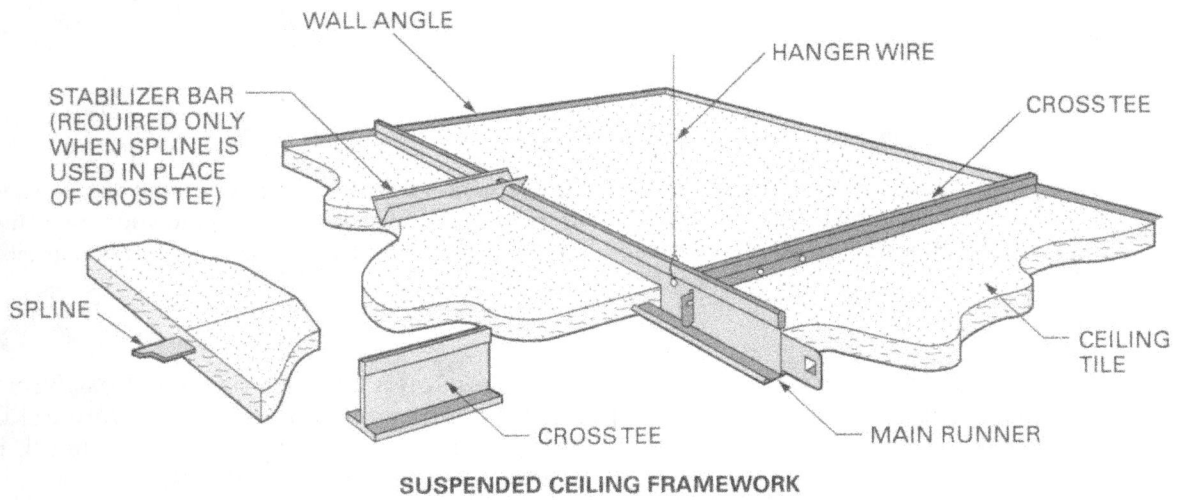

SUSPENDED CEILING FRAMEWORK

STABILIZER BAR
(REQUIRED ONLY
WHEN SPLINE IS
USED IN PLACE
OF CROSS TEE)

WALL ANGLE

HANGER WIRE

CROSS TEE

SPLINE

CROSS TEE

MAIN RUNNER

CEILING
TILE

STEEL I-BEAM HEADER

FINISHED CEILING

THERMAL INSULATION

101F09.EPS

Figure 9 ◆ Other work performed by drywall contractors.

The federal government established registered apprenticeship training via the *Code of Federal Regulations (CFR) 29:29*, which dictates specific requirements for apprenticeship, and *CFR 29:30*, which dictates specific guidelines for recruitment, outreach, and registration into BAT-approved apprenticeship programs.

Compared to the overall employment in the construction industry, the percentage of enrollment in BAT-style programs has been less than 5 percent for the past decade. BAT programs rely upon mandatory classroom instruction and on-the-job training (OJT). The classroom instruction required is 144 hours per year while the OJT requirement is 2,000 hours per year. A typical BAT program requires 8,000 hours of OJT and 576 hours of related classroom training prior to getting the journeyman certificate dispensed by the BAT.

Craft training via the BAT has not been changed for 30 years, which is believed to be one reason for the lack of use of this program in the construction industry today. Education and training throughout the country is undergoing significant change. As education, political, financial, and student factions argue over the direction and future of education, educators and researchers have been learning and applying new techniques to adjust to how today's students learn and apply their education.

NCCER is an independent, private educational foundation founded and funded by the construction industry to solve the training problem plaguing the industry today. The basic idea of the NCCER is to supplant governmental control and credentialing of the construction workforce with industry-driven training and education programs. NCCER departs from traditional classroom learning and has adopted a pure competency-based training regimen. Competency-based training means that instead of requiring specific hours of classroom training and set hours of OJT, you simply have to prove that you know what is required and can demonstrate that you can perform the specific skill. NCCER also uses the latest technology, interactive computer-based training, to deliver the classroom portions of the training. All completion information for every trainee is sent to the NCCER and kept within the Registry. The Registry can then confirm training and skills for workers as they move from company to company, state to state, or even within their own company (see the *Appendix*).

The dramatic shortage of skills within the construction workforce, combined with the shortage of new workers coming into the industry, is forcing the industry to design and implement new training initiatives to combat the problem.

Whether you enroll in a BAT program, an NCCER program, or both, it is critical that you work for an employer who supports a national, standardized training program that includes credentials to confirm your skill development.

4.2.0 Apprenticeship Program

Apprentice training goes back thousands of years; its basic principles have not changed in that time. First, it is a means for individuals entering the craft to learn from those who have mastered the craft. Second, it focuses on learning by doing; real skills versus theory. Although some theory is presented in the classroom, it is always presented in a way that helps the trainee understand the purpose behind the required skill.

4.2.1 Youth Apprenticeship Program

A Youth Apprenticeship Program is also available that allows students to begin their apprentice training while still in high school. A student entering the drywall program in eleventh grade may complete as much as one year of the NCCER Standardized Craft Training two-year program by high school graduation. In addition, the program, in cooperation with local craft employers, allows students to work in the trade and earn money while still in school. Upon graduation, the student can enter the industry at a higher level and with more pay than someone just starting the apprenticeship program.

This training program is similar to the one used by NCCER learning centers, contractors, and colleges across the country. Students are recognized through official transcripts and can enter the second year of the program wherever it is offered. They may also have the option of applying the credits at a two-year or four-year college that offers degree or certificate programs in the construction trades.

4.2.2 Apprenticeship Standards

All apprenticeship standards prescribe certain work-related or on-the-job training. This on-the-job training is broken down into specific tasks in which the apprentice receives hands-on training during the period of the apprenticeship. In addition, a specified number of hours is required in each task. The total number of hours for the drywall apprenticeship program is traditionally 4,000, which amounts to about two years of training. In a competency-based program, it may be possible to shorten this time by testing out of specific tasks through a series of performance exams.

In a traditional program, the required OJT may be acquired in increments of 2,000 hours per year. Layoffs or illness may affect the duration.

The apprentice must log all work time and turn it in to the Apprenticeship Committee (discussed later) so that accurate time control can be maintained. Another important aspect of keeping work records up-to-date is that after each 1,000 hours of related work, the apprentice will receive a pay increase as prescribed by the apprenticeship standards.

The classroom-related instruction and work-related training will not always run concurrently due to such reasons as layoffs, type of work needed to be done in the field, etc. Furthermore, apprentices with special job experience or course-work may obtain credit toward their classroom requirements. This reduces the total time required in the classroom while maintaining the total 4,000-hour on-the-job training requirement. These special cases will depend on the type of program and the regulations and standards under which it operates.

Informal on-the-job training provided by employers is usually less thorough than that provided through a formal apprenticeship program. The degree of training and supervision in this type of program often depends on the size of the employing firm. A small contractor who specializes in home building may provide training in only one area, such as rough framing. In contrast, a large general contractor may be able to provide training in several areas.

For those entering an apprenticeship program, a high school or technical school education is desirable, as are courses in carpentry, shop, mechanical drawing, and general mathematics. Manual dexterity, good physical condition, a good sense of balance, and a lack of fear of working in high places are important. The ability to solve arithmetic problems quickly and accurately and to work closely with others is essential. You must have a high concern for safety.

The prospective apprentice must submit to the apprenticeship committee certain information. This may include the following:

- Aptitude test (General Aptitude Test Battery or GATB Form Test) results (usually administered by the local Employment Security Commission)
- Proof of educational background (candidate should have schools send transcripts to the committee)
- Letters of reference from past employers and friends
- Results of a physical examination

- Proof of age
- If the candidate is a veteran, a copy of Form DD214
- A record of technical training received that relates to the construction industry and/or a record of any pre-apprenticeship training
- High school diploma or General Equivalency Diploma (GED)

The apprentice must:

- Wear proper safety equipment on the job
- Purchase and maintain tools of the trade as needed and required by the contractor
- Submit a monthly on-the-job training report to the committee
- Report to the committee if a change in employment status occurs
- Attend classroom-related instruction and adhere to all classroom regulations such as that for attendance

4.3.0 Responsibilities of the Employee

In order to be successful, the professional must have the skills to use current trade materials, tools, and equipment to produce a finished product of high quality in a minimum period of time. A drywall mechanic must be adept at adjusting methods to meet each situation. The successful drywall specialist must continuously train to remain knowledgeable about the technical advancements in trade materials and equipment and to gain the skills to use them. A professional never takes chances with regard to personal safety or the safety of others.

4.3.1 Professionalism

The word *professionalism* is a broad term that describes the desired overall behavior and attitude expected in the workplace. Most people would argue that it must start at the top in order to be successful. It is true that management support of professionalism is important to its success in the workplace, but it is more important that individuals recognize their own responsibility for professionalism.

Professionalism includes honesty, productivity, safety, civility, cooperation, teamwork, clear and concise communication, being on time and prepared for work, and regard for one's impact on one's co-workers. It can be demonstrated in a variety of ways every minute you are in the workplace. Most important is that you do not tolerate the unprofessional behavior of co-workers. This is not to say that you shun the unprofessional worker; instead, you work to demonstrate the benefits of professional behavior.

Professionalism is a benefit both to the employer and the employee. It is a personal responsibility. Our industry is what each individual chooses to make of it; choose professionalism and the industry image will follow.

4.3.2 Honesty

Honesty and personal integrity are important traits of the successful professional. Professionals pride themselves on performing a job well and on being punctual and dependable. Each job is completed in a professional way, never by cutting corners or reducing materials. A valued professional maintains work attitudes and ethics that protect property such as tools and materials belonging to employers, customers, and other trades from damage or theft at the shop or job site.

Honesty and success go hand-in-hand. It is not simply a choice between good and bad, but a choice between success and failure. Dishonesty will always catch up with you. Whether you are stealing materials, tools, or equipment from the job site or simply lying about your work, it will not take long for your employer to find out. Of course,

you can always go and find another employer, but this option will ultimately run out on you.

If you plan to be successful and enjoy continuous employment, consistency of earnings, and being sought after as opposed to seeking employment, then start out with the basic understanding of honesty in the workplace and you will reap the benefits.

Honesty means more than giving a fair day's work for a fair day's pay; it means carrying out your side of a bargain; it means that your words convey true meanings. Our thoughts as well as our actions should be honest. Employers place a high value on an employee who is strictly honest.

4.3.3 Loyalty

Employees expect employers to look out for their interests, to provide them with steady employment, and to promote them to better jobs as openings occur. Employers feel that they, too, have a right to expect their employees to be loyal to them—to keep their interests in mind; to speak well of them to others; to keep any minor troubles strictly within the plant or office; and to keep absolutely confidential all matters that pertain to the business. Both employers and employees should keep in mind that loyalty is not something to be demanded; rather, it is something to be earned.

INSIDE TRACK   Ethical Principles for Members of the Construction Trades

Honesty: Be honest and truthful in all dealings. Conduct business according to the highest professional standards. Faithfully fulfill all contracts and commitments. Do not deliberately mislead or deceive others.

Integrity: Demonstrate personal integrity and the courage of your convictions by doing what is right even when there is great pressure to do otherwise. Do not sacrifice your principles for expediency, be hypocritical, or act in an unscrupulous manner.

Loyalty: Be worthy of trust. Demonstrate fidelity and loyalty to companies, employers, fellow craftspeople, and trade institutions and organizations.

Fairness: Be fair and just in all dealings. Do not take undue advantage of another's mistakes or difficulties. Fair people display a commitment to justice, equal treatment of individuals, tolerance for and acceptance of diversity, and open-mindedness.

Respect for others: Be courteous and treat all people with equal respect and dignity regardless of sex, race, or national origin.

Law abiding: Abide by laws, rules, and regulations relating to all personal and business activities.

Commitment to excellence: Pursue excellence in performing your duties, be well-informed and prepared, and constantly endeavor to increase your proficiency by gaining new skills and knowledge.

Leadership: By your own conduct, seek to be a positive role model for others.

4.3.4 Willingness to Learn

Every office and plant has its own way of doing things. Employers expect their workers to be willing to learn these ways. Also, it is necessary to adapt to change and be willing to learn new methods and procedures as quickly as possible. Sometimes the installation of a new machine or the purchase of new tools makes it necessary for even experienced employees to learn new methods and operations. It is often the case that employees resent having to accept improvements because of the retraining that is involved. However, employers will no doubt think they have a right to expect employees to put forth the necessary effort. Methods must be kept up-to-date in order to meet competition and show a profit. It is this profit that enables the owner to continue in business and that provides jobs for the employees.

4.3.5 Willingness to Take Responsibility

Most employers expect their employees to see what needs to be done, then go ahead and do it. It is very tiresome to have to ask again and again that a certain job be done. It is obvious that having been asked once, an employee should assume the responsibility from then on. Employees should be alert to see boxes that need to be out of the way, stock that should be stacked, or tools that need to be put away. It is true that, in general, responsibility should be delegated and not assumed; once the responsibility has been delegated, however, the employee should continue to perform the duties without further direction. Every employee has the responsibility for working safely.

4.3.6 Willingness to Cooperate

To cooperate means to work together. In our modern business world, cooperation is the key to getting things done. Learn to work as a member of a team with your employer, supervisor, and fellow workers in a common effort to get the work done efficiently, safely, and on time.

4.3.7 Rules and Regulations

People can work together well only if there is some understanding about what work is to be done, when it will be done, and who will do it. Rules and regulations are a necessity in any work situation and should be so considered by all employees.

4.3.8 Tardiness and Absenteeism

Tardiness means being late for work and absenteeism means being off the job for one reason or another. Consistent tardiness and frequent absences are an indication of poor work habits, unprofessional conduct, and a lack of commitment.

Your work life is governed by the clock. You are required to be at work at a definite time. So is everyone else. Failure to get to work on time results in confusion, lost time, and resentment on the part of those who do come on time. In addition, it may lead to penalties, including dismissal. Although it may be true that a few minutes out of a day are not very important, we must remember that a principle is involved. Our obligation is to be at work at the time indicated. We agree to the terms of work when we accept the job. Perhaps it will help us to see things more clearly if we try to look at the matter from the supervisor's point of view. Supervisors cannot keep track of people if they come in any time they please. It is not fair to others to ignore tardiness. Failure to be on time may hold up the work of fellow workers. Better planning of our morning routine will often keep us from being delayed and so prevent a breathless, late arrival. In fact, arriving a little early indicates your interest and enthusiasm for your work, which is appreciated by employers. The habit of being late is another one of those things that stand in the way of promotion.

It is sometimes necessary to take time off from work. No one should be expected to work when sick or when there is serious trouble at home. However, it is possible to get into the habit of letting unimportant and unnecessary matters keep

Late for Work

us from the job. This results in lost production and hardship on those who try to carry on the work with less help. Again, there is a principle involved. The person who hires us has a right to expect us to be on the job unless there is some very good reason for staying away. Certainly, we should not let some trivial reason keep us home. We should not stay up nights until we are too tired to go to work the next day. If we are ill, we should use the time at home to do all we can to recover quickly. This, after all, is no more than most of us would expect of a person we had hired to work for us, and on whom we depended to do a certain job.

If it is necessary to stay home, then at least phone the office early in the morning so that the boss can find another worker for the day. Some employees remain home without contacting their employer. This is the worst possible way to handle the matter. It leaves those at work uncertain about what to expect. They have no way of knowing whether you have merely been held up and will be in later, or whether immediate steps should be taken to assign your work to someone else. Courtesy alone demands that you let the boss know if you cannot come to work.

The most frequent causes of absenteeism are illness or death in the family, accidents, personal business, and dissatisfaction with the job. Here we see that some of the causes are legitimate and unavoidable, while others can be controlled. One can usually plan to carry on most personal business affairs after working hours. Frequent absences will reflect unfavorably on a worker when promotions are being considered.

Employers sometimes resort to docking pay, demotion, and even dismissal in an effort to control tardiness and absenteeism. No employer likes to impose restrictions of this kind. However, in fairness to those workers who do come on time and who do not stay away from the job, an employer is sometimes forced to discipline those who will not follow the rules.

4.4.0 What You Should Expect from Your Employer

After an applicant has been selected for apprenticeship by the committee, the employer of the apprentice agrees that the apprentice will be employed under conditions that will result in normal advancement. In return, the employer requires the apprentice to make satisfactory progress in on-the-job training and related classroom instruction. The employer agrees that the apprentice will not be employed in a manner that may be considered in violation of the apprenticeship standards. The employer also agrees to pay a prorated share of the cost of operating the apprenticeship program.

4.5.0 What You Should Expect from a Training Program

First and foremost, it is important that the employer you select has a training program. The program should be comprehensive, standardized, and competency-based, not based on the amount of time you spend in a classroom.

When employers take the time and initiative to provide quality training, it is a sign that they are willing to invest in their workforce and improve the abilities of workers. It is important that the training program be national in scope and that transcripts and completion credentials are issued to participants. Construction is unique in that the employers share the workforce. An employee in the trades may work for several contractors throughout their time in the field. Therefore, it is critical that the training program help the worker move from company to company, city to city, or state to state without having to start at the beginning for each move. Ask how many employers in the area use the same program before you enroll. Make sure that you will always have access to transcripts and certificates to ensure your status and level of completion.

Training should be rewarded. The training program should have a well-defined compensation ladder attached to it. Successful completion and mastery of skill sets should be accompanied by increases in hourly wages.

Finally, the curricula should be complete and up-to-date. Any training program has to be committed to maintaining its curricula, developing new delivery mechanisms (CD-ROM, Internet, etc.), and being constantly vigilant for new techniques, materials, tools, and equipment in the workplace.

4.6.0 What You Should Expect from the Apprenticeship Committee

The Apprenticeship Committee is the local administrative body to which the apprentice is assigned and to which the responsibility is delegated for the appropriate training of the individual. Every apprenticeship program, whether state or federal, is covered by standards that have been approved by those agencies. The responsibility of enforcement is delegated to the committee.

The committee is responsible not only for enforcement of standards, but must see to it that proper training is conducted so that a craftsperson graduating from the program is fully qualified in those areas of training designated by the standards.

Among the responsibilities of the committee are the following:

- Screen and select individuals for apprenticeship and refer them to participating firms for training.
- Place apprentices under written agreement for participation in the program.
- Establish minimum standards for related instruction and on-the-job training and monitor the apprentice to see that these criteria are adhered to during the training period.
- Hear all complaints of violations of apprenticeship agreements, whether by employer or apprentice, and take action within the guidelines of the standards.
- Notify the registration agencies of all enrollments, completions, and terminations of apprentices.

5.0.0 ◆ HUMAN RELATIONS

Many people underestimate the importance of working well with others. There is a tendency to pass off human relations as nothing more than common sense. What exactly is involved in human relations? Commonly recognized elements of human relations are being friendly, pleasant, courteous, cooperative, adaptable, and sociable.

5.1.0 Making Human Relations Work

As important as the above-noted characteristics are for personal success, they are not enough. Human relations is much more than just getting people to like you. It is also knowing how to handle difficult situations as they arise.

Human relations is knowing how to work with supervisors who are often demanding and sometimes unfair. It is understanding the personality traits of others as well as yourself. Human relations is building sound working relationships in situations where others are forced on you.

Human relations is knowing how to restore working relationships that have deteriorated for one reason or another. It is learning how to handle frustrations without hurting others. Human relations is building and maintaining relationships with all kinds of people, whether those people are easy to get along with or not.

5.2.0 Human Relations and Productivity

Effective human relations is directly related to productivity. Productivity is the key to business success. Every employee is expected to produce at a certain level. Employers quickly lose interest in an employee who has a great attitude but is able to produce very little. There are work schedules to be met and jobs that must be completed.

All employees, both new and experienced, are measured by the amount of quality work they can safely turn out. The employer expects every employee to do his or her share of the workload.

However, doing one's share in itself is not enough. If you are to be productive, you must do your share (or more than your share) without antagonizing your fellow workers. You must perform your duties in a manner that encourages others to follow your example. It makes little difference how ambitious you are or how capably you perform. You cannot become the kind of employee you want to be, or the type of worker management wants you to be, without learning how to work with your peers.

Employees must sincerely do everything they can to build strong, professional working relationships with fellow employees, supervisors, and clients.

Teamwork

Many of us like to follow all sorts of different teams: racing teams, baseball teams, football teams, and soccer teams. Just as in sports, a job site is made up of a team. As a part of that team, you have a responsibility to your teammates. What does teamwork really mean on the job?

5.3.0 Attitude

A positive attitude is essential to a successful career. First, being positive means being energetic, highly motivated, attentive, and alert. A positive attitude is essential to safety on the job. Second, a positive employee contributes to the productivity of others. Both negative and positive attitudes are transmitted to others on the job. A persistent negative attitude can spoil the positive attitudes of others. It is very difficult to maintain a high level of productivity while working next to a person with a negative attitude. Third, people favor a person who is positive. Being positive makes a person's job more interesting and exciting. Fourth, the kind of attitude transmitted to management has a great deal to do with an employee's future success in the company. Supervisors can determine a subordinate's attitude by their approach to the job, reactions to directives, and the way they handle problems.

5.4.0 Maintaining a Positive Attitude

A positive attitude is far more than a smile, which is only one example of an inner positive attitude. As a matter of fact, some people transmit a positive attitude even though they seldom smile. They do this by the way they treat others, the way they look at their responsibilities, and the approach they take when faced with problems.

Here are a few suggestions that will help you to maintain a positive attitude:

- Remember that your attitude follows you wherever you go. If you make a greater effort to be a more positive person in your social and personal lives, it will automatically help you on the job. The reverse is also true. One effort will complement the other.
- Negative comments are seldom welcomed by fellow workers on the job. Neither are they welcome on the social scene. The solution: Talk about positive things and be complimentary. Constant complainers do not build healthy and fulfilling relationships.

- Look for the good things in people on the job, especially your supervisor. Nobody is perfect, but almost everyone has a few worthwhile qualities. If you dwell on people's good features, it will be easier to work with them.
- Look for the good things where you work. What are the factors that make it a good place to work? Is it the hours, the physical environment, the people, the actual work being done, or is it the atmosphere? Keep in mind that you cannot be expected to like everything. No work assignment is perfect, but if you concentrate on the good things, the negative factors will seem less important and bothersome.
- Look for the good things in the company. Just as there are no perfect assignments, there are no perfect companies. Nevertheless, almost all organizations have good features. Is the company progressive? What about promotional opportunities? Are there chances for self-improvement? What about the wage and benefit package? Is there a good training program? You cannot expect to have everything you would like, but there should be enough to keep you positive. In fact, if you decide to stick with a company for a long period of time, it is wise to look at the good features and think about them. If you think positively, you will act the same way.
- You may not be able to change the negative attitude of another employee, but you can protect your own attitude from becoming negative.

6.0.0 EMPLOYER AND EMPLOYEE SAFETY OBLIGATIONS

An obligation is like a promise or a contract. In exchange for the benefits of your employment and your own well-being, you agree to work safely. In other words, you are obligated to work safely. You are also obligated to make sure anyone you happen to supervise or work with is working safely. Your employer is also obligated to maintain a safe workplace for all employees. Safety is everyone's responsibility (*Figure 10*).

101F10.EPS

Figure 10 ❖ Safety is everyone's responsibility.

Some employers will have safety committees. If you work for such an employer, you are then obligated to that committee to maintain a safe working environment. This means two things:

- Follow the safety committee's rules for proper working procedures and practices.
- Report any unsafe equipment and conditions directly to the committee or your supervisor.

Here is a basic rule to follow every working day:

If you see something that is not safe, REPORT IT! Do not ignore it. It will not correct itself. You have an obligation to report it.

Suppose you see a faulty electrical hookup. You know enough to stay away from it, and you do. But then you forget about it. Why should you worry? It is not going to hurt you. Let somebody else deal with it. The next thing that happens is that a co-worker accidentally touches the live wire.

In the long run, even if you do not think an unsafe condition affects you—it does. Report what is not safe. Do not think your employer will be angry because your productivity suffers while the condition is corrected. On the contrary, your employer will be more likely to criticize you for not reporting a problem.

Your employer knows that the short time lost in making conditions safe again is nothing compared with shutting down the whole job because of a major disaster. If that happens, you are out of work anyway. Do not ignore an unsafe condition. In fact, Occupational Safety and Health Administration (OSHA) regulations require you to report hazardous conditions.

This applies to every part of the construction industry. Whether you work for a large contractor or a small subcontractor, you are obligated to report unsafe conditions. The easiest way to do this is to tell your supervisor. If that person ignores the unsafe condition, report it to the next highest supervisor. If it is the owner who is being unsafe, let that person know your concerns. If nothing is done about it, report it to OSHA. If you are worried about your job being on the line, think about it in terms of your life, or someone else's, being on the line.

The U.S. Congress passed the *Occupational Safety and Health Act* in 1970. This act also created OSHA. It is part of the U.S. Department of Labor. The job of OSHA is to set occupational safety and health standards for all places of employment, enforce these standards, ensure that employers provide and maintain a safe workplace for all employees, and provide research and educational programs to support safe working practices.

OSHA requires each employer to provide a safe and hazard-free working environment. OSHA also requires that employees comply with OSHA rules and regulations that relate to their conduct on the job. To gain compliance, OSHA can perform spot inspections of job sites, impose fines for violations, and even stop work from proceeding until the job site is safe.

According to OSHA standards, you are entitled to on-the-job safety training. As a new employee, you must be:

- Shown how to do your job safely
- Provided with the required personal protective equipment
- Warned about specific hazards
- Supervised for safety while performing the work

THINK ABOUT IT

Drugs and Alcohol

When people use drugs and alcohol, they are putting both themselves and the people around them at serious risk. A construction site can be a dangerous environment, and it is important to be alert at all times. Using drugs and alcohol on the job is an accident waiting to happen. You have an obligation to yourself, your employer, and your fellow employees to work safely. What should you do if you discover someone abusing drugs and/or alcohol at work?

OSHA was adopted with the stated purpose "to assure as far as possible every working man and woman in the nation safe and healthful working conditions and to preserve our human resources."

The enforcement of the *Occupational Safety and Health Act* is provided by the federal and state safety inspectors who have the legal authority to make employers pay fines for safety violations. The law allows states to have their own safety regulations and agencies to enforce them, but they must first be approved by the U.S. Secretary of Labor. For states that do not develop such regulations and agencies, federal OSHA standards must be obeyed.

These standards are listed in *OSHA Safety and Health Standards for the Construction Industry (29 CFR, Part 1926)*, sometimes called *OSHA Standards 1926*. Other safety standards that apply to the trade are published in *OSHA Safety and Health Standards for General Industry (29 CFR, Parts 1900 to 1910)*.

The most important general requirements that OSHA places on employers in the construction industry are:

- The employer must perform frequent and regular job site inspections of equipment.
- The employer must instruct all employees to recognize and avoid unsafe conditions, and to know the regulations that pertain to the job so they may control or eliminate any hazards.
- No one may use any tools, equipment, machines, or materials that do not comply with *OSHA Standards 1926*.
- The employer must ensure that only qualified individuals operate tools, equipment, and machines.

1. A competency-based training program is one that requires the student to _____.
   a. receive at least four years of classroom training
   b. receive on-the-job training for at least two years
   c. demonstrate the ability to perform specific job-related tasks
   d. pass a series of written tests

2. The common dimensions of a drywall panel are _____.
   a. 3' × 3'
   b. 4' × 8'
   c. 5' × 8'
   d. 5' × 10'

3. The position that follows an apprenticeship is _____.
   a. architect
   b. journeyman
   c. general contractor
   d. estimator

4. The _____ is most likely to handle the day-to-day operations on the job site.
   a. supervisor
   b. project manager
   c. architect
   d. owner

5. The purpose of the Youth Apprenticeship Program is to _____.
   a. make sure all young people know how to use basic carpentry tools
   b. provide job opportunities for people who quit high school
   c. allow students to start in an apprenticeship program while still in high school
   d. make sure that people under 18 have proper supervision on the job

6. The combined total of on-the-job and classroom training needed for a drywall apprentice to advance to journeyman is _____ hours.
   a. 2,000
   b. 4,000
   c. 6,000
   d. 8,000

7. Which of the following is true with respect to honesty?
   a. It is okay to borrow tools from the job site as long as you return them before anyone notices.
   b. You are doing your company a favor by using lower-grade materials than those listed in the specifications.
   c. It is okay to take materials or tools from your employer if you feel the company owes you for past efforts.
   d. Being late and not making up the time is the same as stealing from your employer.

8. If one of your co-workers complains about your company, you should _____.
   a. contribute your own complaints to the conversation
   b. agree with the person to avoid conflict
   c. suggest that the person look for another job
   d. find some good things to say about the company

9. If you see an unsafe condition on the job, you should _____.
   a. ignore it because it is not your job
   b. tell a co-worker
   c. call OSHA
   d. report it to a supervisor

10. The purpose of OSHA is to _____.
    a. catch people breaking safety regulations
    b. make rules and regulations governing all aspects of construction projects
    c. ensure that the employer provides and maintains a safe workplace
    d. assign a safety inspector to every project

Summary

There are many job and career opportunities for skilled drywall mechanics. An apprenticeship program that combines competency-based, hands-on training with classroom instruction has proven to be the most effective means for a person to learn and advance in the craft. Developing job skills is only part of the solution; it is just as important to learn good work habits, convey a positive, cooperative attitude to those around you, and practice good safety habits every day.

Notes

David E. Ackerman

Technical Instructor – Building Construction
Greenville Technical College
Greenville, SC

David Ackerman's experience in construction served as a springboard to elevate him to a college-level teaching position and subsequently provided an opportunity to serve on the industry committee overseeing the development of this curriculum. His commitment to education was recognized in 2006 when he received his division's Faculty of the Semester award.

How did you choose a career in the construction field?

Once I was exposed to construction work, it seemed like a natural extension of my personality. It's a field that demands careful, precise work, but every structure you work on has an artistic component to it. It's a trade that allows you to use both your creative and physical energies.

What types of training have you had?

I received an Associates Degree in Occupational Technology with a Carpentry minor focusing on Masonry. I am currently working on a Bachelor of Science Degree in Applied Management in Building Construction. I also have some training in accounting, which helps in areas such as business management and estimating.

What types of work have you done?

Before I started teaching, I worked in both residential and commercial construction. My experience includes finish carpentry, as well as cabinetry. I have also done accounting work.

What do you like about your present job?

I really enjoy teaching building construction because I can see that I'm making a difference in the lives of my students. I'm able to see the effects of my work because the job placement rate for our graduates is exceptionally high. In addition, our program directly helps the community through our involvement with Habitat for Humanity. Our administration is very supportive, which helps make the job more enjoyable.

What factors have contributed most to your success?

I have benefited greatly from the support and mentoring of my department head at Greenville Technical College. I also credit my ability to make a commitment to learning and sticking with it. I like taking courses and I am committed to learning through continuing education.

What advice would you give to those who are new to the field?

Be patient; it takes time and experience to become proficient in a craft. Listen to those with experience, and don't pass up an opportunity to work with someone who is good at the craft, even though they may be demanding. It's a great way to learn.

Trade Terms Introduced in This Module

Firestopping: A special putty or mechanical device that is used to plug openings in fire-rated structures such as wall and floors.

Gypsum: A chalky type of rock that serves as the basic ingredient of plaster and gypsum wallboard.

Joint: The place where two pieces of material meet. For example, the space between two drywall panels.

Joists: Equally-spaced framing members that support floors and ceilings.

Lath: Thin, narrow strips of wood used as a base for plaster.

Plaster: A compound consisting of lime, sand, and water used to cover walls and ceilings.

Studs: The vertical support members for walls.

Samples of NCCER
Training Credentials

Student 7 Samp
Certified Plus
4671784
Sample Student
2781481

NCCER

The Standard for Developing Craft Professionals

This is to certify that

Steven Whitaker

has fulfilled the requirements for

Carpentry Level One

*in NCCER's standardized training curriculum
on this Sixteenth day of September, 2012*

_____
Donald E. Whyte
President, NCCER

Official Transcript

January 17, 2012

NCCER Card #: 1720726
Trainee Name: John Q Smith
Sponsor: Austin Industrial Incorporated
Address: 2801 E 13th St
La Porte, TX 77571

Current Employer/School:
Solomon Plumbing Company

Module	Description	Instructor	Training Location	Date Completed
00101-04	Basic Safety	Kevin Jenkins	Solomon Plumbing Company	2/20/2008
00102-04	Introduction to Construction Math	Dave Buck	Building Trades Institute, LLC	8/8/2008
00103-04	Introduction to Hand Tools	Kevin Jenkins	Solomon Plumbing Company	1/1/2008
00104-04	Introduction to Power Tools	Dave Buck	Building Trades Institute, LLC	8/8/2008
00105-04	Introduction to Blueprints	Kevin Jenkins	Solomon Plumbing Company	3/20/2008
00106-04	Basic Rigging	Dave Buck	Building Trades Institute, LLC	8/8/2008
00108-04	Basic Employability Skills	Rod Blackburn	Utility Contractors, Inc.	3/15/2009
02101-05	Introduction to the Plumbing Profession	Kevin Jenkins	Solomon Plumbing Company	3/22/2008
26101-02	Electrical Safety	Don Whyte	National Center for Construction Education &	7/29/2002
26102-02	Hand Bending	Don Whyte	National Center for Construction Education &	7/29/2002
26103-02	Fasteners and Anchors	Don Whyte	National Center for Construction Education &	7/29/2002
26104-02	Electrical Theory One	Don Whyte	National Center for Construction Education &	7/29/2002
26105-02	Electrical Theory Two	Don Whyte	National Center for Construction Education &	7/29/2002
26106-02	Electrical Test Equipment	Don Whyte	National Center for Construction Education &	7/29/2002
26107-02	Introduction to the National Electrical Code	Don Whyte	National Center for Construction Education &	7/29/2002
26108-02	Raceways, Boxes, and Fittings	Don Whyte	National Center for Construction Education &	7/29/2002
26109-02	Conductors	Don Whyte	National Center for Construction Education &	7/29/2002

President, NCCER

Additional Resources and References

Additional Resources

This module is intended to be a thorough resource for task training. The following reference work is suggested for further study. This is optional material for continued education rather than for task training.

Gypsum Construction Handbook. Chicago, IL: United States Gypsum Company, 2000.

Figure Credits

NCCER CURRICULA — USER UPDATE

NCCER makes every effort to keep its textbooks up-to-date and free of technical errors. We appreciate your help in this process. If you find an error, a typographical mistake, or an inaccuracy in NCCER's curricula, please fill out this form (or a photocopy), or complete the online form at **www.nccer.org/olf**. Be sure to include the exact module ID number, page number, a detailed description, and your recommended correction. Your input will be brought to the attention of the Authoring Team. Thank you for your assistance.

Instructors – If you have an idea for improving this textbook, or have found that additional materials were necessary to teach this module effectively, please let us know so that we may present your suggestions to the Authoring Team.

NCCER Product Development and Revision
13614 Progress Blvd., Alachua, FL 32615

Email: curriculum@nccer.org
Online: www.nccer.org/olf

❏ Trainee Guide   ❏ Lesson Plans   ❏ Exam   ❏ PowerPoints   Other _____

Craft / Level: _____ Copyright Date: _____

Module ID Number / Title: _____

Section Number(s): _____

Description: _____

_____

_____

_____

Recommended Correction: _____

_____

_____

_____

Your Name: _____

Address: _____

Email: _____ Phone: _____

45102-07

Construction Materials
and Methods

45102-07
Construction Materials and Methods

Topics to be presented in this module include:

Overview

Whether you are installing drywall or suspended ceiling grids, or erecting steel stud walls, it is essential that you be familiar with the construction methods used in the industry and the materials used in the various types of structures. This module introduces drywall mechanics to the types of structures and building materials they are likely to encounter in their work.

Objectives

When you have completed this module, you will be able to do the following:

1. Describe the composition and uses of the common types of residential building materials.
2. Identify the major structural components of a residential building.
3. Describe the composition and uses of the common types of commercial building materials.
4. Describe common methods of residential and commercial construction.
5. State the major steps in the construction of a frame residence.
6. Explain common terms used in construction.
7. Identify various types of suspended ceilings.
8. Identify the various types of gypsum board and their applications.
9. Describe types of firestopping systems.
10. Describe the construction of walls to meet code requirements for fire and sound ratings.

Trade Terms

Admixture
APA-rated
Blocking
Bridging
Cantilever
Corrugated
Cripple stud
Dimension lumber
Dormer
Double top plate
Fire rating
Firestop
Firestopping
Footing
Furring strips
Gable
Girder
Green concrete
Gypsum
Gypsum wallboard
Header
Kerf
Millwork
Oriented strand board (OSB)
Plastic concrete

Plenum
Post-tensioned concrete
Pre-stressed concrete
Rabbeted
Rafter
Reinforced concrete
Ribband
Shakes
Sheathing
Shiplap
Sill plate
Soleplate
Striated
Stringer
Strongback
Stucco
Subfloor
Substrate
Top plate
Trimmer joist
Trimmer stud
Truss
Underlayment
Vaulted ceiling
Veneer

Required Trainee Materials

1. Pencil and paper
2. Appropriate personal protective equipment

Prerequisites

Before you begin this module, it is recommended that you successfully complete *Core Curriculum* and *Drywall Level One*, Module 45101-07.

This course map shows all of the modules in the first level of the *Drywall* curriculum. The suggested training order begins at the bottom and proceeds up. Skill levels increase as you advance on the course map. The local Training Program Sponsor may adjust the training order.

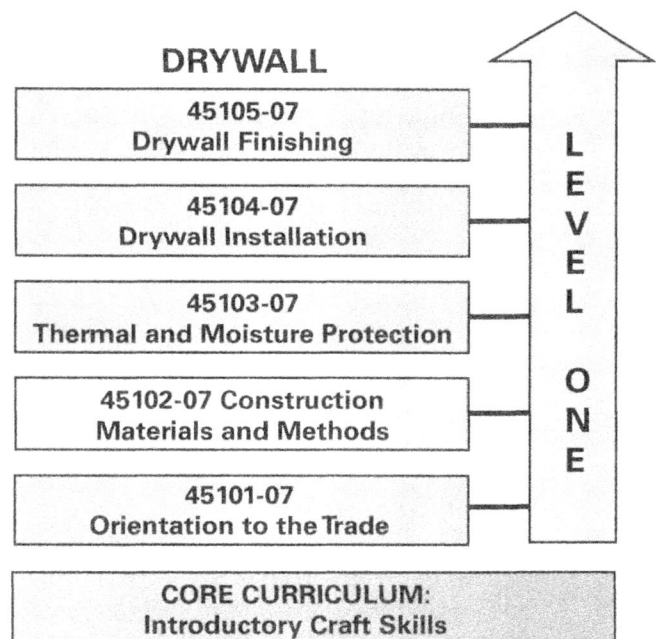

102CMAP.EPS

1.0.0 ◆ INTRODUCTION

This module provides an overview of the various types of building materials used in residential and commercial construction.

Residential and commercial (including multi-family) construction methods are very different, so they will be covered separately. The term *residential* refers to single-family and two-family dwellings. Keep in mind, however, that some small commercial buildings, such as apartment buildings and townhouses, may use the same construction techniques and materials as those used in residential buildings.

2.0.0 ◆ BUILDING MATERIALS

Many different materials are used in the construction of a building. Wood frame construction is most common in residential work. Concrete block and brick are also used in residential and light commercial construction.

The construction of large commercial buildings such as office buildings, warehouses, apartment buildings, and parking garages generally involves the use of a steel or concrete support structure and walls made of concrete or steel and glass.

2.1.0 Lumber

The framework of a single-family or two-family dwelling is usually built from lumber, which is divided into five categories:

- *Boards* – Members up to 1½" thick and 2" wide or wider.
- *Light framing (L.F.)* – Members 2" to 4" thick and 2" to 4" wide.
- *Joists and planks (J&P)* – Members 2" to 4" thick and 6" wide or wider.
- *Beams and stringers (B&S)* – Members 5" and thicker by 8" and wider.
- *Posts and timbers (P&T)* – Members 5" × 5" and greater, approximately square.

The vast majority of lumber used in framing a house is softwood such as pine or fir. Hardwoods such as oak and maple are used primarily in furniture and decorative pieces.

Light framing lumber, studs, joists, and planks are all classified as dimension lumber.

You are probably familiar with the terms 2 × 4, 1 × 6, and so on. These numbers represent the nominal (rough) size of the lumber in inches. Once the lumber is dressed (finished) at the lumber mill, it is somewhat smaller, typically ½" to ¾" less than the nominal size in each dimension.

Table 1 shows the final dimensions for some standard sizes of softwood dimension lumber. Note that these dressed dimensions apply only to softwoods; hardwoods have different conversion tables.

2.1.1 Pressure-Treated Lumber

Pressure-treated lumber is softwood lumber protected by chemical preservatives forced deep into the wood through a vacuum-pressure process. Pressure-treated lumber has been used for many years in on-ground and below-ground applications such as landscape timbers, sill plates, and foundations. In some parts of the country, it is also used extensively in the building of decks, porches, docks, and other outdoor structures. It is popular for these uses in areas where structures are exposed to snow for several months of the year. A major advantage of pressure-treated lumber is its relatively low price in comparison with

Table 1 Nominal and Dressed Sizes of Dimension Lumber (in inches)

Nominal	Dressed
2 × 2	1½ × 1½
2 × 4	1½ × 3½
2 × 6	1½ × 5½
2 × 8	1½ × 7¼
2 × 10	1½ × 9¼
2 × 12	1½ × 11¼

102T01.EPS

redwood and cedar. When natural woods such as these are used, only the more expensive heartwood will resist decay and insects.

Because the chemicals used in pressure-treated lumber present some hazards to people and the environment, special precautions apply to its use:

- When cutting pressure-treated lumber, always wear eye protection and a dust mask.
- Wash any skin that is exposed while cutting or handling the lumber.
- Wash clothing that is exposed to sawdust separately from other clothing.
- Do not burn pressure-treated lumber, as the ash poses a health hazard. Bury it or put it in with the trash.
- Be sure to read and follow the manufacturer's safety instructions as defined in the material safety data sheet (MSDS).

One place to look for pressure-treated lumber is any location where wood comes into contact with the ground, or outdoors where the wood is exposed to moisture.

2.2.0 Plywood

Plywood is made by gluing together thin layers of wood known as veneers. Plywood can have three or more plies (layers). These are bonded together at right angles with glue and heat under tremendous pressure. Putting the plies together at right angles increases the strength; also, the more plies there are, the greater the strength. The ply that is in the center is called the core and each of the exposed plies is called a veneer or face (*Figure 1*). All other plies between the core and veneer are called the crossbands. Constructing the plywood with the grain of adjacent plies running at right angles reduces the possibility of warping.

The average or standard size of plywood is 4'-0" × 8'-0". A few companies produce plywood from 6' to 8' widths and up to 16' in length. Sheathing-grade plywood is nominally sized by the manufacturer to allow for expansion; for example, 4'-0" × 8'-0" is really 47¾" × 95¾".

The thickness of plywood will vary from ³⁄₁₆" to 1¼". The common sizes are ¼", ½", and ¾" for finish paneling and ⅜", ½", ⅝", and ¾" for structural purposes.

Plywood is rated by the American Plywood Association (APA) for interior or exterior use. Exterior-rated plywood is used for sheathing, siding, and other applications where there may be exposure to moisture or wet weather conditions. Exterior plywood panels are made of high-grade veneers bonded together with a waterproof glue that is as strong as the wood itself.

Interior plywood uses lower grades of veneer for the back and inner plies. Although the plies may be bonded with a water-resistant glue, waterproof glue is normally used. The lower-grade veneers reduce the bonding strength, however, which means that interior-rated panels are not suitable for exterior use.

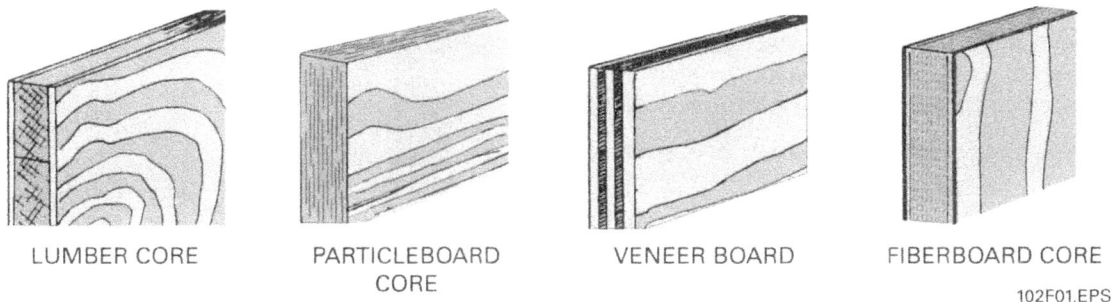

LUMBER CORE   PARTICLEBOARD CORE   VENEER BOARD   FIBERBOARD CORE

102F01.EPS

Figure 1 Types of plywood.

Plywood

Plywood that is expressly manufactured for either interior or exterior use may be used for other purposes in certain situations. Some local codes may require the use of pressure-treated plywood for exterior construction, in bathrooms, or in other high-moisture areas of a house. Pressure-treated plywood can withstand moisture better than interior plywood.

INSIDE TRACK

2.3.0 Building Boards

The ingenuity and technology that helped develop the plywood industry also assisted in the development of other materials in sheet form. The main ingredients for these products, known as building boards, are vegetable or mineral fibers. After mixing these ingredients with binder, the mixture becomes very soft.

At this point, the mixture passes through a press, which uses heat and pressure to produce the required thickness and density of the finished board.

Sawdust, wood chips, and wood scraps are the major waste materials at sawmills. These scrap materials are softened with heat and moisture, mixed with a binder and other ingredients, and then run through presses that produce the desired density and thickness.

The finished wood products that come off the presses are classified as hardboard, particleboard, or oriented strand board (OSB).

2.3.1 Hardboard

Hardboard is a manufactured building material, sometimes called tempered board or pegboard. Hardboards are water-resistant and extremely dense. The common thicknesses for hardboards are $\frac{3}{16}$", $\frac{1}{4}$", and $\frac{5}{16}$". The standard sheet size for hardboards is 4'-0" × 8'-0". However, they can be made in widths up to 6' and lengths up to 16' for specialized uses.

These boards are susceptible to breaking at the edges if they are not properly supported. Holes must be predrilled for nailing; direct nailing into the material will cause it to fracture.

Three grades of hardboard are manufactured:

- *Standard* – Suitable only for interior use, such as cabinets.
- *Tempered* – The same as standard grade except that it is denser, stronger, and more brittle. Tempered hardboard is suitable for either interior or exterior uses such as siding, wall paneling, and other decorative purposes.
- *Service* – Not as dense, strong, or heavy as standard grade. It can be used for basically everything for which standard or tempered hardboard is used. Service grade hardboard is manufactured for items such as cabinets, parts of furniture, and perforated hardboard.

2.3.2 Particleboard

The main composition of this type of material is small particles or flakes of wood. Particleboard is pressed under heat into panels. The sheets range in size from $\frac{1}{4}$" to $1\frac{1}{2}$" in thickness and from 3' to 8' in width. There are also thicknesses of 3" and lengths ranging up to 24' for special purposes. Particleboard has no grain, is smoother than plywood, is more resilient, and is less likely to warp.

Some types of particleboard can be used for underlayment if permitted by the local building codes. If particleboard is used as underlayment, it is laid with the long dimension across the joists and the edges staggered. Particleboard can be nailed, although some types will crumble or crack when nailed close to the edges.

2.3.3 Oriented Strand Board (OSB)

Oriented strand board (OSB) is a manufactured structural panel used for wall and roof sheathing and single-layer floor construction (*Figure 2*). OSB consists of compressed wood strands arranged in three perpendicular layers and bonded with phenolic resin. Some of the qualities of OSB are dimensional stability, stiffness, fastener holding capacity, and no voids in the core material. Before cutting into OSB, be sure to check the applicable MSDS for safety hazards. The MSDS is the most reliable source of safety information.

2.3.4 Mineral Fiberboards

The building boards just covered are classified as vegetable fiberboards. Mineral fiberboards fall into the same category as vegetable fiberboards. The main difference is that they will not sup-

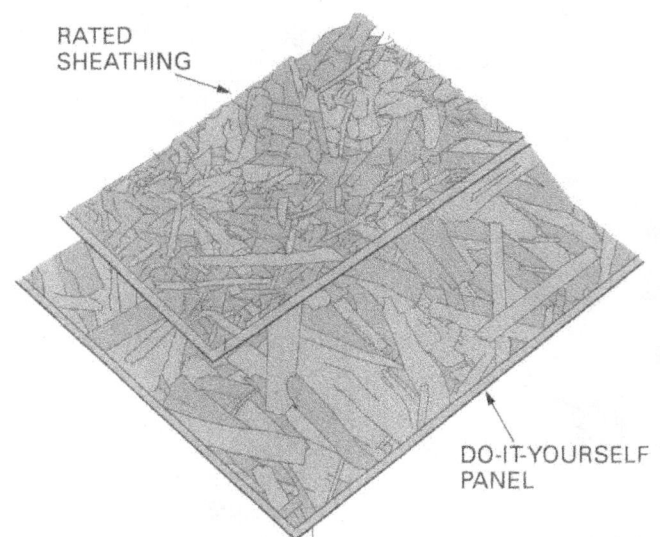

RATED SHEATHING

DO-IT-YOURSELF PANEL

102F02.EPS

Figure 2 ◈ OSB panels.

port combustion. Glass and gypsum rock are the most common minerals used in the manufacture of these fiberboards. Fibers of glass or gypsum powder are mixed with a binder and pressed or sandwiched between two layers of asphalt-impregnated paper, producing a rigid insulation board.

Some types of chemical foam mixed with glass fibers will also make a good, rigid insulation. However, this mineral insulation will crush and should not be used when it must support a heavy load.

WARNING!

Whenever working with older materials that may be made with asbestos, contact your supervisor for the company's policies on safe handling of the material. State and federal regulations require specific procedures to follow prior to removing, cutting, or disturbing any suspect materials. Also, some materials emit a harmful dust when cut. Check the MSDS before cutting. Asbestos can be found in structures built before 1978. It was used in ceiling tiles, siding, floor coverings, shingles, and pipe insulation.

2.3.5 High-Density Overlay (HDO) and Medium-Density Overlay (MDO) Plywood

High-density overlay (HDO) plywood panels have a hard, resin-impregnated fiber overlay heat-bonded to both surfaces. HDO panels are resistant to both abrasion and moisture and can be used for concrete forms, cabinets, countertops, and similar high-wear applications. HDO also resists damage from chemicals and solvents. HDO is available in four common thicknesses: ⅜", ½", ⅝", and ¾".

Medium-density overlay (MDO) panels are coated on one or both surfaces with a smooth, opaque overlay. MDO accepts paint well and is suitable for use as structural siding, exterior decorative panels, and soffits. MDO panels are available in eight common thicknesses ranging from ¹¹⁄₃₂" to ²³⁄₃₂".

Both HDO and MDO panels are manufactured with waterproof adhesive and are suitable for exterior use. If MDO panels are to be used outdoors, however, the panels should be edge-sealed with one or two coats of a good-quality exterior housepaint primer. This is easier to do when the panels are stacked.

2.4.0 Engineered Wood Products

In the past, the primary source of structural beams, timbers, joists, and other weight-bearing lumber was old-growth trees. These trees, which need more than 200 years to mature, are tall and thick and can produce a large amount of high-quality, tight-grained lumber. Extensive logging of these trees to meet demand resulted in higher prices and conflict with forest conservation interests.

The development of wood laminating techniques by lumber producers has permitted the use of younger-growth trees in the production of structural building materials. These materials are given the general classification of engineered lumber products.

Engineered wood products fall into five categories: laminated veneer lumber (LVL), parallel strand lumber (PSL), laminated strand lumber (LSL), wood I-beams, and glue-laminated lumber or glulam (*Figure 3*).

Engineered wood products provide several benefits:

- They can be made from younger, more abundant trees.
- They can increase the yield of a tree by 30 to 50 percent.
- They are stronger than the same size of structural lumber. Therefore, the same size piece of engineered lumber can bear more weight than that of solid lumber. Or, looked at another way, a smaller piece of engineered lumber can bear equal weight.

INSIDE TRACK

Use of Engineered Wood Products

Engineered wood products are used in a wide array of applications that were once exclusively served by cut lumber. For example, PSL is used for columns, ridge beams, and headers. LVL is also used for form headers and beams. Wood I-beams are used to frame roofs as well as floors. An especially noteworthy application is the use of LSL studs, top plates, and soleplates in place of lumber to frame walls.

PSL

LVL

LSL

WOOD I-BEAMS

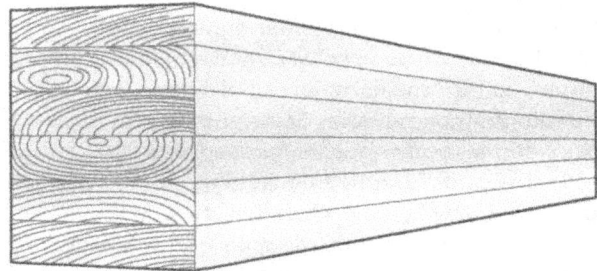
GLULAM BEAM

102F03.EPS

Figure 3 ◆ Examples of engineered wood products.

- Greater strength allows the engineered lumber to span a greater distance.
- A length of engineered wood is lighter than the same length of solid lumber. It is therefore easier to handle.

LVL is used for floor and roof beams and for headers over windows and doors. It is also used in scaffolding and concrete forms. No special cutting tools or fasteners are required.

PSL is used for beams, posts, and columns. It is manufactured in thicknesses up to 7". Columns can be up to 7" wide, and beams range up to 18" in width.

LSL is used for millwork such as doors and windows, and any other product that requires high-grade lumber. However, LSL will not support as much of a load as a comparable size of PSL because PSL is made from stronger wood.

Wood I-beams consist of a web with flanges bonded to the top and bottom. This arrangement, which mimics the steel I-beam, provides exceptional strength. The web can be made of OSB or plywood. The flanges are grooved to fit over the web. Wood I-beams are used as floor joists, rafters, and headers. Because of their strength, wood I-beams can be used in greater spans than a comparable length of dimension lumber. Lengths of up to 80' are available.

Glulam is made from several lengths of solid lumber that have been glued together. It is popular in architectural applications where exposed beams are used (*Figure 4*). Because of its exceptional strength and flexibility, glulam can be used in areas subject to high winds or earthquakes.

Glulam beams are available in widths from 2½" to 8¾". Depths range from 5½" to 28½". They are available in lengths up to 40'. They are used for many purposes, including ridge beams; basement beams; headers of all types; stair treads, supports, and stringers. They are also used in cantilever and vaulted ceiling applications.

102F04.EPS

Figure 4 ◆ Glulam beam application.

Fire-Retardant Building Materials

Lumber and sheet materials are sometimes treated with fire-retardant chemicals. The lumber can either be coated with the chemical in a non-pressure process or impregnated with the chemical in a pressure-treating process. Fire-retardant chemicals react to extreme heat, releasing vapors that form a protective coating around the outside of the wood. This coating, known as char, delays ignition and inhibits the release of smoke and toxic fumes.

2.5.0 Gypsum Board

Gypsum wallboard, also known as gypsum drywall, is one of the most popular and economical methods of finishing the interior walls and ceilings of wood-framed and metal-framed buildings. Properly installed and finished, drywall can give a wall or ceiling made from many panels the appearance of being made from one continuous sheet (*Figure 5*).

Gypsum board is a generic name for products consisting of a noncombustible core. It is rated as limited combustible because of the paper. This product is made primarily of gypsum with a paper surfacing covering the face, back, and long edges. It is also called plaster board, gypsum drywall, and Sheetrock®, a trade name of USG Corporation.

2.5.1 Types of Gypsum Products

Many types of gypsum board are available for a variety of building needs (*Table 2*). Gypsum board panels are mainly used as the surface layer for interior walls and ceilings; as a base for ceramic, plastic, and metal tile; for exterior soffits; for elevator and other shaft enclosures; and to provide fire protection for architectural elements.

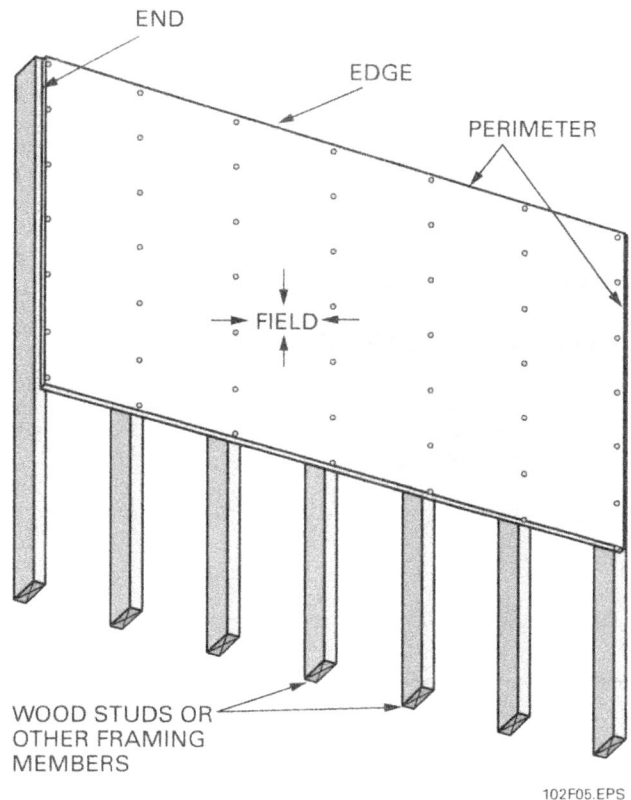

102F05.EPS

Figure 5 Typical gypsum wallboard application.

The Way It Was

Until the 1930s, walls were typically finished by installing thin, narrow strips of wood or metal known as lath between studs, and then coating the lath with wet plaster. Skilled plasterers could produce a very smooth wall finish, but the process was time-consuming and messy. In the early 1930s, paper-bound gypsum board was introduced and soon came into widespread use as a replacement for the tedious lath and plaster process.

Table 2 Types and Uses of Gypsum Wallboard

Type	Thickness	Sizes	Use
Regular, paper faced	¼" ⅜" ½", ⅜"	4' × 8' and 10' 4' × 8' to 14' 4' × 8' to 14'	Recovering old gypsum walls Double-layer installation Standard single-ply installation
Regular with foil back	½", ⅝", ⅜"	4' × 8' to 14'	Use as a vapor barrier or radiant heat retarder
Type X, fire-retardant	⅜", ½", ⅝"	4' × 8' to 14'	Use in garages, workshops, and kitchens, as well as around furnaces, fireplaces, and chimney walls; ⅝" is ⅜-hour fire rated.
Water-resistant	½", ⅝"	4' × 8' to 12'	For tile backing in areas not exposed to constant moisture. Use in kitchens, bathrooms and in utility rooms.
Architectural panels	5⁄16"	4' × 8' to 10'	Any room in the house
Gypsum lath	⅜", ½", ⅝"	16" × 4' 2' × 8' to 12	Use as a base for plaster and sound resistance Use ⅜" for 16" OC stud spacing ½" or ⅝" for 24" OC stud spacing
Gypsum coreboard	1"	2' × 8' to 12'	Shaft liner Laminated partitions
Stretch gypsum wallboard	½"	54" × 8' to 14'	Use for interior, 9' walls with extremely smooth finishes.
Exterior gypsum soffit (ceiling) board	½", ⅝"	4' × 8' to 12'	Use in metal and wood-framed soffits, on exteriors with indirect weather exposure.
Mold-resistant	½", ⅝", 1" (core)	4' × 8' to 14'	Use in places where there is increased moisture, but can be installed anywhere in interior or exterior.
Abuse-resistant	½", ⅝"	4' × 8' to 14'	Use in environments where heavy impacts to the walls are expected.

102T02.EPS

Gypsum board products are available with reflective aluminum foil backing, which provides an effective vapor barrier for exterior walls. When applied with the foil surface against the framing, with a minimum of ¾" enclosed air space adjacent to the foil, additional insulating efficiency is achieved.

This combination effectively reduces radiant heat loss in the cold season and radiant heat gain in the warm season. However, foil-backed gypsum board is not used as a backing material for tile, as a second face ply on a two-ply system, in conjunction with heating cables, or when laminating directly to masonry, ceiling, and roof assemblies.

Various thicknesses of gypsum wallboard are available in regular, Type X, water-resistant, and predecorated boards.

- *¼" gypsum board* – A lightweight, low-cost board used as a base in a multi-layer application for improving sound control, to cover existing walls and ceilings in remodeling, for curved walls, and for barrel ceilings.
- *5⁄16" gypsum board* – A lightweight board developed for use in manufactured construction, primarily mobile homes.
- *⅜" gypsum board* – A lightweight board principally applied in a double-layer system over wood framing and as a face layer in repair and remodeling.
- *½" gypsum board* – Generally used for single-layer wall and ceiling construction in residential work and in double-layer systems for greater sound and fire ratings.

- ⅝" gypsum board – Used in quality single-layer and double-layer wall systems. The greater thickness provides additional fire resistance, higher rigidity, and better impact resistance. It is also used to separate occupied and unoccupied areas, such as a house from a garage or an office from a warehouse.
- 1" gypsum board (either a single 1" board or two ½" factory-laminated boards) – Used as a liner or as a core board in shaft walls and in semi-solid or solid gypsum board partitions. It is also known as coreboard.

Standard gypsum boards are 4' wide and 8', 10', or 14' long. The width is compatible with the standard framing of studs or joists spaced 16" or 24" on center.

Regular gypsum board is used as a surface layer on walls and ceilings. Type X gypsum board is available in ½" and ⅝" thicknesses and has an improved fire resistance made possible by the use of special core additions. It is also available with a predecorated finish. Type X gypsum board is used in most fire-rated assemblies.

Foil-back gypsum boards have aluminum foil backings applied to the back surface of regular or fire-rated gypsum panels. The backing acts as a vapor retarder in cold climates, and it helps prevent moisture from getting into walls and ceilings. It can be used alone or as a base in multi-layered drywall systems. Foil-back gypsum is less effective in areas with high temperatures and humidity, however. In these climates, vapor control must be determined by a mechanical engineer.

Water-resistant gypsum board, also known as green board or MR board, has a water-resistant gypsum core and water-repellent paper. The facing typically has a light green color. It is available with a regular or Type X core and in ½" or ⅝" thicknesses. Water-resistant gypsum board is not recommended for use in tubs and shower enclosures and other areas exposed to constant water.

A special type of wallboard is replacing water-resistant gypsum wallboard as a backing for tile in damp areas such as baths and shower stalls. One type is known as cement board. It is made from a slurry of portland cement mixed with glass fibers. It is colored light blue for easy recognition. These backer boards, in addition to their use as a tile backer, can be used as a floor underlayment, countertop base, heat shield for stoves, and as a base for exterior finishes such as stucco and brick veneer. They are available in 4" × 8" and 3" × 5" panels. Common thicknesses are ¼", ⁷⁄₁₆", and ½".

Architectural (predecorated) gypsum board has a decorated surface that does not require further treatment. The surfaces may be coated, printed, or have a vinyl film. Textured patterns are also available. It requires additional trim, dividers, and corners.

Drywall manufacturers have also improved the capability of gypsum board to resist mold, now that mold growth has been shown to be a severe environmental and health hazard. Mold can grow just about on any surface, but there are gypsum boards that reduce absorption of moisture and slow down the growth of mold. Mold-resistant gypsum comes both as a 1" core board, and as ½" and ⅝" exterior boards. Some manufacturers chemically treat the papering on the gypsum to prevent mold growth. These include Sheetrock® Brand MoldTough™, National Gypsum's Gold Bond Brand XP, and Temple-Inland's Silent-Guard™ TS gypsum shaftliner. Rather than treating the paper, other manufacturers replace it with another mold-resistant material, such as the glass mat facings in Georgia Pacific's Dens-Armor® Plus. FiberRoc® AquaTough™, developed by U.S Gypsum, uses a gypsum-cellulose combination in their panels. The sizes and installations of mold-resistant gypsum board are similar to those of regular board.

WARNING!

When handling mold-resistant gypsum board, be sure to wear proper PPE, including hand, face, and respiratory protection. The chemical treatments and cellulose fibers can get on your skin or be inhaled, and may cause irritation or health problems.

Gypsum core board is available as a 1"-thick solid core board or as a factory-laminated board composed of two ½" boards. It is used in shaft walls and laminated gypsum partitions with additional layers of gypsum board applied to the core board to complete the wall assembly. It is available in a width of 24" and with a variety of edges (square and tongue-and-groove are the most common).

Exterior gypsum ceiling board, or exterior gypsum soffit board, is a weather-resistant panel installed on the soffit side of exteriors indirectly exposed to weather. Applications include ceilings of canopies, carports, or other types of outdoor ceilings. The panels come with regular and fire-rated cores, in ½" or ⅝" thicknesses, and they are designed to prevent ceiling sag. Lighter-weight ceiling panels are used for interiors (½" thick) and for interior ceiling finishing.

Also known as Stretch 54® board, these panels come 6 inches wider than the standard 4'. It is designed for backing the thinner wallpapers and new no-texture finishes done in newer construction. Since two standard-width boards come up one foot short when hanging drywall on nine-foot walls, Stretch 54® boards are wide enough to reduce the number of drywall joints, and make a smoother surface.

Gypsum sheathing is used as a protective, fire-resistant membrane under exterior wall surfacing materials such as wood siding, masonry veneer, stucco, and shingles. It also provides protection against the passage of water and wind and adds structural rigidity to the framing system. The noncombustible core is surfaced with firmly bonded, water-repellent paper. In addition, a water-repellent material may be incorporated in the core. It is available in 2' and 4' widths, and ½" to ⅝" thicknesses. The ⅝"-thick boards are available with Type X cores.

Gypsum board substrate for floor or roof assemblies has a ½"-thick Type X core and is available in 24" or 48" widths. It is used under combustible roof coverings to protect the structure from fires originating on the roof. It can also serve as an underlayment when applied to the top surfaces of floor joists and under the subfloor. It may also be used as a base for built-up roofing applied over steel decks.

Abuse-resistant panels are designed to withstand impacts against the wall, and to prevent indentations and damage from objects. They have strong face paper and backing, and/or can be made with a gypsum-cellulose fiber core. Abuse-resistant gypsum panels are more rigid and have higher structural integrity.

Gypsum base for veneer plaster is used as a base for thin coats of hard, high-strength gypsum veneer plaster. This board is colored light blue.

2.6.0 Masonry Materials

For the purposes of this module, the term *masonry* includes construction using stone, brick, concrete block, and poured concrete. These materials are used extensively in residential and commercial construction. Special tools and fasteners are used with these materials.

2.6.1 Concrete

Concrete is a mixture of four basic materials: portland cement, fine aggregates, coarse aggregates, and water. When first mixed, concrete is in a semi-liquid state and is referred to as plastic concrete. When the concrete hardens, but has not yet gained structural strength, it is called green concrete. After the concrete has hardened and gained its structural strength, it is called cured concrete. Various types of concrete can be obtained by varying the basic materials and/or by adding other materials to the mix. These added materials are called admixtures.

Special-Use Wallboard

INSIDE TRACK

Regular ½" and ⅝" gypsum wallboard are the most common types. There are, however, several types of gypsum wallboard designed for special applications. These include:

- Type X gypsum wallboard provides improved fire ratings because its core material is mixed with fire-retardant additives. Type X is often used on walls that separate occupancies. Examples are walls and ceilings between apartments or a wall separating a garage from the living area of a house. Use of Type X is normally specified by local building codes for protection of occupants.

- Flexible ¼" drywall panels have a heavy paper face and are designed to bend around curved surfaces.

- Special high-strength drywall panels are made for ceiling applications. The core of these panels is specially treated to resist sagging.

- A weather-resistant drywall panel is available for installation on soffits, porch ceilings, and carport ceilings.

Gypsum sheathing panels are used in cases where the required fire rating of exterior walls exceeds that available with OSB, plywood, or other types of sheathing. Gypsum sheathing panels have a water-resistant core covered on both sides with water-repellent paper. Gypsum sheathing panels are widely used in commercial construction.

The desirable properties of concrete in the plastic state are as follows:

- *Moldability* – Plastic concrete may be molded by forms into almost any shape. This is often used to obtain a decorative effect.
- *Portability* – Plastic concrete may be moved in mixing trucks, motorized buggies, wheelbarrows, or by belt conveyors or hydraulic pumps.

The desirable properties of cured concrete are:

- *High structural strength* – Unreinforced concrete has great compressive strength. Reinforced concrete, pre-stressed concrete, or post-tensioned concrete has high structural strength under compression, tension, and lateral pressure.
- *Watertightness* – Although water is used to prepare concrete and concrete can harden under water, properly proportioned and mixed concrete is virtually watertight in most cases.
- *Durability* – Properly mixed and placed concrete usually continues to gain strength for several years and becomes almost as durable and abrasion-resistant as the hardest natural stone.

Portland cement is a finely ground powder consisting of varying amounts of lime, silica, alumina, iron, and other trace components. While dry, it may be moved in bulk or can be bagged in moisture-resistant sacks and stored for relatively long periods of time. Portland cement is a hydraulic cement because it will set and harden by reacting with water with or without the presence of air. This chemical reaction is called hydration and can occur even when the concrete is submerged in water. The reaction creates a calcium silicate hydrate gel and releases heat. This reaction begins the instant water is mixed with the cement and continues as the mixture hardens and cures. The reaction occurs rapidly at first, depending on how finely the cement is ground and what admixtures are present. Then, after its initial cure and strength are achieved, a cement mixture continues to slowly cure over a longer period of time until its ultimate strength is attained.

Because it is in a semi-liquid form when poured, concrete is placed in reinforced forms made of wood, metal, or other materials (*Figure 6*). Concrete floors, walls, and columns can be poured on site. Walls and other structural concrete components are sometimes prefabricated off site and moved to the site on a truck. They are then lifted into place with cranes.

In residential construction, concrete may be used in foundation walls and footings, basement floors, or as the foundation slab if the house has no basement.

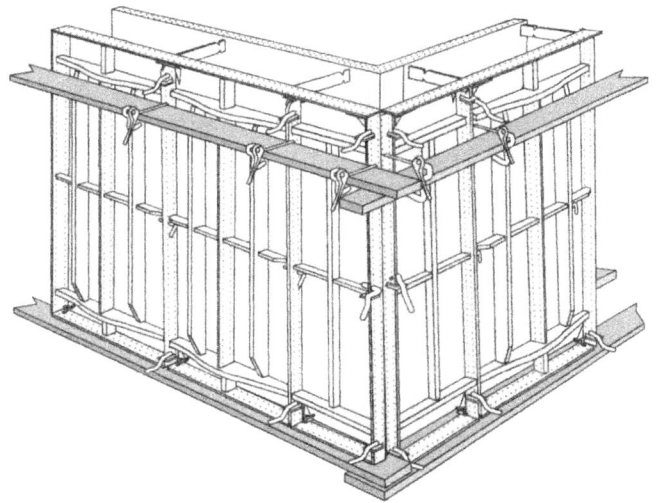

102F06.EPS

Figure 6 Wall form made from EFCO Hand-E-Form® panel system components.

In commercial construction, the entire structure, including floors, walls, and support columns, may be made of concrete. Walls can be anywhere from a few inches to several feet thick.

The ratio of basic ingredients in concrete is determined by a number of variables, such as the application or weather conditions. A common mix for do-it-yourself applications is 3:2:1—one part portland cement, two parts sand, three parts aggregate, with enough water to make the mix workable. In the construction trades, the correct ratio for a given situation is determined much more scientifically, and is usually done by an engineer. Admixtures may be added to affect drying time, increase strength, and add color.

WARNING!

Those working with cement should be aware that it is harmful. Dry cement dust can enter open wounds and cause blood poisoning. When the cement dust comes in contact with body fluids, it can cause chemical burns to the membranes of the eyes, nose, mouth, throat, or lungs. Wet cement or concrete can also cause chemical burns to the eyes and skin. Make sure that appropriate personal protective equipment is worn when working with dry cement or wet concrete. If wet concrete enters waterproof boots from the top, remove the boots and rinse your legs, feet, boots, and clothing with clear water as soon as possible. Repeated contact with cement or wet concrete can cause an allergic reaction in certain individuals.

2.6.2 Concrete Masonry Units (CMUs)

Commonly known as concrete block, concrete masonry units (CMUs) are one of the most common building materials in both residential and commercial construction (*Figure 7*). They are made from a mixture of portland cement, aggregates such as sand and gravel, and water.

Hollow concrete block is used in all kinds of residential and commercial applications. Residential basement walls are usually made of concrete block and it is often used as a base for finish materials such as brick and stucco.

The typical size of a concrete block used in loadbearing construction is 7⅝" wide, 7⅝" high, and 15⅝" long. This is known as an 8" × 8" × 16" unit because it is designed for a ⅜" mortar joint. Mortar is a bonding agent made of cement, fine aggregate such as sand, and water. It is used to provide a watertight bond between blocks. In some cases, CMU construction is reinforced by placing rebar in the openings and then filling the openings with a thinned mortar known as grout.

2.6.3 Brick

Brick is commonly used as a veneer for residential and commercial buildings. Brick is made from pulverized clay that is mixed with water and then molded into various shapes, primarily rectangular. Once the brick hardens and dries, it is fired in a furnace to provide the necessary hardness. Although there are many sizes available, a standard brick is 2¼" × 3¾" × 8".

Like cement block, bricks are bonded together with mortar. Brick is typically laid against a supporting structure such as a concrete block wall or a frame wall sheathed with plywood (*Figure 8*). An air space is maintained between the two walls to allow moisture to escape. A weep hole is provided to drain condensation that develops in the air space. The main difference between conventional frame construction and brick facing is that the foundation wall is extended to provide support for the brick.

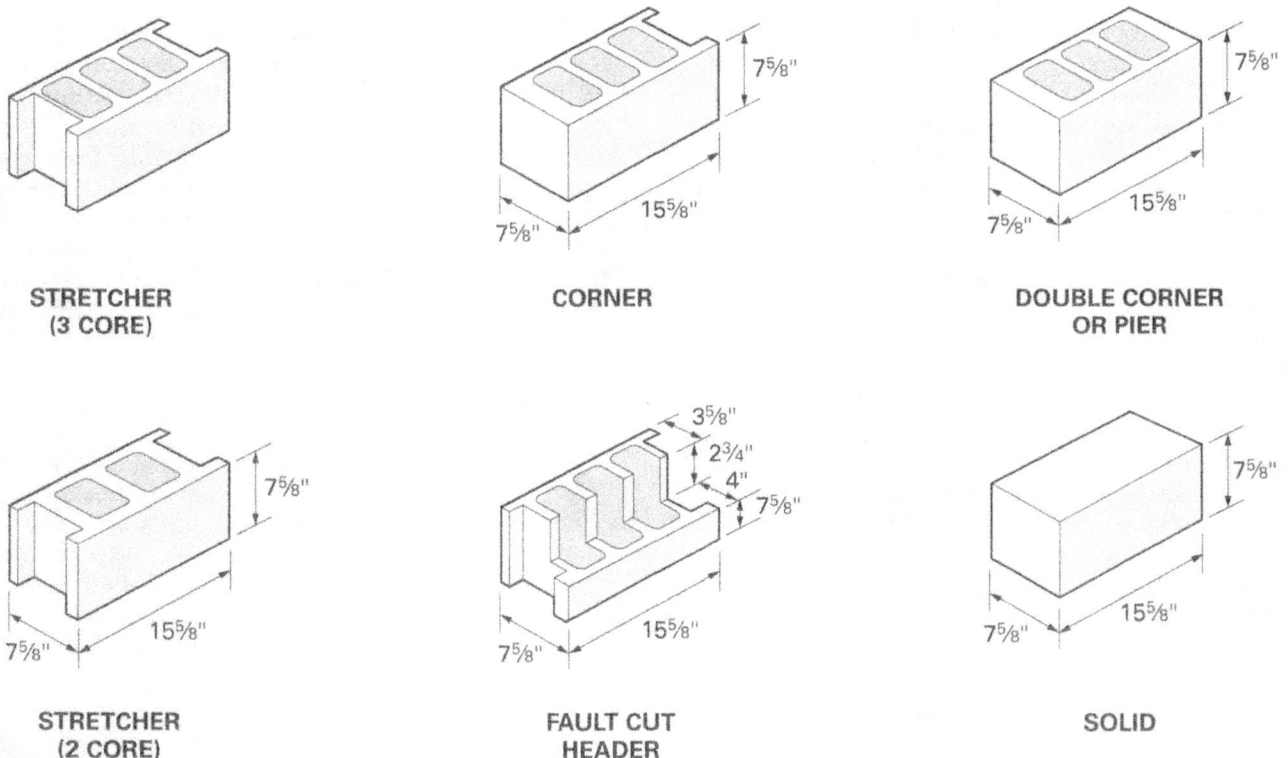

STRETCHER (3 CORE)   **CORNER**   **DOUBLE CORNER OR PIER**

STRETCHER (2 CORE)   **FAULT CUT HEADER**   **SOLID**

102F07.EPS

Figure 7 ❖ Examples of concrete blocks.

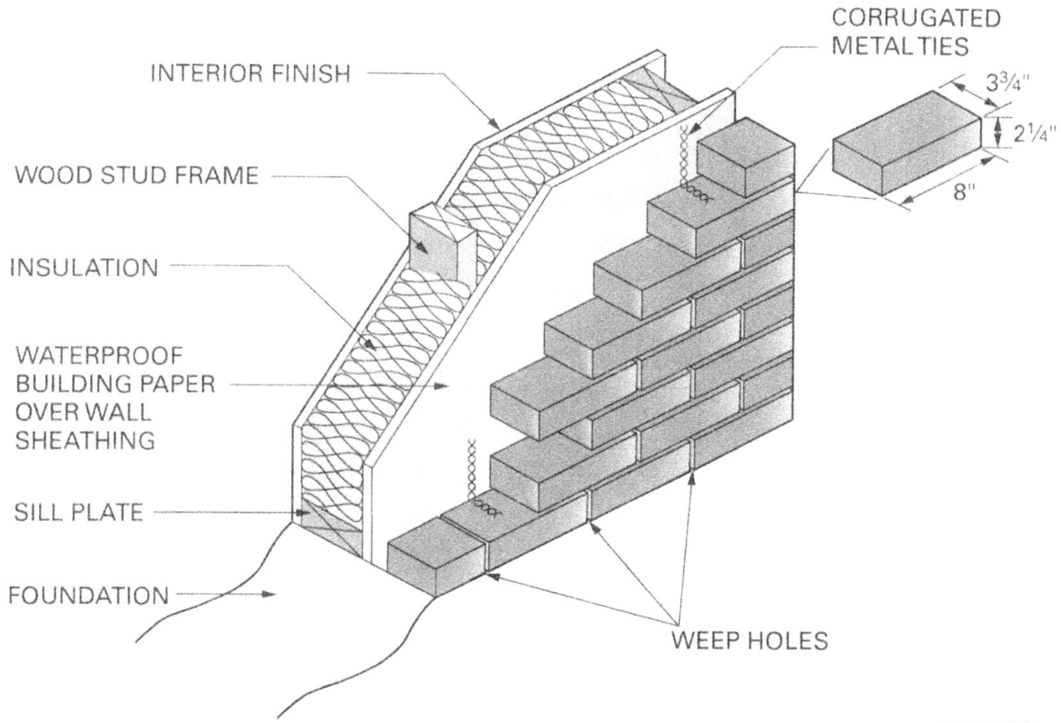

INTERIOR FINISH

WOOD STUD FRAME

INSULATION

WATERPROOF
BUILDING PAPER
OVER WALL
SHEATHING

SILL PLATE

FOUNDATION

CORRUGATED
METAL TIES

3¾"
2¼"
8"

WEEP HOLES

102F08.EPS

Figure 8 Brick veneer wall.

2.6.4 Stone

Like brick, stone is used primarily as a facade over block or frame walls. Stone used for this purpose can be as much as 6" thick. However, in renovating very old homes, you may find stone foundations and walls a foot or more thick, and they are very difficult to drill through. *Figure 9* shows a stone veneer wall with the stone laid in a random pattern.

102F09.EPS

Figure 9 Stone wall.

2.6.5 Metal

Metals have a variety of applications, especially in commercial construction. Lightweight steel and aluminum studs are used in framing walls, floors, and roofs. Metal sheet material is common in walls and roofs of commercial buildings. Corrugated steel decking is used as a base for poured concrete floors in multi-story commercial buildings.

Heavy-gauge structural steel girders and beams are used as the horizontal and vertical support members in many commercial buildings. Steel reinforcing bars and mesh are used to strengthen poured concrete in all applications.

3.0.0 RESIDENTIAL FRAME CONSTRUCTION

Wood frame construction (*Figure 10*) has been in common use since the 1800s. Frame construction begins by building a foundation, which usually consists of a poured concrete footing. In cold climates, the footing must be built below the frost line to prevent it from cracking. If there is no basement, a short foundation wall of poured concrete is set onto poured concrete footings.

Figure 10 ◆ Example of rough carpentry (western platform framing).

It is common to use wooden forms to shape the poured concrete footing (*Figure 11*), and then to use edge forms to make the slab floor. In some cases, the two pours are combined to make a monolithic pour (*Figure 12*). Reinforcing bars are embedded in the concrete both to reinforce it and to provide a connection to the adjoining concrete.

If the house is to have a basement, the basement walls, usually made of concrete block, are set onto the footings (*Figure 13*), which are below ground level.

Reinforced polystyrene foam wall forms (*Figure 14*) have become popular for basement walls because they are easy to build and can be left in place after the wall is poured. In addition, they provide substantial insulation. The concrete walls made with these forms can range from 3" or 4" to 10" thick.

INSIDE TRACK **Framing Methods**

Western platform framing is a method of construction in which a first floor deck is built on top of the foundation walls. Then, the first floor walls are erected on top of the platform. Upper floor platforms are built on top of the first floor walls, and upper floor walls are erected on top of the upper floor platforms. In balloon framing, which is a method seldom used today, the studs extend from the sill plate to the rafter plate. Balloon framing requires the use of much longer studs.

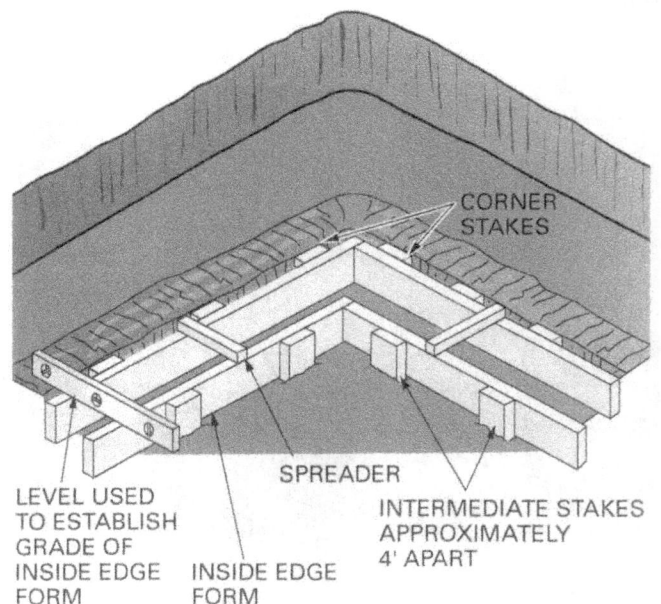

Figure 11 ◆ Foundation form.

SLAB WITH FOUNDATION

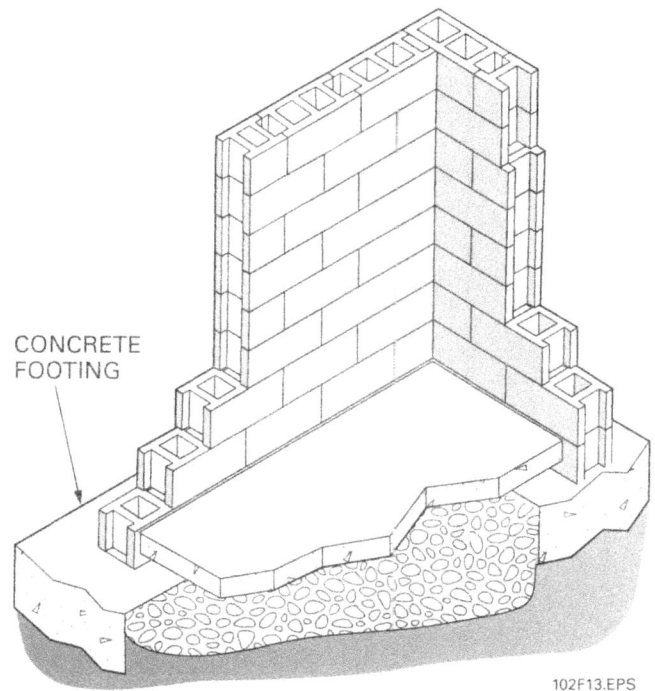

CONCRETE
FOOTING

102F13.EPS

Figure 13 Concrete block basement wall.

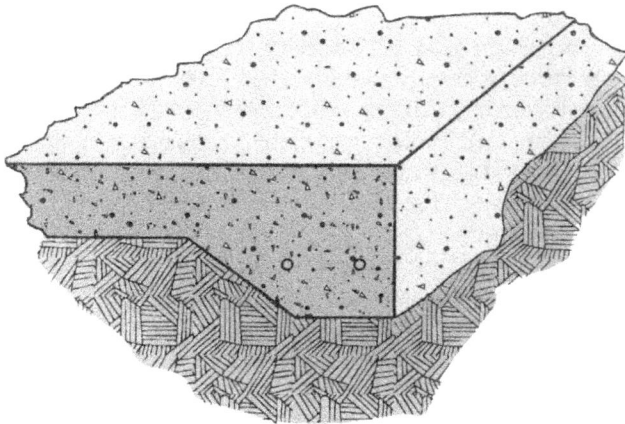

SLAB WITH THICKENED EDGE

102F12.EPS

Figure 12 Types of slabs.

102F14.EPS

Figure 14 Polystyrene form system.

A sill plate, which acts as the anchor for the wood framing, is installed onto the foundation wall (*Figure 15*). Anchor bolts or straps embedded in the concrete are used to attach the sill plate to the foundation. The sills are often made from pressure-treated lumber. Otherwise, a vapor barrier must be placed between the sill and the foundation.

The sills provide a means of leveling the top of the foundation wall and also prevent the other wood framing lumber from making contact with the concrete or masonry, which can cause the lumber to rot.

Today, sills are normally made using a single layer of 2 × 6 lumber. Local codes normally require that pressure-treated lumber and/or foundation-grade redwood lumber be used for the sill whenever it comes into direct contact with any type of concrete. However, where codes allow, untreated softwood can be used.

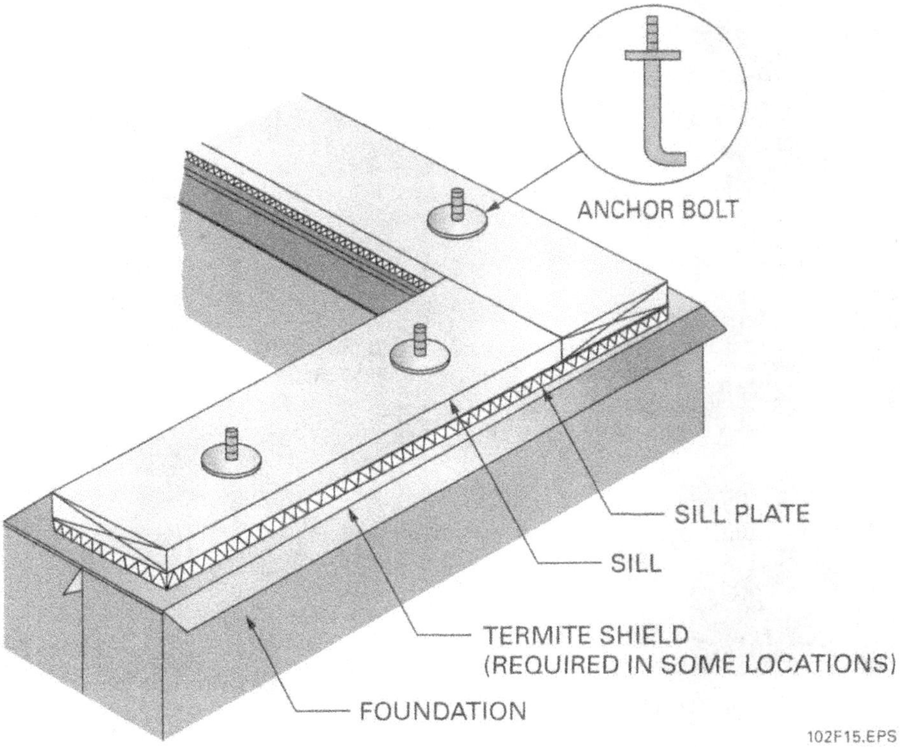

ANCHOR BOLT

SILL PLATE

SILL

TERMITE SHIELD
(REQUIRED IN SOME LOCATIONS)

FOUNDATION

102F15.EPS

Figure 15 ✦ Typical sill installation.

3.1.0 Floor Construction

Floor systems provide a base for the remainder of the structure to rest on. They transfer the weight of people, furniture, and materials from the subfloor, to the floor framing, to the foundation wall, to the footing, then finally to the earth. Floor systems are built over basements or crawl spaces. Single-story structures built on slabs do not have floor systems; however, multi-level structures may have both a slab and a floor system (*Figure 16*).

3.1.1 Girders

Floor joists rest on the sill and provide the support for the floor, as well as an attaching surface for the ceiling of the floor below, if applicable.

The distance between two outside walls is frequently too great to be spanned by a single joist. When two or more joists are needed to cover the span, support for the inboard joist ends must be provided by one or more beams, commonly called girders. Girders carry a very large portion of the weight of the building. They must be well designed, rigid, and properly supported at the foundation walls and on the supporting posts or columns. They must also be installed so that they will properly support the floor joists. Girders

may be made of solid timbers, built-up lumber, engineered lumber, or steel beams. In some instances, precast reinforced concrete girders may be used.

Girders and beams must be properly supported at the foundation walls, and at the proper intervals in between, either by supporting posts, columns, or piers (*Figure 17*). Solid or built-up wooden posts installed on pier blocks are commonly used to support floor girders, especially for floors built over a crawl space.

Four-inch round steel columns filled and reinforced with concrete, called lally columns, are commonly used as support columns in floors built over basements. Some types of lally columns must be cut to the required height, while others have a built-in jack screw that allows the column to be adjusted to the proper height. Metal plates are installed at the top and bottom of the column to distribute the load over a wider area. The plates normally have predrilled holes so that they may be fastened to the girder.

3.1.2 Floor Joists

Floor joists are a series of parallel, horizontal framing members that make up the body of the floor frame (*Figure 16*). They rest on and transfer the building load to the sills and girders.

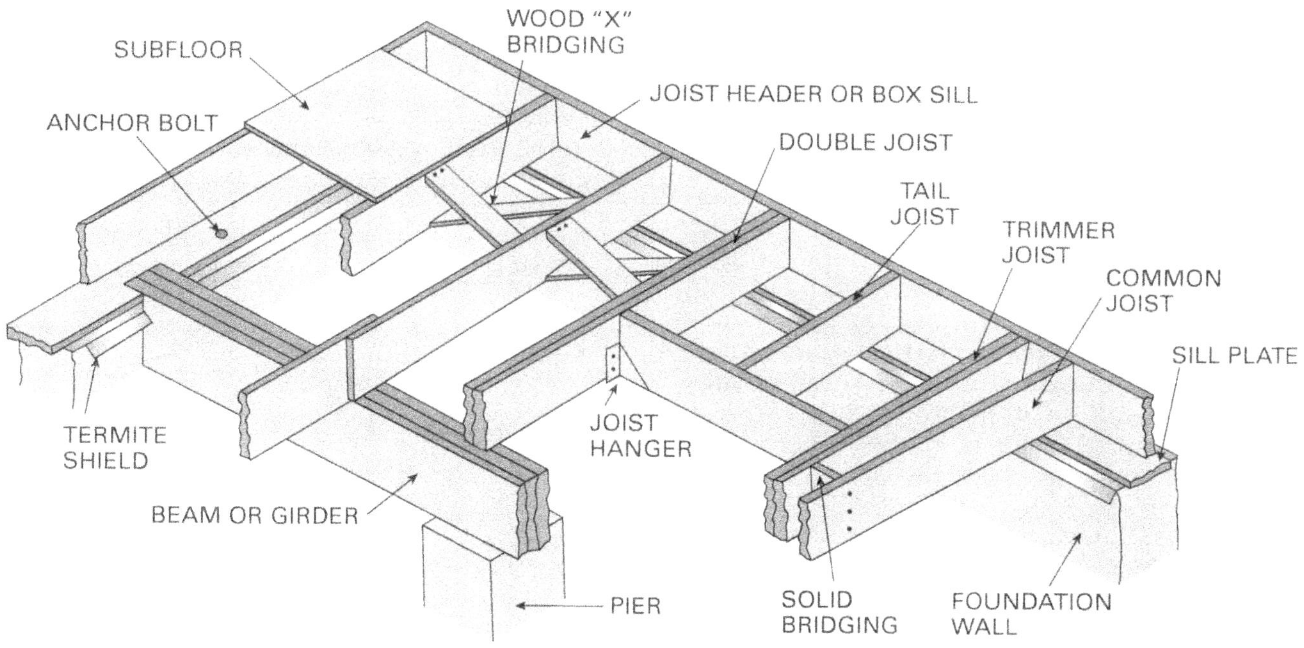

Figure 16 Typical platform frame floor system.

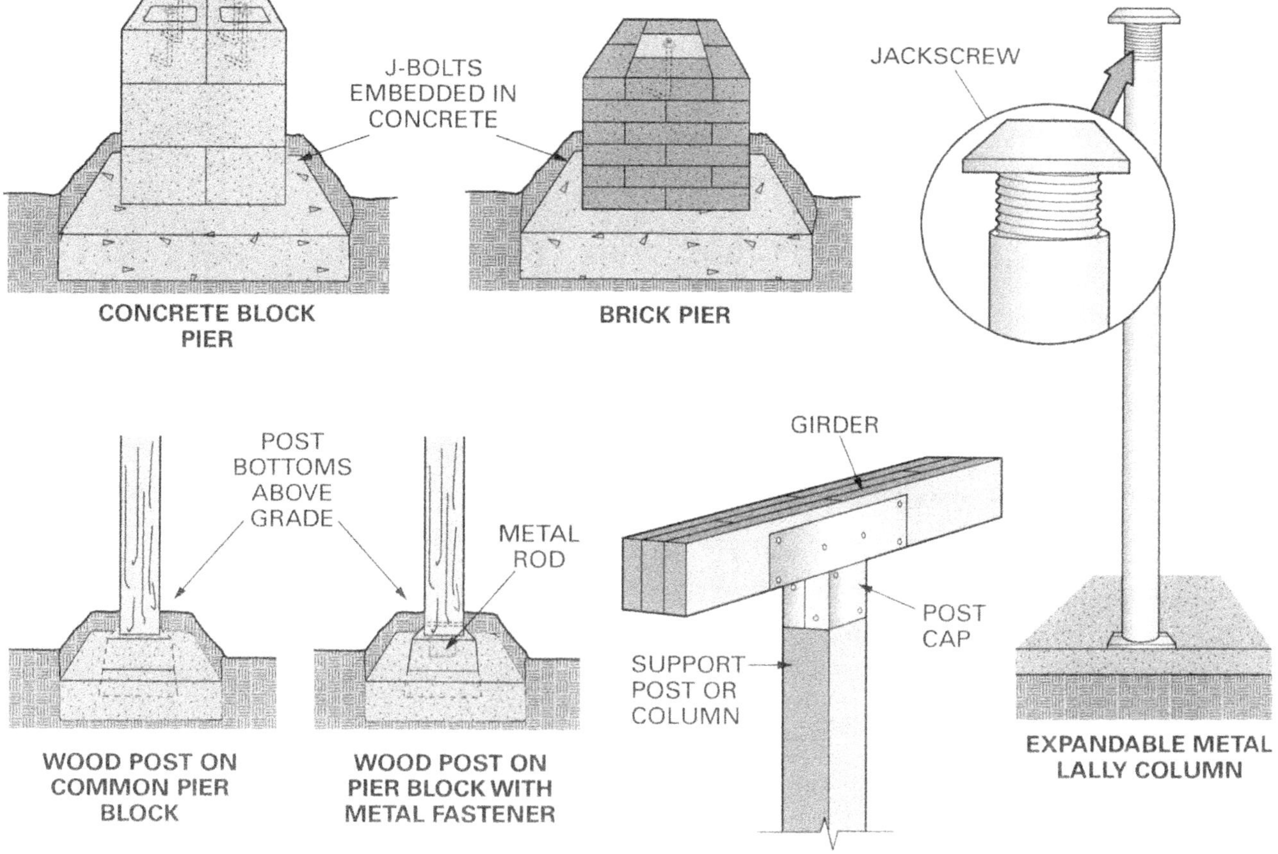

Figure 17 Typical methods of supporting girders.

Joists are normally placed 16" on center (OC). However, there are applications when joists can be set as close as 12" OC or as far apart as 24" OC. These distances are used because they accommodate 4' × 8' subfloor panels and provide a nailing surface where two panels meet. Joists can be supported by the top of the girder or may be framed to the side (*Figure 18*). If joists are lapped over the girder, the minimum amount of lap is 4" and the maximum amount of lap is 12".

There are many different types of joist hangers that can be used to fasten joists to girders and other support framing members. Joist hangers are used where the bottom of the girder must be flush with the bottoms of the joists. At the sill end of the joist, the joist should rest on at least 1½" of wood. In platform construction, the ends of all the joists are fastened to a header joist, also called a band joist or rim joist, to form the box sill.

Joists are doubled where extra loads need to be supported. When a partition runs parallel to the joists, a double joist is placed underneath. Joists must also be doubled around all openings in the floor frame for stairways, chimneys, etc., to reinforce the rough opening in the floor. These additional joists used at such openings are called trimmer joists.

In residential construction, floors traditionally have been built using wooden joists. However, the use of prefabricated engineered wood products such as wood I-beams and various types of trusses is also becoming common.

3.1.3 Wood I-Beams

Wood I-beam joists are typically manufactured with 1½" diameter, pre-stamped knockout holes in the web about 12" OC that can be used to accommodate wiring. Other holes or openings can be cut into the web, but these can only be of a certain size and at the locations specified by the I-beam manufacturer. Under no circumstances should the flanges of I-beam joists be cut or notched because it will weaken the joist.

3.1.4 Trusses

Trusses are manufactured joist assemblies made of wood or a combination of steel and wood (*Figure 19*). Solid light-gauge steel and open-web steel trusses are also made, but these are used mainly in commercial construction. Like the

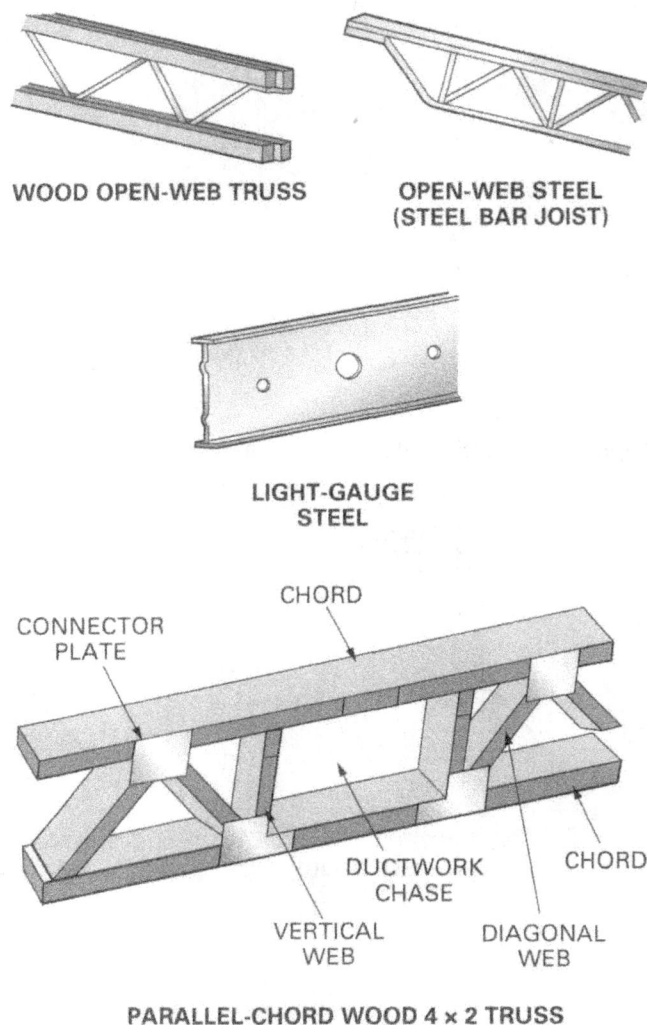

WOOD OPEN-WEB TRUSS

OPEN-WEB STEEL
(STEEL BAR JOIST)

LIGHT-GAUGE
STEEL

CONNECTOR PLATE
CHORD
DUCTWORK CHASE
CHORD
VERTICAL WEB
DIAGONAL WEB

PARALLEL-CHORD WOOD 4 × 2 TRUSS

102F19.EPS

Figure 19 ♦ Typical floor trusses.

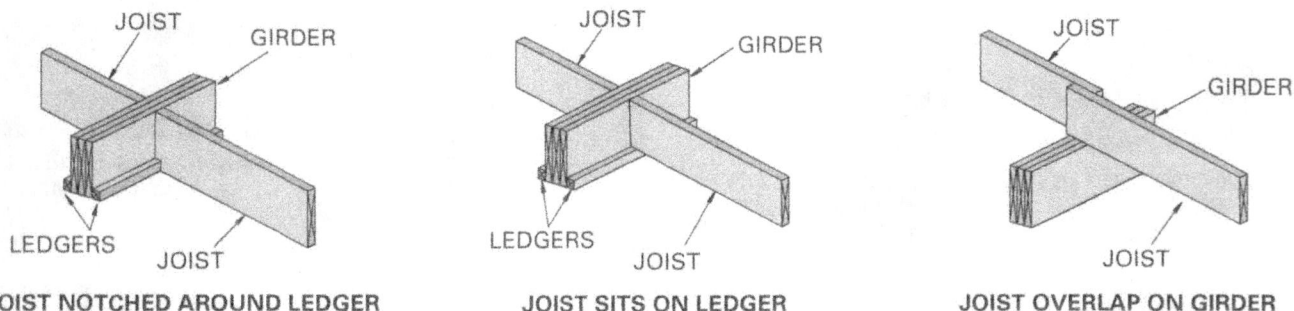

JOIST
GIRDER
LEDGERS
JOIST

JOIST NOTCHED AROUND LEDGER

JOIST
GIRDER
LEDGERS
JOIST

JOIST SITS ON LEDGER

JOIST
GIRDER
JOIST

JOIST OVERLAP ON GIRDER

102F18.EPS

Figure 18 ♦ Methods of joist framing at a girder.

wood I-beams, trusses are stronger than comparable lengths of dimension lumber, allowing them to be used over longer spans. Longer spans allow more freedom in building design because interior load bearing walls and extra footings can often be eliminated. Trusses are generally faster and easier to erect, with no need for trimming or cutting in the field. They also provide the additional advantage of permitting ductwork, plumbing, and wiring to be run easily between the open webs (*Figure 20*).

3.1.5 Notching and Drilling of Wooden Joists

When it is necessary to notch or drill through a floor joist, most building codes will specify how deep a notch can be made. For example, the *International Building Code (IBC) or NFPA 5000 Building*

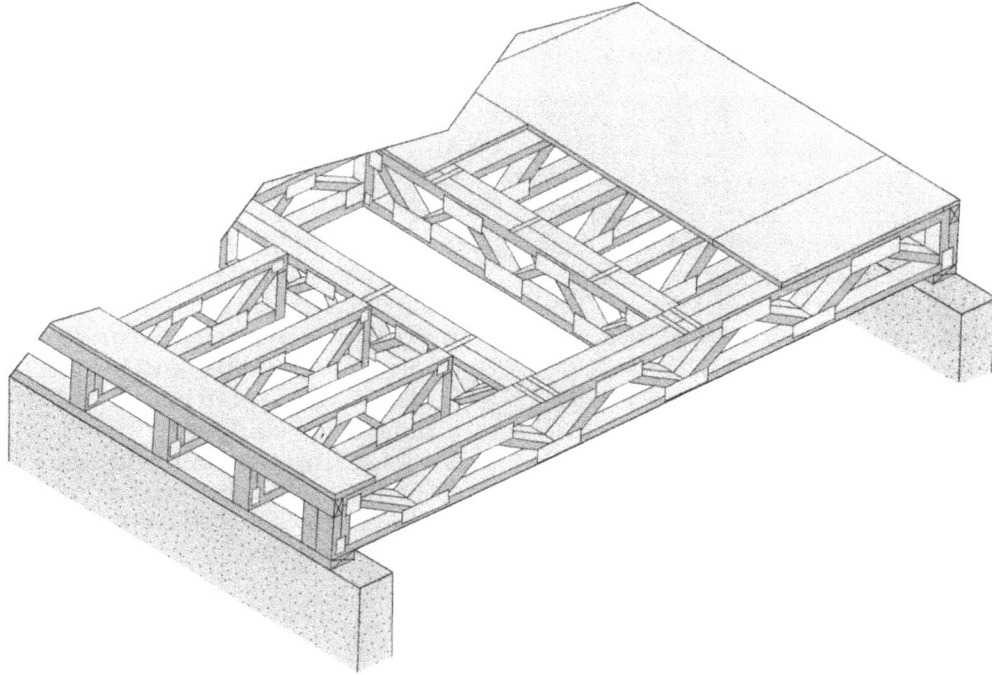

102F20.EPS

Figure 20 ♦ Typical floor system constructed with trusses.

I-Beams

The first plywood I-beam was created in 1969. In 1977, the first I-beam was created using LVL. This new construction offered superior strength and stability. In 1990, OSB web material, constructed of interlocking fibers, began to be used in I-beams, as shown here. OSB is less expensive than plywood and is not as prone to warping or cracking. Engineered wood products were once only available through a handful of companies that pioneered the industry. Today, engineered lumber and lumber systems are offered by a wide variety of companies.

102SA01.EPS

Code specifies that notches on the ends of joists shall not exceed one-fourth the depth. Therefore, in a 2 × 10 floor joist, the notch could not exceed 2½" (*Figure 21*).

This code also states that notches for pipes in the top or bottom shall not exceed one-sixth the depth, and shall not be located in the middle third of the span. Therefore, when using a 2 × 10 floor joist, a notch cannot be deeper than 1⅜". This notch can be made either in the top or bottom of the joist, but it cannot be made in the middle third of the span. This means that if the span is 12', the middle span from 4' to 8' may not be notched.

3.1.6 Bridging

Bridging is used to stiffen the floor frame and to enable an overloaded joist to receive some support from the joists on either side. Most building codes require that bridging be installed in rows between the floor joists, at intervals of not more than 8'. For example, floor joists with spans of 8' to 16' need one row of bridging in the center of the span.

Three types of bridging (*Figure 22*) are commonly used: wood cross-bridging, solid wood bridging, and metal cross-bridging. Wood and metal cross-bridging are composed of pieces of wood or metal set diagonally between the joists to form an X. Wood cross-bridging is typically 1 × 4 lumber placed in double rows that cross each other in the joist space.

Metal cross-bridging is installed in a similar manner. Metal cross-bridging comes in a variety of styles and different lengths for use with a particular joist size and spacing. It is usually made of 18-gauge steel and is ¾" wide. Solid bridging, also called blocking, consists of solid pieces of lumber (usually the same size as the floor joists) installed between the joists. The bridging pieces are offset from one another to enable end nailing.

3.1.7 Subflooring

Subflooring consists of panels or boards laid directly on and fastened to floor joists (*Figure 23*) in order to provide a base for underlayment and/or the finish floor material. Underlayment is a material, such as particleboard or plywood, laid on top of the subfloor to provide a smoother surface for finish flooring. The subfloor adds rigidity to the structure and provides a surface upon which walls and other framing can be laid out and constructed. Subfloors also act as a barrier to cold and dampness, thus keeping the building warmer and drier in winter. Subflooring can be constructed of plywood, OSB or other manufactured board panels, or common wooden boards.

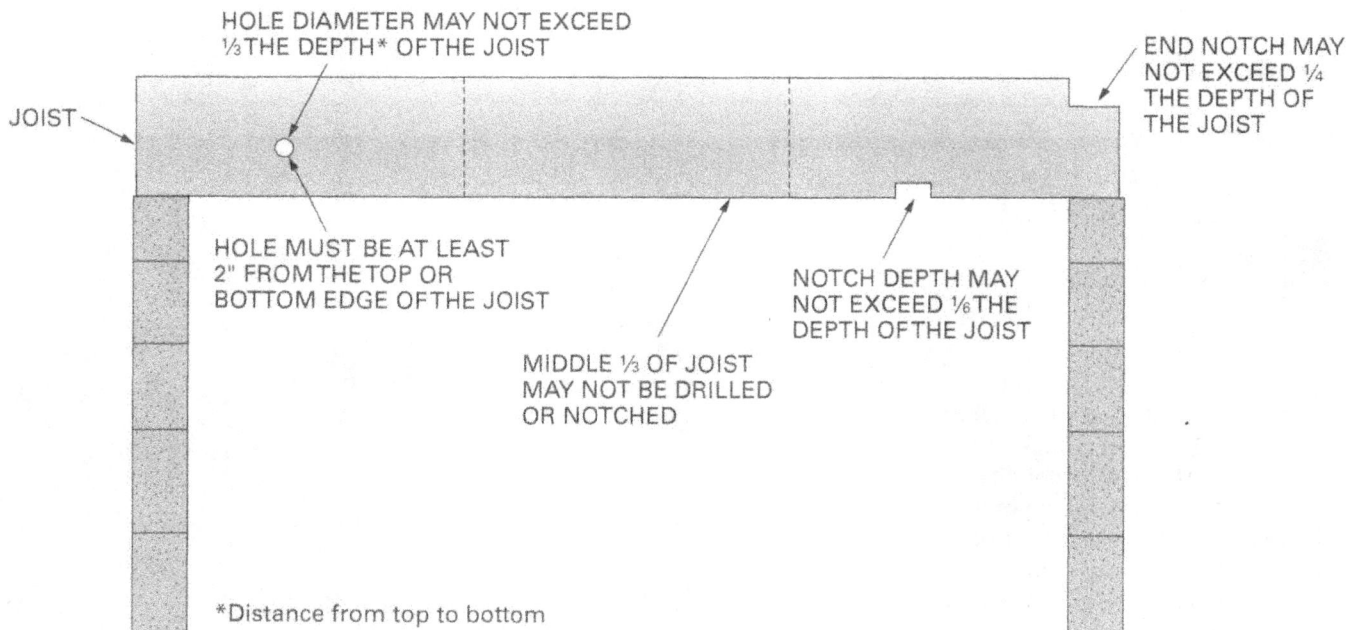

HOLE DIAMETER MAY NOT EXCEED ⅓ THE DEPTH* OF THE JOIST

END NOTCH MAY NOT EXCEED ¼ THE DEPTH OF THE JOIST

JOIST

HOLE MUST BE AT LEAST 2" FROM THE TOP OR BOTTOM EDGE OF THE JOIST

NOTCH DEPTH MAY NOT EXCEED ⅙ THE DEPTH OF THE JOIST

MIDDLE ⅓ OF JOIST MAY NOT BE DRILLED OR NOTCHED

*Distance from top to bottom

102F21.EPS

Figure 21 ❖ Notching and drilling of wooden joists.

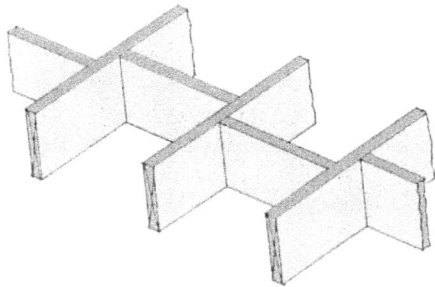

WOOD CROSS-BRIDGING

SOLID WOOD BRIDGING

STEEL CROSS-BRIDGING

102F22.EPS

Figure 22 Types of bridging.

3.2.0 Wall Construction

Wall framing is generally done with 2 × 4 studs spaced 16" OC. In many cases, 24" spacing is used on interior walls. Some codes permit 24" spacing on exterior walls for one-story buildings. If 24" spacing is used in a two-story building, the lower floor must be framed with 2 × 6 lumber.

Figure 24 identifies the structural members of a wood frame wall. Each of the members shown on the illustration is described here.

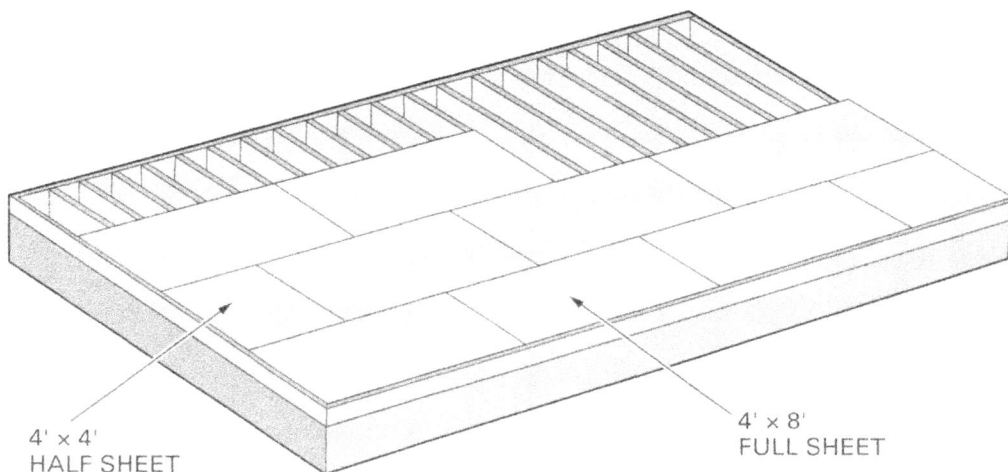

4' × 4'
HALF SHEET

4' × 8'
FULL SHEET

102F23.EPS

Figure 23 Subflooring installation.

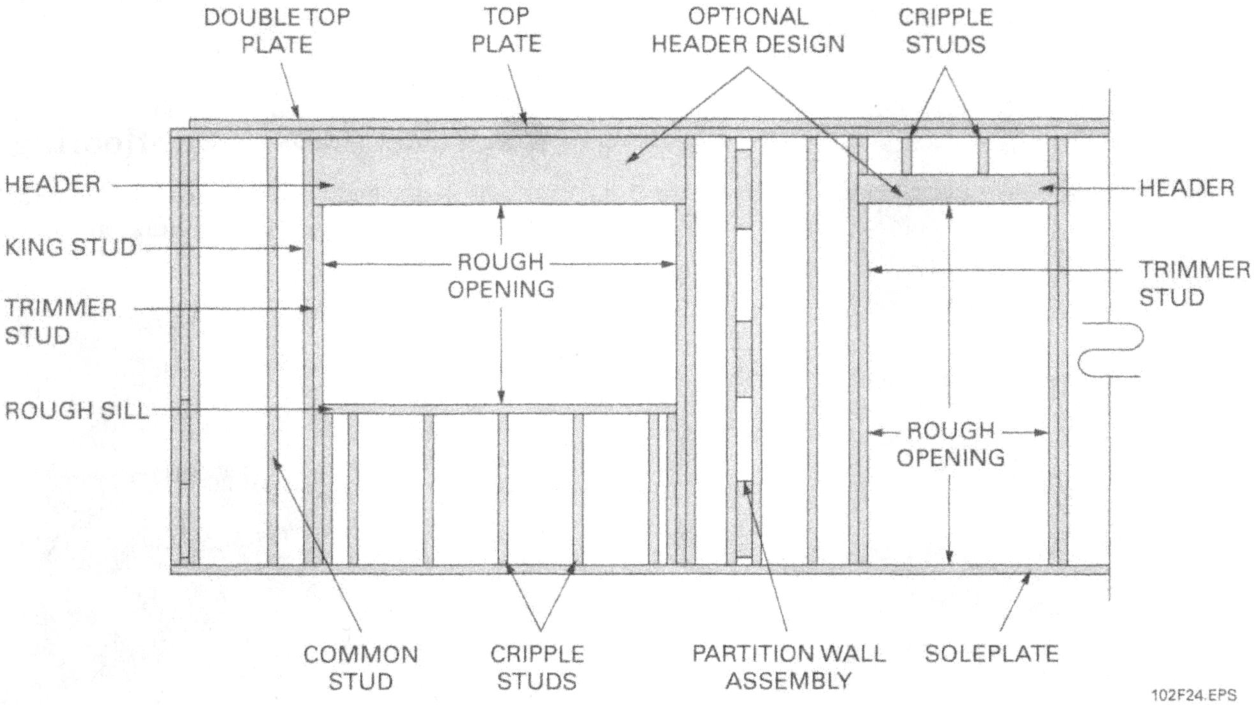

Figure 24 ❖ Wall and partition framing members.

> **NOTE**
>
> Codes may require variations in structure for seismic and other hazards.

- *Blocking (spacer)* – A wood block that is used as a filler piece and support between framing members. Blocking also provides a surface for attaching equipment, etc.
- *Cripple stud* – In wall framing, this is a short framing stud that fills the space between a header and a top plate or between the sill and the soleplate.
- *Double top plate* – This is a plate made of two members to provide better stiffening of a wall. It is also used for connecting splices, corners, and partitions that are at right angles (perpendicular) to the wall.
- *Header (lintel)* – This is a horizontal structural member that supports the load over an opening such as a door or window.
- *King stud* – This is the full-length stud next to the trimmer stud in a wall opening.
- *Partition* – This is a wall that subdivides space within a building. A bearing partition or wall is one that supports the floors and roof directly above in addition to its own weight.
- *Rough opening* – This is an opening in the framing formed by framing members, usually for a window or a door.

- *Rough sill* – This is the lower framing member attached to the top of the lower cripple studs to form the base of a rough opening for a window.
- *Soleplate* – This is the lowest horizontal member of a wall or partition to which the studs are nailed. It rests on the rough floor.
- *Stud* – The main vertical framing member in a wall or partition.
- *Top plate* – The upper horizontal framing member of a wall used to carry the roof trusses or rafters.
- *Trimmer stud* – The vertical framing member that forms the sides of rough openings for doors and windows. It provides stiffening for the frame and supports the weight of the header.

3.2.1 Corners

A wall must have solid corners that can take the weight of the structure. In addition to contributing to the strength of the structure, corners must provide a good nailing surface for sheathing and interior finish materials. Building contractors generally select the straightest, least defective studs for corner framing.

There are many methods for constructing corners (*Figure 25*). Some builders will construct the corner in place, then plumb and brace it before raising the wall frames. This approach makes it easier to plumb and brace the frame, but it prevents installation of the sheathing before the

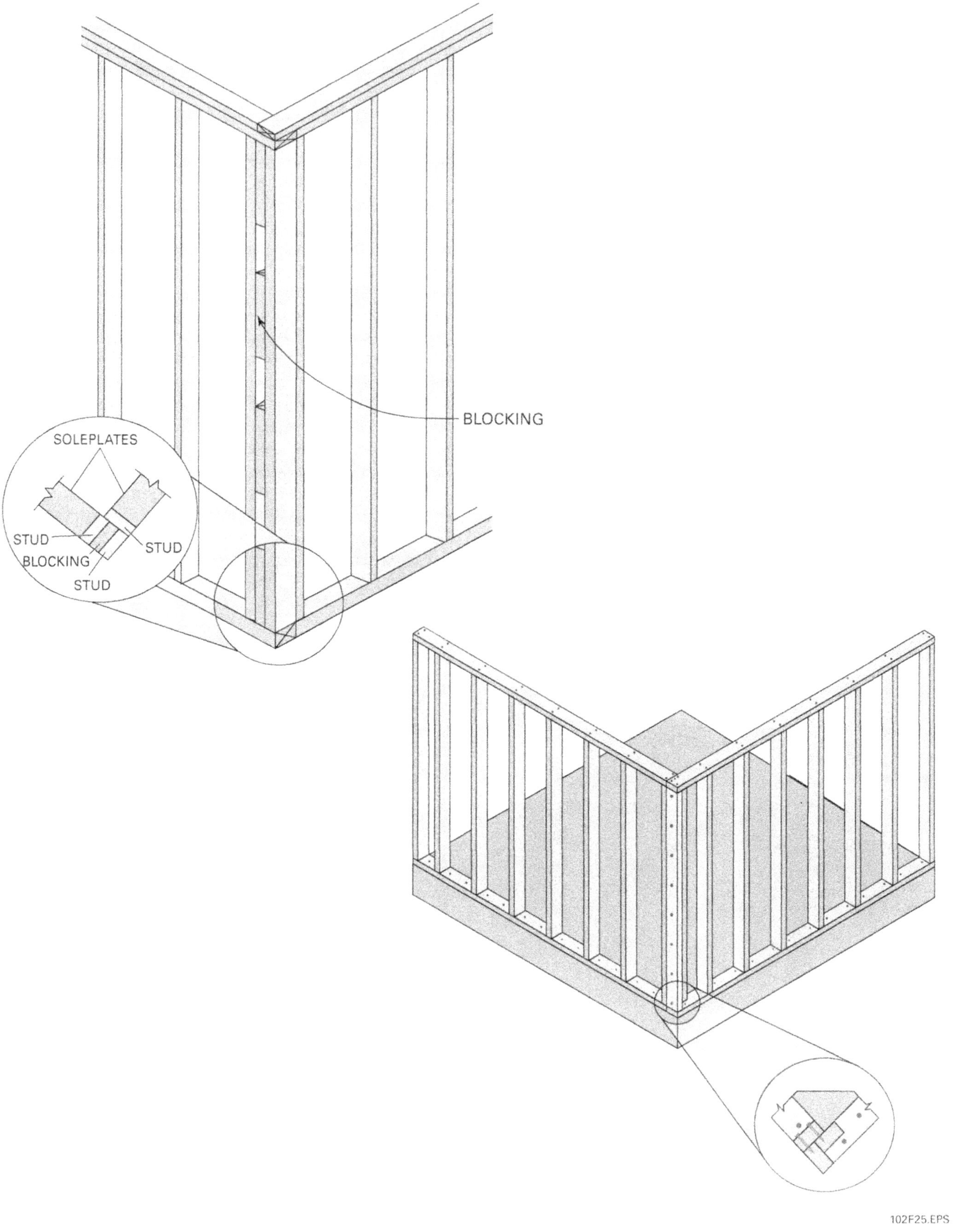

BLOCKING

SOLEPLATES

STUD

STUD

BLOCKING

STUD

102F25.EPS

Figure 25   Corner construction.

frame is erected. If the corners are included in the frame, then a portion of the corner is included with each of the mating frame sections.

3.2.2 Partition Intersections

Interior partitions must be securely fastened to outside walls. For that to happen, there must be a solid nailing surface where the partition intersects the exterior frame. There are several methods used to construct framing for partition Ts (*Figure 26*).

3.2.3 Window and Door Openings

When wall framing is interrupted by an opening such as a window or door, a method is needed to distribute the weight of the structure around the opening. This is done by the use of a header. The header is placed so that it rests on the trimmer studs, which transfer the weight to the soleplate or subfloor and then to the foundation.

Headers are made of solid or built-up lumber. Laminated lumber and beams have become popular as header material, especially where the load is heavy.

Built-up headers are usually made from 2" lumber separated by ½" plywood spacers (*Figure 27*).

A full header is used for large openings and fills the area from the rough opening to the bottom of the top plate. A small header with cripple studs is suitable for average-size windows and doors and is usually made from 2 × 4 or 2 × 6 lumber. Built-up headers are sometimes made by gluing and nailing ½" plywood the entire length of the header, instead of inserting plywood blocks. This method allows the framing crew to make a long section (16') of built-up header, then cut what they need for each opening from that section. The crew may use the same header for all openings. This saves time because it eliminates the need for cutting and installing cripple studs.

Truss headers are used when the load is especially heavy or the span is extra wide. The design of the trusses is generally included in the architect's plans.

Other types of headers used for heavy loads are wood or steel I-beams and box beams. The latter are made of plywood webs connected by lumber flanges in a box configuration.

The width of a header is equal to the rough opening plus the thickness of the trimmer studs. For example, if the rough opening for a 3'-wide window is 38" and the thickness of the trimmer studs measures 3", the width of the header would be 41".

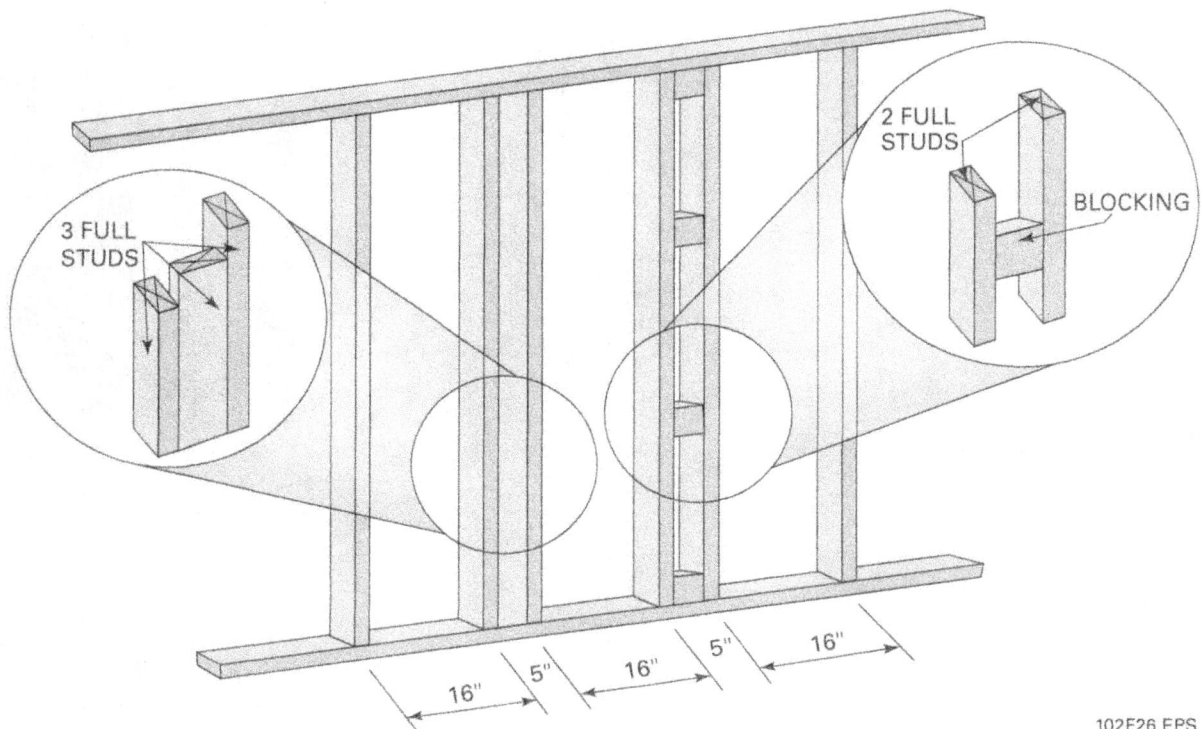

102F26.EPS

Figure 26 ◆ Constructing nailing surfaces for partitions.

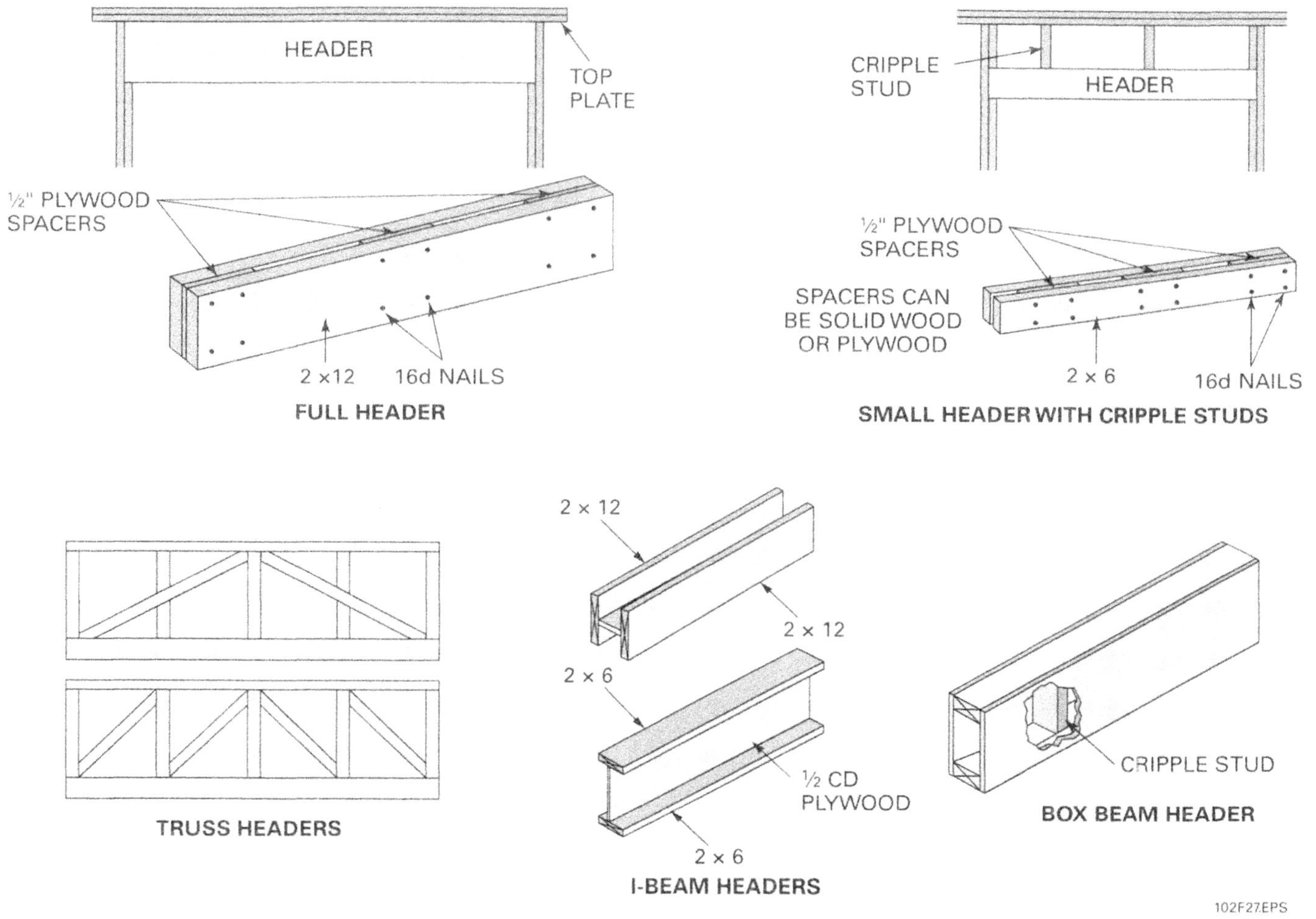

HEADER
TOP PLATE
½" PLYWOOD SPACERS
2 × 12   16d NAILS
FULL HEADER

CRIPPLE STUD
HEADER
½" PLYWOOD SPACERS
SPACERS CAN BE SOLID WOOD OR PLYWOOD
2 × 6   16d NAILS
SMALL HEADER WITH CRIPPLE STUDS

TRUSS HEADERS

2 × 12
2 × 12
2 × 6
½ CD PLYWOOD
2 × 6
I-BEAM HEADERS

CRIPPLE STUD
BOX BEAM HEADER

102F27.EPS

Figure 27 Types of headers.

Figure 28 shows cross sections of typical wood-framed walls.

3.2.4 Firestops

In some areas, local building codes may require firestops. Firestops are short pieces of 2 × 4 blocking (or 2 × 6 pieces if the wall is framed with 2 × 6 lumber) that are nailed between studs (*Figure 29*).

Without firestops, the space between the studs will act like a flue in a chimney. Any holes drilled through the soleplate and top plate create a draft, and air will rush through the space. In a fire, air, smoke, gases, and flames can race through the chimney-like space.

The installation of firestops has two purposes. First, it slows the flow of air, which feeds a fire through the cavity. Second, it can temporarily block flames from traveling up through the cavity.

If the local code requires firestops, it may also require that holes through the soleplate and top plate (for plumbing or electrical runs) be plugged with a firestopping material to prevent airflow.

3.2.5 Bracing

Bracing is important in the construction of exterior walls. Many local building codes require bracing if certain types of sheathing are used. In areas where high winds or earthquakes are a hazard, lateral bracing may be required even when ½" plywood is used as the sheathing.

Several methods of bracing have been used since the early days of construction. One method is to cut a notch (let-in) for a 1×4 or 1×6 at a 45-degree angle on each corner of the exterior walls. Another method is to cut 2×4 braces at a 45-degree angle for each corner. Still another type of bracing (used where permitted by the local code), is metal strap bracing (*Figure 30*). This product is made of galvanized steel.

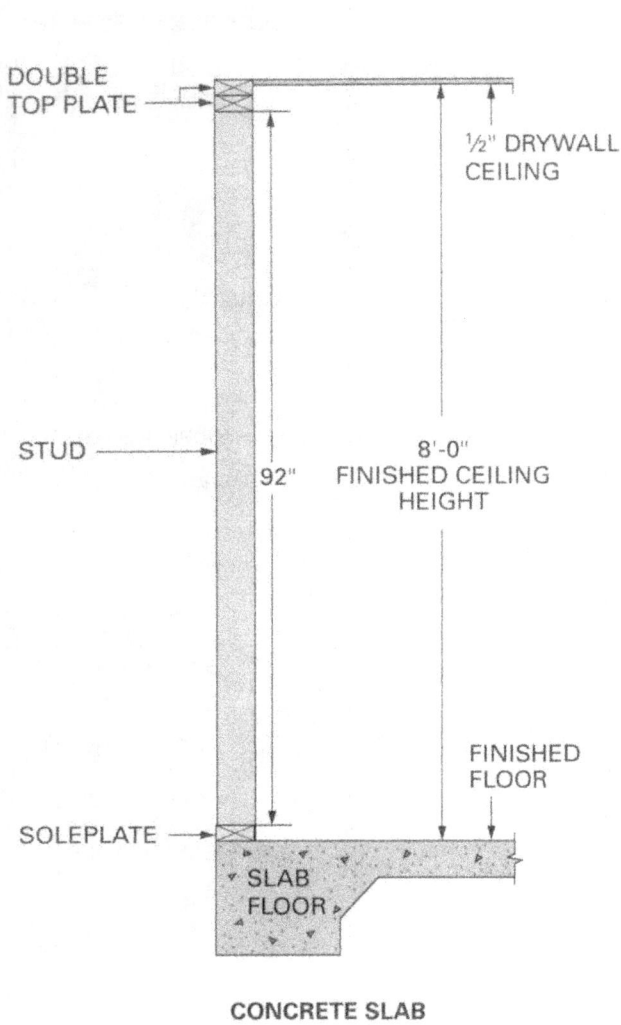

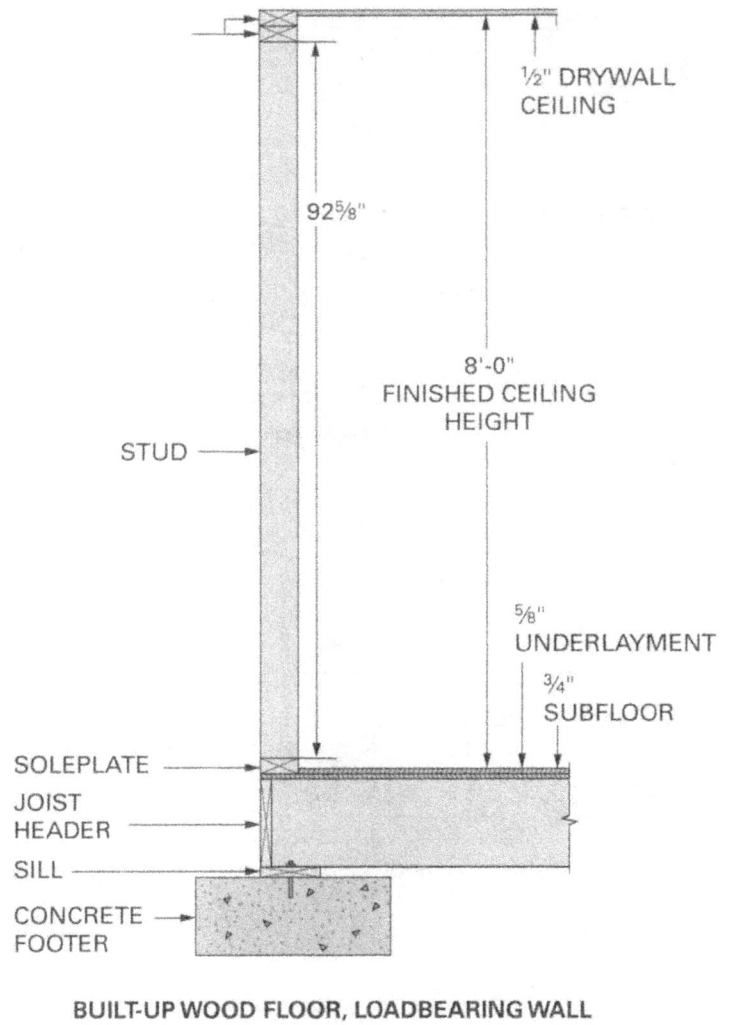

Figure 28 ◆ Cross sections of wood-framed walls.

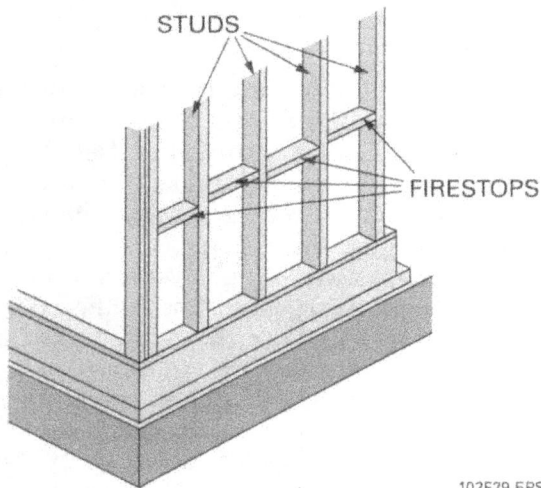

Concrete Slab diagram labels:
- DOUBLE TOP PLATE
- STUD
- SOLEPLATE
- 92"
- ½" DRYWALL CEILING
- 8'-0" FINISHED CEILING HEIGHT
- FINISHED FLOOR
- SLAB FLOOR
- CONCRETE SLAB

Built-up Wood Floor, Loadbearing Wall diagram labels:
- ½" DRYWALL CEILING
- 92⅝"
- STUD
- 8'-0" FINISHED CEILING HEIGHT
- ⅝" UNDERLAYMENT
- ¾" SUBFLOOR
- SOLEPLATE
- JOIST HEADER
- SILL
- CONCRETE FOOTER
- BUILT-UP WOOD FLOOR, LOADBEARING WALL

102F28.EPS

Figure 29 ◆ Firestops.

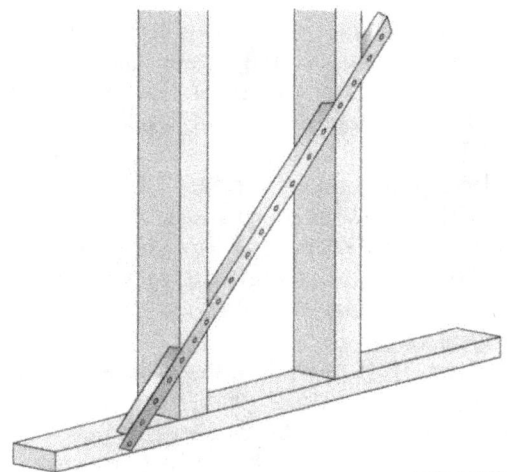

- STUDS
- FIRESTOPS

102F29 EPS

Figure 30 ◆ Metal bracing.

102F30.EPS

Headers

Headers can be constructed in many ways, some of which are shown here.

HEADER WITH CRIPPLE STUDS

SOLID HEADER

GARAGE DOOR HEADER USING
ENGINEERED LUMBER

102SA03.EPS

Metal strap bracing is easier to use than let-in wood bracing. Instead of notching out the studs for a 1 × 4 or 2 × 4, a circular saw is used to make a diagonal groove in the studs, top plate, and sole-plate for the rib of the bracing strap. The strap is then nailed to the framing.

With the introduction of plywood, some areas of the country have done away with corner bracing. However, along with plywood came different types of sheathing that are byproducts of the wood industry and do not have the strength to withstand wind pressures. When these are used, permanent bracing is needed. Building codes in some areas will allow a sheet of ½" plywood to be used on each corner of the structure in lieu of diagonal bracing when the balance of the sheathing is fiberboard. In other areas, the use of bracing is required regardless of the type of sheathing used. Always check local codes.

3.2.6 Sheathing

Sheathing is the material used to close in the walls. APA-rated material, such as plywood and non-veneer panels, are generally used for sheathing.

When plywood is used, the panels will range from ⁵⁄₁₆" to ¾" thick. A minimum thickness of ⅜" is recommended when siding is to be applied.

The higher end of the range is recommended when the sheathing acts as the exterior finish surface. The panels may be placed with the grain running horizontally or vertically. If they are placed with the grain running horizontally, local building codes may require that blocking be used along the top edges.

Typical nailing requirements call for 6d (6 penny) nails for panels ½" thick or less and 8d nails for thicker panels. Nails are spaced 6" apart at the panel edges and 12" apart at intermediate studs.

Other materials that are sometimes used as sheathing are fiberboard (insulation board), rigid foam sheathing, and exterior-rated gypsum wallboard. A major disadvantage of these materials is that siding cannot be nailed to them. It must either be nailed to the studs or special fasteners must be used.

When material other than rated panels is used as sheathing, rated plywood panels may be installed vertically at the corners to eliminate the need for corner bracing.

3.3.0 Ceiling Construction

Ceiling joists are usually laid across the width of a building at the same positions as the wall studs.

The length of a joist is the distance from the outside edges of the double top plates. The ends of the joists are cut to match the rafter pitch so that the roof sheathing will lay flush on the framing (*Figure 31*). If the joist exceeds the allowable span, two pieces of joist material must be spliced over a bearing wall or partition. *Figure 32* shows two splicing methods. There should be a minimum overlap of 6". Another method of splicing is to place the two joists on either side of the rafter with a piece of blocking between the joists at the splice.

If the spacing is the same as that of the wall studs, the joists are nailed directly above the studs. This makes it easier to run ductwork, piping, flues, and wiring above the ceiling. Metal joist hangers can be used in place of nailing.

After the joists are installed, a ribband or strongback is nailed across them to prevent twisting or bowing (*Figure 33*). The strongback is used for larger spans. In addition to holding the joists in line, it provides support for the joists at the center of the span.

(A) CEILING JOISTS LAPPED OVER BEARING PARTITION

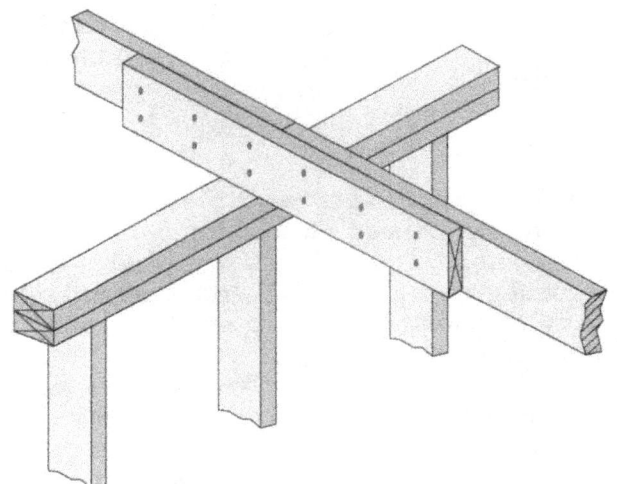

HEIGHT OF BACK OF RAFTER ABOVE PLATE

102F31.EPS

Figure 31 ♦ Cutting joist ends to match the roof pitch.

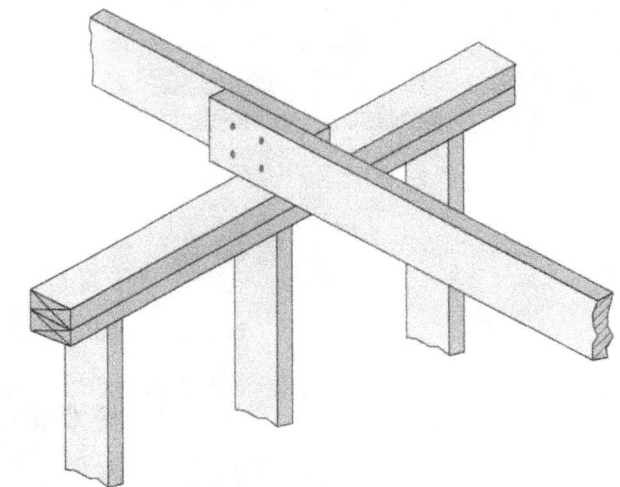

(B) CEILING JOISTS BUTTED OVER BEARING PARTITION

102F32.EPS

Figure 32 ♦ Spliced ceiling joists.

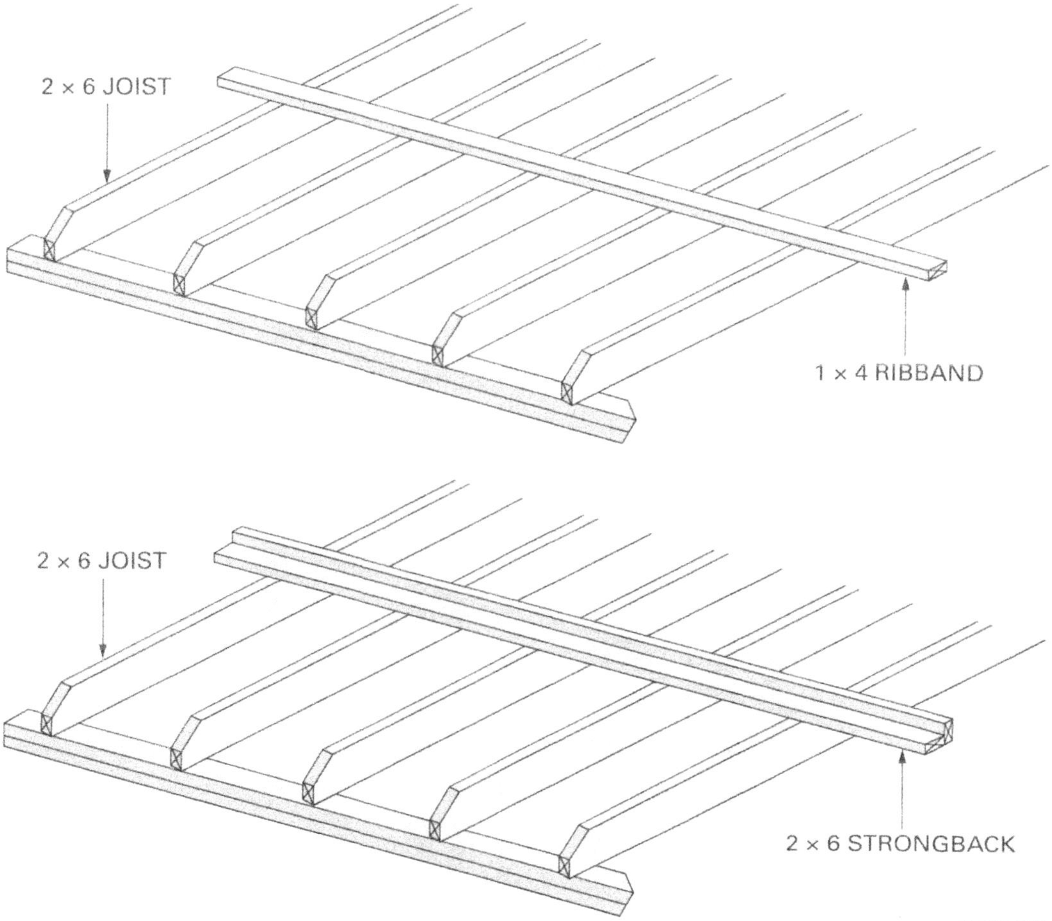

Figure 33   Reinforcing ceiling joists.

3.4.0 Roof Construction

The most common types of roofs used in residential construction (*Figure 34*) are described here.

- *Gable roof* – A gable roof has two slopes that meet at the center (ridge) to form a gable at each end of the building. It is the most common type of roof because it is simple, economical, and can be used on any type of structure.
- *Hip roof* – A hip roof has four sides or slopes running toward the center of the building. Rafters at the corners extend diagonally to meet at the ridge. Additional rafters are framed into these rafters.
- *Gable and valley roof* – This roof consists of two intersecting gable roofs. The part where the two roofs meet is called a valley.
- *Mansard roof* – The mansard roof has four sloping sides, each of which has a double slope. As compared with a gable roof, this design provides more available space in the upper level of the building.

INSIDE TRACK   **Mansard Roof**

Mansard roofs are popular in the design of many fast-food restaurants and small office buildings.

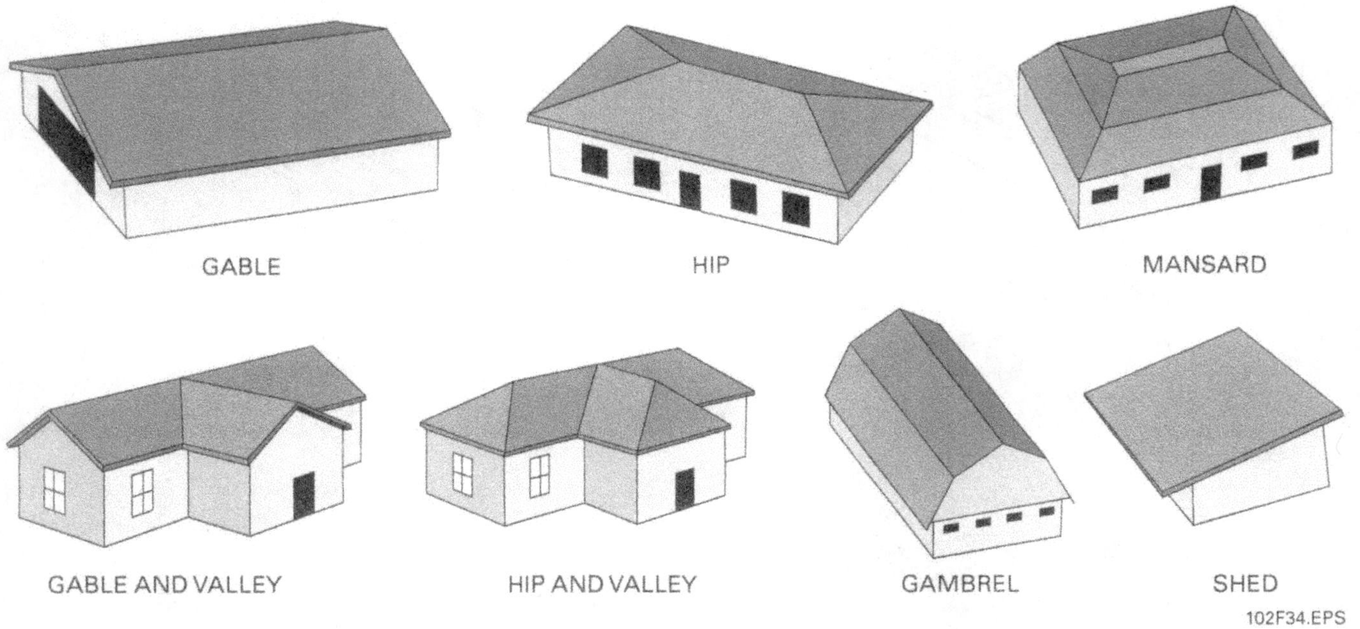

GABLE   HIP   MANSARD

GABLE AND VALLEY   HIP AND VALLEY   GAMBREL   SHED

102F34.EPS

Figure 34 ◆ Types of roofs.

- *Gambrel roof* – The gambrel roof is a variation on the gable roof in which each side has a break, usually near the ridge. The gambrel roof provides additional space in the upper level.
- *Shed roof* – Also known as a lean-to roof, the shed roof has a flat, sloped construction. It is commonly used in high-ceiling contemporary residences and for additions.

There are two basic roof framing systems. In stick-built framing, ceiling joists and rafters are laid out and cut by the builder on the site and the frame is constructed one stick at a time.

In truss-built construction, the roof framework is prefabricated off site. The truss contains both the rafters and the ceiling joist. Trusses and truss construction will be discussed later in this module.

3.4.1 Roof Components

Rafters and ceiling joists provide the framework for all roofs. The main components of a roof (*Figure 35*) are described here.

- *Ridge (ridgeboard)* – The highest horizontal roof member. It helps to align the rafters and tie them together at the upper end. The ridgeboard is one size larger than the rafters.
- *Common rafter* – A structural member that extends from the top plate to the ridge in a direction perpendicular to the wall plate and ridge. Rafters often extend beyond the roof plate to form the overhang (eaves) that protect the side of the building.

- *Hip rafter* – A roof member that extends diagonally from the corner of the plate to the ridge.
- *Valley rafter* – A roof member that extends diagonally from the plate to the ridge along the lines where two roofs intersect.
- *Jack rafter* – A roof member that does not extend the entire distance from the ridge to the top plate of a wall. Hip jacks and valley jacks are shown in *Figure 35*. A rafter fitted between a hip rafter and a valley rafter is called a cripple jack. It touches neither the ridge nor the plate.
- *Plate* – The wall framing member that rests on top of the wall studs. It is sometimes called the rafter plate because the rafters rest on it. It is also referred to as the top plate.

On any pitched roof, rafters rise at an angle to the ridgeboard. Therefore, the length of the rafter is greater than the horizontal distance from the plate to the ridge. In order to calculate the correct rafter length, the builder must factor in the slope of the roof (*Figure 36*).

- *Span* – The horizontal distance from the outside of one exterior wall to the outside of the other exterior wall.
- *Run* – The horizontal distance from the outside of the top plate to the center line of the ridgeboard (usually one-half of the span).
- *Rise* – The total height of the rafter from the top plate to the ridge. This is stated in inches per foot of run.

Figure 35 ◇ Roof framing members.

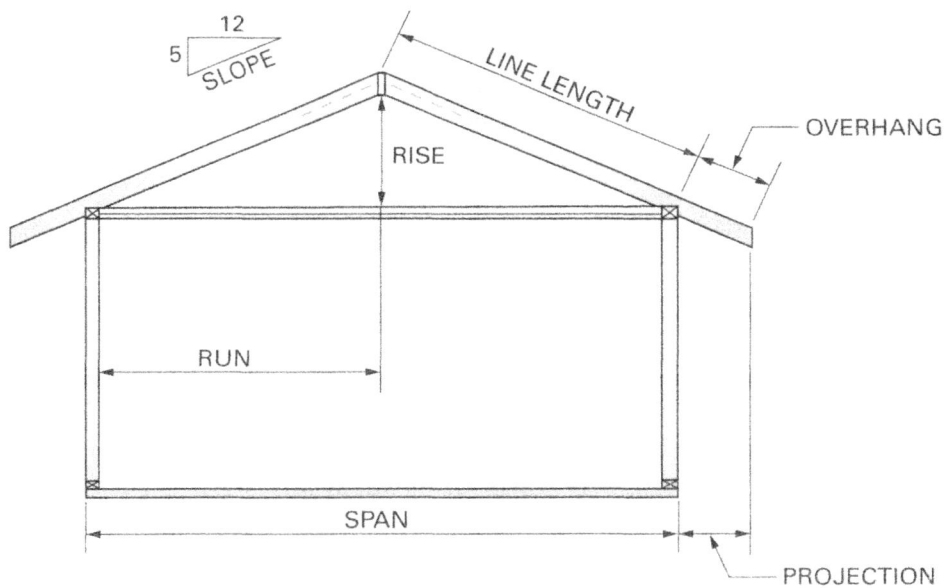

Figure 36 ◇ Roof layout factors.

- *Pitch* – The angle or degree of slope of the roof in relation to the span. Pitch is expressed as a fraction. For example, if the total rise is 6' and the span is 24', the pitch would be ¼ (6 over 24).

- *Slope* – The inclination of the roof surface expressed as the relationship of rise to run. It is stated as a unit of rise to so many horizontal units. For example, a roof that has a rise of 5" for each foot of run is said to have a 5 in 12 slope (*Figure 36*). The roof slope is sometimes referred to as the roof cut.

Pitch and Slope

The terms *pitch* and *slope* may be used inter-
changeably on the job site, but the two terms
actually refer to two different concepts. Slope is
the amount of rise per foot of run and is always
referred to as a number in 12. For example, a roof
that rises 6" for every foot of run has a 6 in 12 slope
(the 12 simply refers to the number of inches in a
foot). Pitch, on the other hand, is the ratio of rise to
the span of the roof and is expressed as a fraction.
For example, a roof that rises 8' over a 32' span is
said to have a pitch of ¼ (⁸/₃₂ = ¼).

3.4.2 Roof Sheathing

Sheathing is applied as soon as the roof framing
is finished. It provides additional strength to the
structure and a base for the roofing material. Some
of the materials commonly used for sheathing are
plywood, OSB, waferboard, shiplap, and common
boards. When composition shingles are used, the
sheathing must be solid. If wood shakes are used,
the sheathing boards may be spaced. When solid
sheathing is used, a ⅛" space is left between panels to allow for expansion.

Once the sheathing has been installed, an underlayment of asphalt-saturated felt or other specified material is installed to keep moisture out until the shingles are laid. For roofs with a slope of 4"
or more, 15-pound roofer's felt is commonly used.

The underlayment is applied horizontally with
a 2" top lap and a 4" side lap, as shown in *Figure
37*. A 6" lap should be used on each side of the center line of hips and valleys. A metal drip edge is
installed along the rakes and eaves to keep out
wind-driven moisture.

In climates where snow accumulates, a waterproof underlayment should be used at roof edges
and around chimneys, skylights, and vents. This
underlayment has an adhesive backing that adheres to the sheathing. It protects against water
damage that can result from melting ice and snow
that backs up under the shingles. Sheet metal or
other material is used at roof intersections and
around chimneys, vents, and skylights to prevent
water from entering. In snowy climates, sheet
metal eave flashing is often installed at the edge
of a roof to prevent ice dams from forming.

UNDERLAYMENT OVERLAP

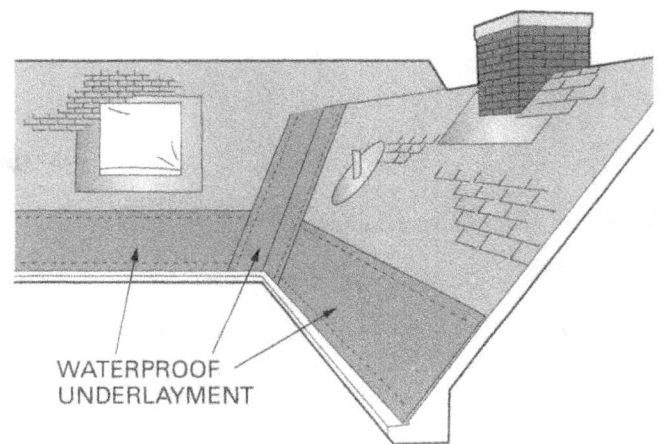

UNDERLAYMENT LOCATIONS

102F37.EPS

Figure 37 ◆ Underlayment installation.

3.4.3 Truss Construction

In most cases, it is much faster and more economical to use prefabricated trusses in place of rafters
and joists. Even if a truss costs more to buy than
the comparable framing lumber (and this is not
always the case), it takes significantly less labor
than stick framing. Another advantage is that a
truss will span a greater distance without a bearing wall than stick framing. Almost any type of
roof can be framed with trusses. Some of the special terms used to identify the members of a truss
are shown in *Figure 38*.

A truss is a framed or jointed structure (*Figure
39*). It is designed so that when a load is applied
at any intersection, the stress in any member is in
the direction of its length.

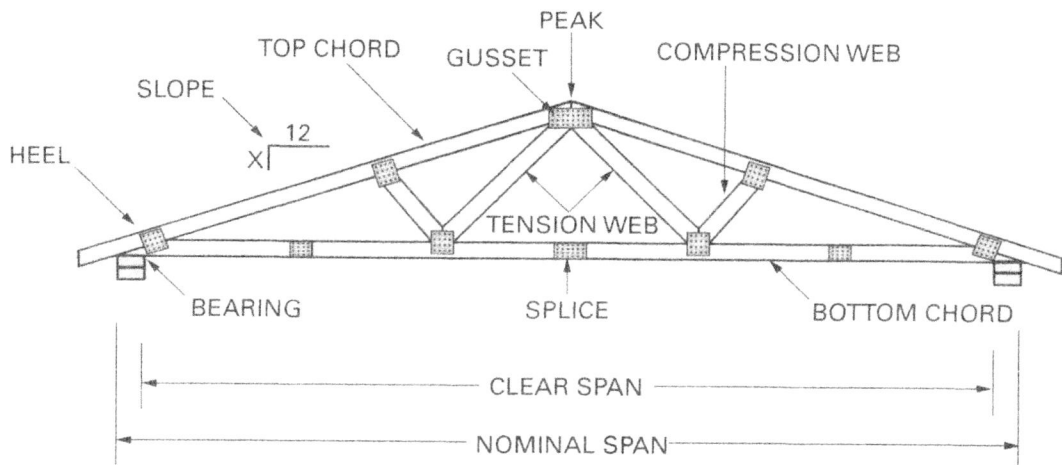

Figure 38 Components of a truss.

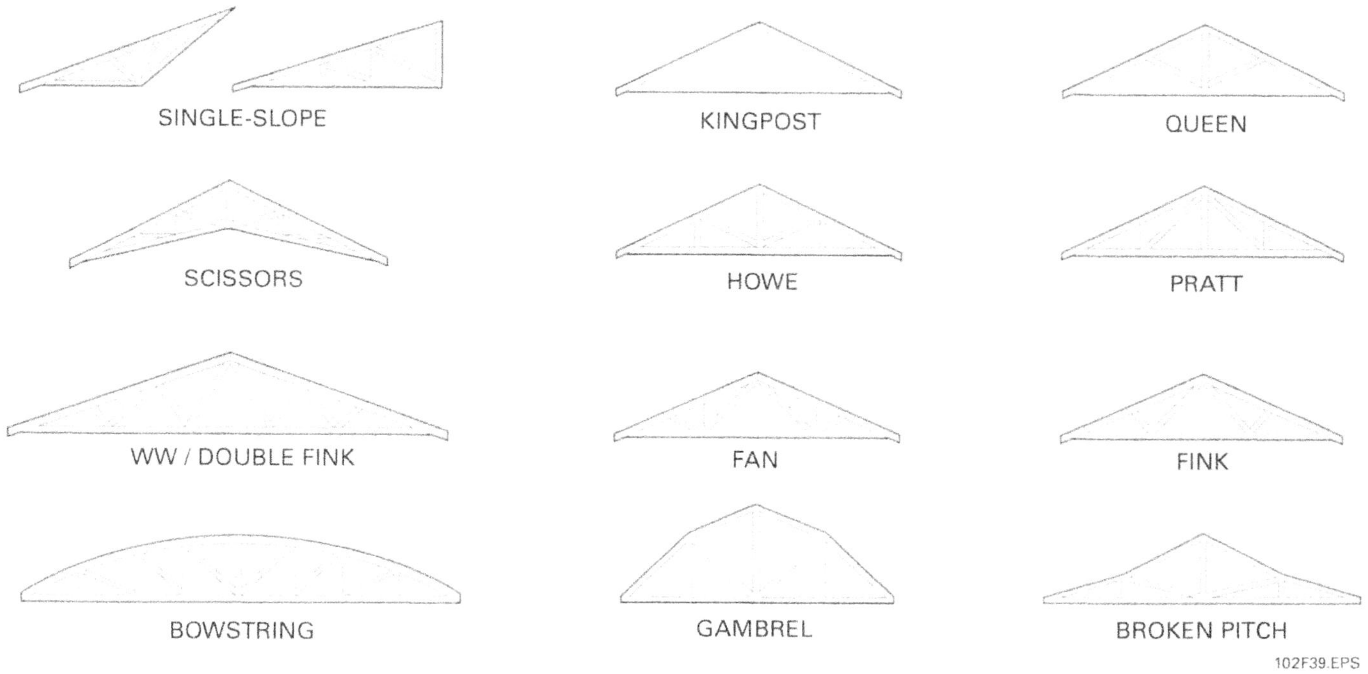

Figure 39 Types of trusses.

Even though some trusses look nearly identical, there is some variation in the interior (web) pattern. Each web pattern distributes weight and stress a little differently, so different web patterns are used to deal with different loads and spans. The decision of which truss to use for a particular application is made by the architect or engineer and is shown on the blueprints.

3.4.4 Dormers

A dormer is a framed structure that projects out from a sloped roof. A dormer provides additional space and is often used in a Cape Cod style home, which is a single-story dwelling in which the attic is often used for sleeping rooms.

A shed dormer (*Figure 40*) is a good way to obtain a large amount of additional living space. If it is added to the rear of the house, it will not affect the appearance of the house from the front.

A gable dormer (*Figure 41*) serves as an attractive addition to a house. It provides a little extra space, as well as some light and ventilation. They are sometimes used over garages to provide a small living area or studio.

Figure 40 ◆ Shed dormer.

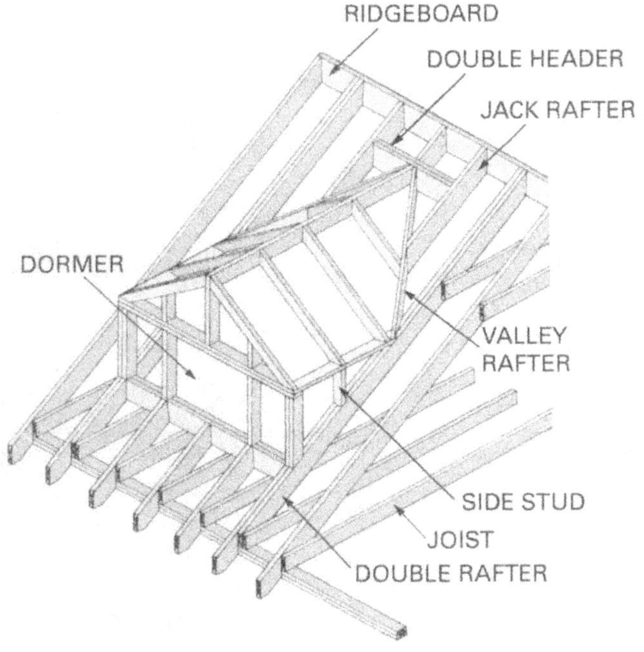

Figure 42 ◆ Example of post and beam framing.

Figure 41 ◆ Gable dormer framing.

3.5.0 Plank-and-Beam Framing

Plank-and-beam framing, also known as post-and-beam framing (*Figure 42*), employs much sturdier framing members than common framing. It is often used in framing roofs for luxury residences, churches, and lodges, as well as other public buildings where a striking architectural effect is desired.

Because the beams used in this type of construction are very sturdy, wider spacing may be used. Vertical supports are typically spaced 48" OC, as compared with 16" OC used in conventional framing. When plank-and-beam framing is used for a roof, the beams and planking can be finished and left exposed. The underside of the planking takes the place of an installed ceiling.

In lighter construction, solid posts or beams such as 4 × 4s are used. In heavier construction, laminated beams made of glulam, LVL, and PSL are used.

In post-and-beam framing, plank subfloors or roofs are usually of 2" nominal thickness, supported on beams spaced up to 8' apart. The ends of the beams are supported on posts or piers. Wall spaces between posts are provided with supplementary framing as required for attachment of exterior and interior finishes. This additional framing also provides lateral bracing for the building.

If local building codes allow end joints in the planks to fall between supports, planks of random lengths may be used and the beam spacing adjusted to fit the house dimensions. Windows and doors are normally located between posts in the exterior walls, eliminating the need for headers over the openings. The wide spacing between posts permits ample room for large glass areas.

A combination of conventional framing with post-and-beam framing is sometimes used where the two adjoin each other.

Where a post-and-beam floor or roof is supported on a stud wall, a post is usually placed under the end of the beam to carry a conventional load. A conventional roof can be used with post-and-beam construction by installing a header between posts to carry the load from the rafters to the posts.

3.6.0 Framing in Masonry

In order to work in buildings constructed of masonry, you must be aware of the methods used in furring masonry walls. As a general rule, furring of masonry walls is done on 16" centers. Some contractors will apply 1 × 2 furring strips 24" OC. This may save material, but it does not provide the same quality as a wall done on 16" centers.

In addition to the furring strips, 1×4 and 1×6 stock is used. All material that comes in contact with concrete or masonry must be pressure-treated.

Backing for partitions against a masonry block wall is done using one of the methods shown in *Figure 43*.

When preparing the corners of the block wall to receive the furring strips, enough space is left for the drywall to slip by the furring strips (*Figure 44*).

A 1 × 4 is used at floor level to receive the baseboard. Either a narrow or wide baseboard can be used. Some builders will install a simple furring strip at floor level and depend on the vertical strips for baseboard nailing. Once the drywall has been installed, it is difficult to find the strips when nailing the baseboard.

3.7.0 Walls Separating Occupancies

Every wall, floor, and ceiling in a building is rated for its fire resistance, as established by building codes. The fire rating is stated in terms of hours; for example, one-hour wall, two-hour wall, and so on. The rating denotes the length of time an assembly can withstand fire and give protection from it as determined under laboratory conditions. The greater the fire rating, the thicker the wall is likely to be. This subject is covered further in the section on commercial construction.

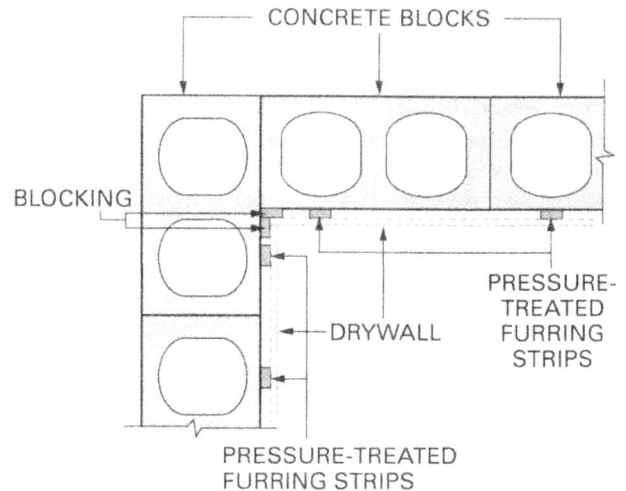

102F44.EPS

Figure 44 Corner construction.

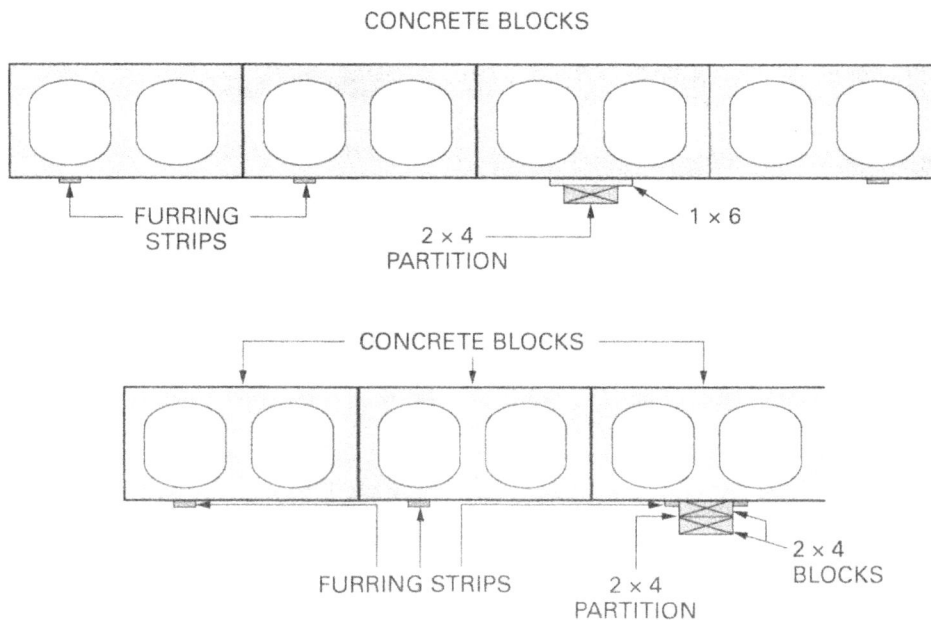

102F43.EPS

Figure 43 Partition backing on a block wall.

In multi-family residential construction, such as apartments and townhouses, the walls and ceilings dividing the occupancies must meet special fire and soundproofing requirements. The code requirements will vary from one location to another and may even vary within areas of a jurisdiction. That is, dwellings in high-risk areas may have stiffer standards than those in other areas of the same city or county. In some cases, the code may require a masonry wall between occupancies. This masonry wall may even be required to penetrate the roof of the building so that if a fire occurs, it is contained within the unit in which it started, because it is unable to travel through the walls or across the attic space.

There are many different construction methods for multi-dwelling (party) walls. Each is designed to meet different fire and soundproofing standards. The wall is likely to be more than 3" thick and contain several layers of gypsum wallboard and insulation.

4.0.0 ◆ COMMERCIAL CONSTRUCTION METHODS

The structural framework of large buildings, such as office buildings, hospitals, apartment houses, and hotels, is usually made from concrete or structural steel. The exterior finish is often concrete panels that are either prefabricated and raised into place or poured into forms built at the site. Floors are usually made of concrete that is poured at the site using wood, metal, or fiberglass forms. Exterior walls (curtain walls) may also be made of glass in a metal or concrete framework (*Figure 45*).

In some buildings, the framework is made of structural steel. Panels fabricated off site are lifted into place and bolted or welded to the steel (*Figure 46*).

102F45.EPS

Figure 45 ◆ Installing a corner curtain wall panel.

102F46.EPS

Figure 46 ◆ Curtain wall under construction.

INSIDE TRACK   **Party Walls**

When gypsum drywall is used in a party wall, the architect's plans must be followed precisely. If the inspector finds flaws in the construction, a certificate of occupancy will not be issued.

This photo illustrates why building codes place so much emphasis on properly constructed party walls.

102SA05.EPS

Masonry Curtain Wall

This masonry curtain wall panel combines 2"-thick architectural precast concrete brick with a heavy-gauge stainless steel frame and insulated stainless steel anchor doors.

102SA06.EPS

Figure 47 shows the structure of a building in which all the structural framework is made of concrete poured at the site. Each component of the structure requires a different type of form. In this case, the floor and beams were made in a single pour using integrated floor and beam forms, which were removed once the concrete hardened.

In some specialized commercial applications, tilt-up concrete construction is used. In tilt-up construction, the wall panels are usually poured on the concrete floor slab, then tilted into place on the footing using a crane (*Figure 48*). The panels are welded together.

The main difference between tilt-up and other types of large commercial construction is that there is no steel or concrete framework in tilt-up construction. The walls and floor slab bear the entire load. Tilt-up is most common in one- or

Tilt-Up Records

The Tilt-Up Concrete Association (**www.tilt-up.org**) keeps records of tilt-up projects. As of 2002, several amazing statistics had been recorded, including:

- Heaviest tilt-up panel – 310,000 pounds
- Largest tilt-up panel by area – 1,815 square feet, including window openings (30' × 60'-6")
- Tallest tilt-up panel – 91'-7¼"
- Largest number of panels in a single building – 1,310
- Largest tilt-up building (floor area) – 1,750,000 square feet

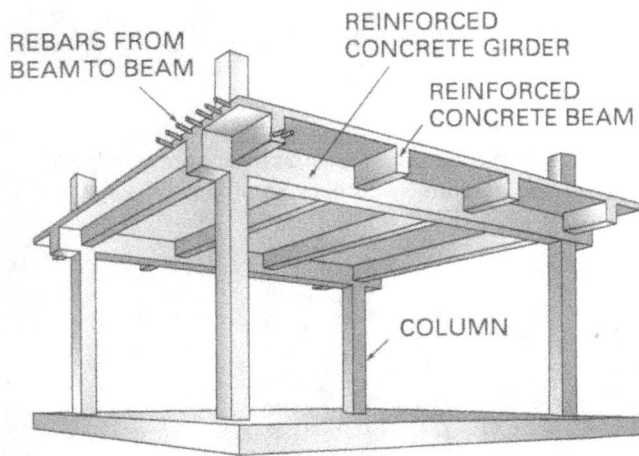

Figure 47 ◆ Concrete structure.

Figure 48 ◆ Tilt-up panel being lifted into place.

Figure 49 ◆ Corrugated steel forms.

two-story buildings with a slab at grade (no below-grade foundation). It is popular for warehouses, low-rise offices, churches, and a variety of other commercial and multi-family residential applications. Tilt-up panels of 50' in height are not uncommon. They typically range from 5" to 8" thick, but thicker walls can be obtained when using lighter-weight concrete.

4.1.0 Floors

Once the framework is in place, the concrete floors are poured using deck forms. Shoring is placed under the form to support it until the concrete hardens. *Figure 49* shows cellular floors poured over corrugated steel forms, which remain in place, providing channels through which cabling can be run.

A section of metal, plastic, or fiber sleeve is often inserted vertically into the form before the concrete is poured to allow for electrical, communications, and other cabling to pass through the floor.

In some installations, underfloor duct systems are embedded in the concrete floor and are used to provide horizontal distribution of cables. Vertical access ports (handholes) are embedded in the form so that cable can be fished to various locations in the space. *Figure 50* shows a single-level feeder duct system. In two-level systems, one level carries electrical power cables and the other carries low-voltage cables.

Trench ducts (*Figure 51*) are metal troughs that are embedded in the concrete floor and used as feeder ducts for electrical power and telecommunication lines.

Access floors consist of modular floor panels supported by pedestals. They may or may not have horizontal bracing in addition to the pedestals. This type of structure is used in computer rooms, intensive care facilities, and other areas where a lot of cabling is required. In some applications, such as a factory, a trench may be formed in the concrete floor to accommodate cabling and other services.

4.2.0 Exterior Walls

When walls are formed of concrete, openings for doors and windows are made by inserting wooden or metal bucks in the form (*Figure 52*). Openings for services such as piping and cabling are accommodated with fiber, plastic, or metal tubes inserted into the form.

4.3.0 Interior Walls and Partitions

The construction of walls and partitions in commercial applications is driven by the fire and soundproofing requirements specified in local building codes. In some cases, a frame wall with ½" gypsum drywall on either side is satisfactory. In extreme cases, such as the separation between offices and manufacturing space in a factory, it

102F50.EPS

Figure 50 Flushduct underfloor system.

102F52.EPS

Figure 52 Framing openings in concrete walls.

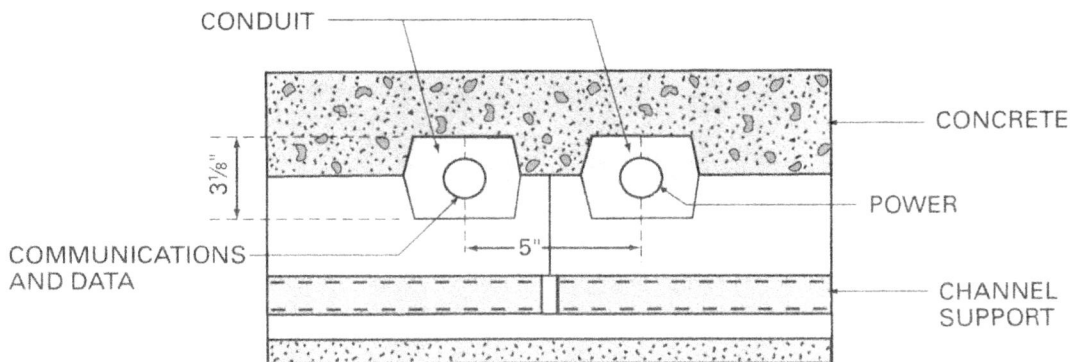

102F51.EPS

Figure 51 Trench duct in a cellular floor.

may be necessary to have a concrete block wall combined with fire-resistant gypsum wallboard, along with rigid and/or fiberglass insulation (*Figure 53*). This is especially true if there is any explosion or fire hazard.

While they are sometimes used in residential construction, steel studs are the standard for framing walls and partitions for commercial construction. Once the studs are installed, one or more layers of gypsum wallboard and insulation are applied. The type and thickness of the wallboard and insulation depend on the fire rating and soundproofing requirements. Soundproofing needs vary from one use to another and are often based on the amount of privacy required for the intended use. For example, executive and physician offices, high-rise condos, hospitals, and houses for the elderly may require more privacy than general offices.

The requirements for sound reduction and fire resistance can significantly affect the thickness of a wall. For example, a wall with a high sound transmission class (STC) and fire resistance might have a total thickness of nearly 6½", while a low-rated wall might have a thickness of only 3½" (*Figure 54*).

As discussed previously, the fire rating specified by the applicable building code determines the types and amount of material used in a wall or partition. As shown in *Figure 54*, a one-hour rated wall might be made of single sheets of ⅝" gypsum wallboard on wooden or 25-gauge metal studs. A two-hour rated wall requires metal studs and two layers of fire-resistant gypsum wallboard.

Steel thickness was traditionally referenced in terms of gauge numbers, but that has given way to designating thickness in terms of mils. *Table 3* lists common base steel thicknesses and equivalent gauge identifiers.

4.3.1 Metal Framing Materials

Metal framing components include metal studs and, in some cases, metal joists and metal roof trusses (*Figure 55*). The vertical and horizontal framing members serve as structural load-carrying components for a variety of low- and high-rise structures. Metal stud framing is compatible with all types of surfacing materials that can be screw-applied.

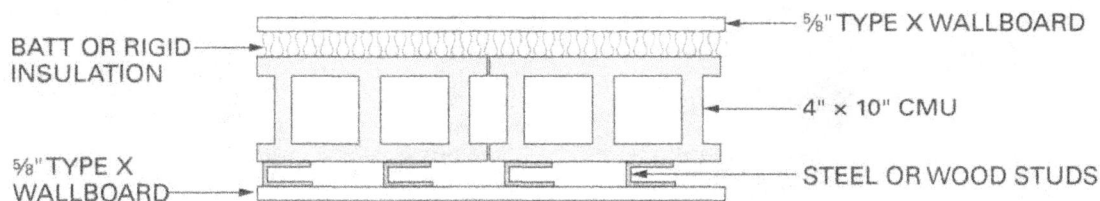

102F53.EPS

Figure 53 ◆ High fire/noise resistance partition.

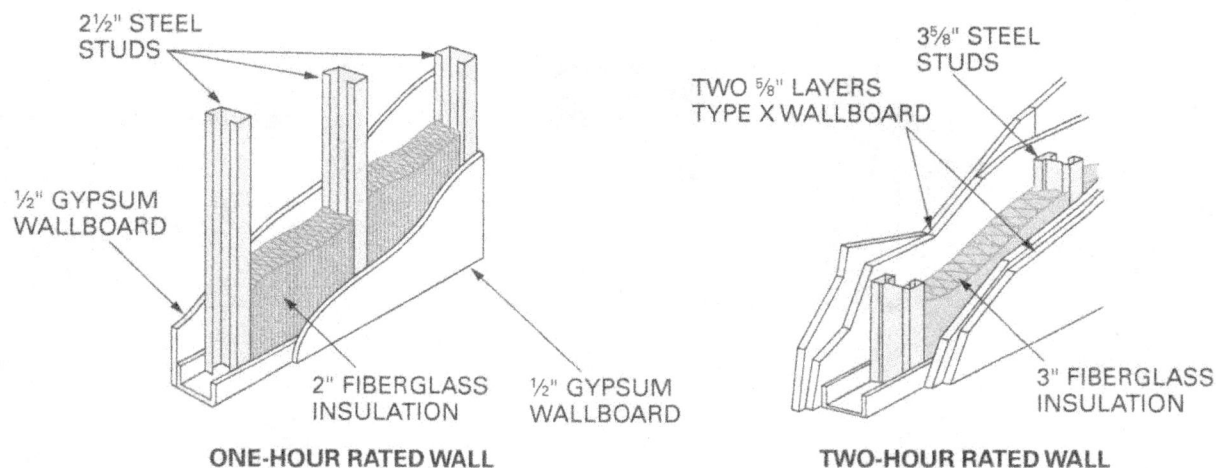

102F54.EPS

Figure 54 ◆ Partition wall examples.

Table 3 Minimum Base Steel Thickness of Cold-Formed Steel Members

Designation (Thickness in mils)	Minimum Base Steel [Thickness in inches (mm)][1]	Old Reference Gauge Number[2]
18	0.0179 (0.455)	25
27	0.0269 (0.683)	22
30	0.0296 (0.752)	20 – drywall[3]
33	0.0329 (0.836)	20 – structural[3]
43	0.0428 (1.09)	18
54	0.0538 (1.37)	16
68	0.0677 (1.72)	14
97	0.0966 (2.45)	12
118	0.1180 (3.00)	10

[1] Design thickness shall be the minimum base steel thickness divided by 0.95.

[2] Gauge thickness is an obsolete method of specifying sheet and strip steel thickness. Gauge numbers are only a very rough approximation of steel thickness and shall not be used to order, design or specify any sheet or strip steel product.

[3] Historically, 20 gauge material has been furnished in two different thicknesses for structural and drywall (non-structural) applications.

102T03.EPS

Figure 55   Metal trusses.

The advantages of metal framing include non-combustibility, uniformity of dimension, lightness of weight, freedom from rot and moisture, and ease of construction. The components of metal frame systems are manufactured to fit together easily. There are a variety of metal framing systems, both loadbearing and nonbearing types. Some nonbearing partitions are designed to be demountable or moveable and still meet the requirements of sound insulation and fire resistance when covered with the proper gypsum system.

When metal studs are used for drywall framing systems, the channel stock features knurled sides for positive screw settings and comes in two grades. The first grade is the standard drywall stud (*Figure 56*). Standard studs come in widths of 1⅝" to 6". The flanges are 1¼" with a ¼" return lip. Lengths of 6' to 16' are commercially available. The standard drywall stud is 18-mil (25-gauge) steel (the higher the gauge, the lighter the metal). Depending on the product number, lengths range from 8' to 12'; other lengths are available by special order.

> **NOTE**
>
> The term *gauge* has traditionally been used to state the weight of steel studs. In recent years, however, the industry has converted to using mils, which represents the thickness of the steel in millimeters.

The second grade is the extra heavy drywall stud (*Figure 57*). These studs have knurled flanges for positive screw settings. They also have cutouts and utility knockout holes 12" from each end and at the mid-point of the stud.

The width of the extra heavy studs will also vary from 1⅝" to 6". The flanges are 1¼" with a ¼" return lip. The extra heavy drywall stud is 20 gauge (30 mils). This type of stud can be ordered in any length that is needed.

102F56.EPS

Figure 56   Standard metal stud stock.

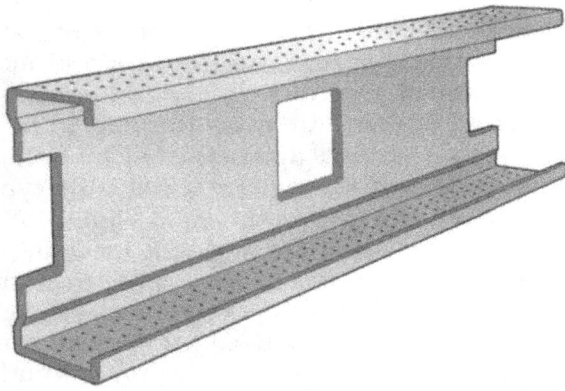

Figure 57 ◆ Heavy-duty stud stock.

102F57.EPS

Metal studs are also available in greater strengths of 18 gauge, 16 gauge, 14 gauge, and 12 gauge. These are typically used on exterior work. These strengths are classified as structural steel studs. They are available in widths of 2½" to 14". The flanges are 1⅝", 2", or 2½". Return lips vary from ¼" to 2". They can be ordered in whatever length is needed.

Although different materials are used, the general approach to framing with metal studs is the same as that used for wooden studs. In fact, metal studs can be used with either wood plates or metal runners.

Like wood framing, metal framing is installed 12", 16", or 24" OC, openings are framed with headers and cripples, and special framing is needed for corners and partition Ts.

Depending on the load, reinforcement may be needed when framing openings. Bracing of walls to keep them square and plumb is also required. The illustrations in this section show examples of common framing techniques. *Table 4* shows the framing spacing for various gypsum drywall applications.

The erection of metal studs typically starts by laying metal tracks in position on the floor and ceiling and securing them (*Figure 58*). If the tracks are being applied to concrete (*Figure 59*), a low-velocity powder-actuated fastener is generally used. If the tracks are being applied to wood joists, such as in a residence, screws can be driven with a screw gun. Always check the shop drawings for such details.

> **WARNING!**
> The use of a powder-actuated fastener requires special training and certification. The use of these tools may be prohibited by local codes because of safety and seismic concerns.

Table 4 Maximum Framing Spacing

	Single-Ply Gypsum Board Thicknesses	Application to Framing	Maximum OC Spacing of Framing
Ceilings	⅜"	Perpendicular or Parallel	16"
	½"	Parallel	16"
	½	Perpendicular	24"
	⅝"	Perpendicular	24"
Sidewalls	½" or ⅝"	Perpendicular or Parallel	24"

Fasteners Only – No Adhesive Between Plies				
	Multi-Ply Gypsum Board Thicknesses		Application to Framing		Maximum OC Spacing of Framing

	Base	Face	Base	Face	
Ceilings	⅜"	⅜"	Perpendicular	Perpendicular*	16"
	⅜"	⅜"	Parallel	Perpendicular	24"
Sidewalls	½" or ⅝"	½" or ⅝"	Perpendicular	Perpendicular	24"
	⅝" or ½"	⅝" or ½"	Perpendicular	Perpendicular	24"

*Must use adhesive

102T04.EPS

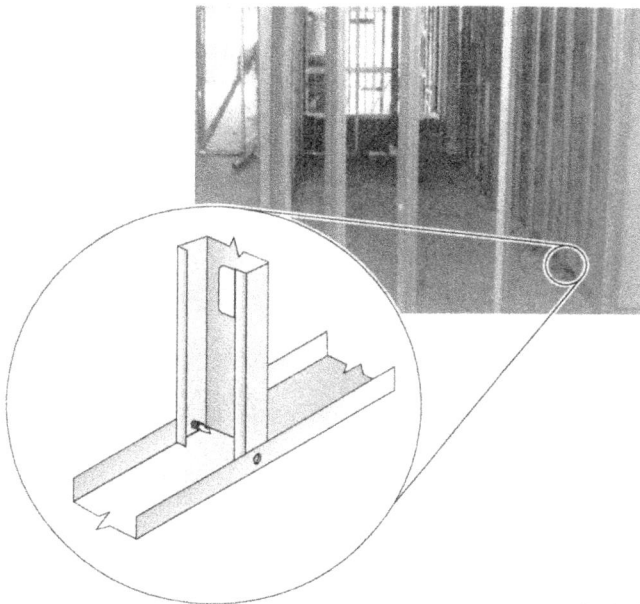

Figure 58 Metal framing.

102F58.EPS

STUD

BOTTOM
TRACK

TOP
TRACK

DEEP LEG

STUD

BOTTOM
TRACK

ALLOWING FOR
VERTICAL MOVEMENT

102F59.EPS

Figure 59 Metal studs with concrete floors and ceiling.

Once the tracks are in place, the studs and openings are laid out in the same way as a wood frame wall. The studs may be secured to the tracks with screws or they may be welded. In some cases, the entire wall will be laid out on the floor, then raised and secured. When heavy-gauge walls are used, they may be assembled and welded in a shop and brought to the site.

There are some differences between installing metal nonbearing partitions and wooden nonbearing partitions. When building with wood, all partitions must be nailed together. With metal studs, this is not required.

As shown in *Figure 60*, the partitions are held back from the other partitions so that the drywall will slide past. Note the conduit fed through the openings in the stud.

When metal studs are used to frame around steel beams, the metal studs are secured to the metal beam with powder-actuated fasteners (if allowed) and the support members are screwed to the metal studs (*Figure 61*). Note the wire hanger for the suspended ceiling at the right of the picture.

When the metal studs are installed against metal channels or flanges, the studs are secured to the channel with scrap pieces of metal studs (*Figure 62*).

4.3.2 Bracing Walls

Different forms of bracing are used to support metal stud walls. Lateral bracing using continuous metal strapping is always recommended as the minimum support for metal stud walls (*Figure 63*).

102F60.EPS

Figure 60 Partition held back to allow drywall to slide by.

102F61.EPS

Figure 61 ❖ Plate attached to a beam.

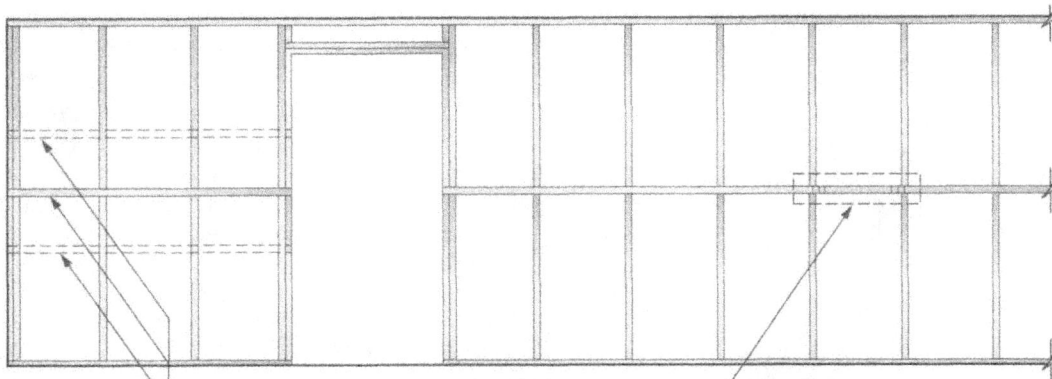

102F62.EPS

Figure 62 ❖ Studs secured to a channel.

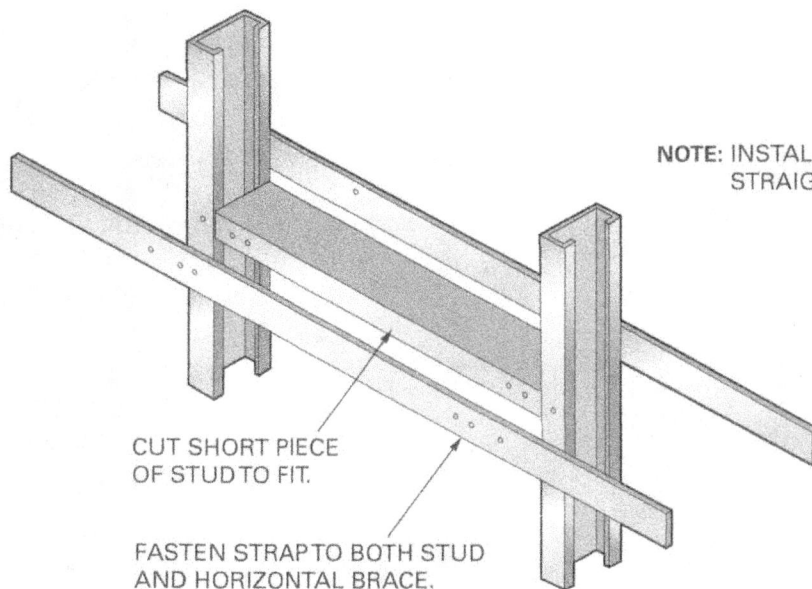

INSTALL HORIZONTAL STRAPS
AT MIDPOINT OR THIRD POINTS
AS REQUIRED BY LOAD OR
ENGINEERED TABLES.

STRAPPING SHOULD NOT SPAN
MORE THAN 8'-0" WITHOUT BEING
FASTENED TO A JAMB STUD OR
BRACED AS SHOWN.

NOTE: INSTALL STRAPPING
STRAIGHT AND TAUT.

CUT SHORT PIECE
OF STUD TO FIT.

FASTEN STRAP TO BOTH STUD
AND HORIZONTAL BRACE.

102F63.EPS

Figure 63 ❖ Lateral strapping for stud walls.

Diagonal bracing using metal strapping is sometimes required (*Figure 64*). This is done with 20-gauge, 2" wide metal straps placed close to the end of the wall. Lateral and diagonal braces can be screwed and/or welded to the studs.

For wider studs (4" and wider), steel channel threaded through the openings in the studs and welded or screwed to angle clips is sometimes required (*Figure 65*). Flat strap screw-applied or welded can also be used.

4.3.3 Metal Joists and Roof Trusses

In commercial work, metal studs are commonly used to frame interior nonbearing walls and partitions. In residential work, an entire house can be framed with steel studs, joists, and roof trusses.

Steel joists are available in the same sizes as wood joists. Joists can rest directly on concrete or masonry or they can be attached to a wood sill plate or top plate (*Figure 66*).

One method of installing floor joists in a poured concrete foundation wall is to form slots in the wall to accept the joists (*Figure 67*).

Metal roofs are framed with prefabricated trusses in which the framing members are welded together.

4.4.0 Ceilings

Although suspended ceilings are sometimes found in residential applications, they are most commonly used in commercial construction. Suspended ceilings have a number of advantages in commercial work:

- They provide excellent noise suppression.
- They provide an area in which horizontal runs of cabling, piping, heating and cooling ducts, and other services can be readily accessed.

102F64.EPS

Figure 64 Diagonal strapping for shear walls.

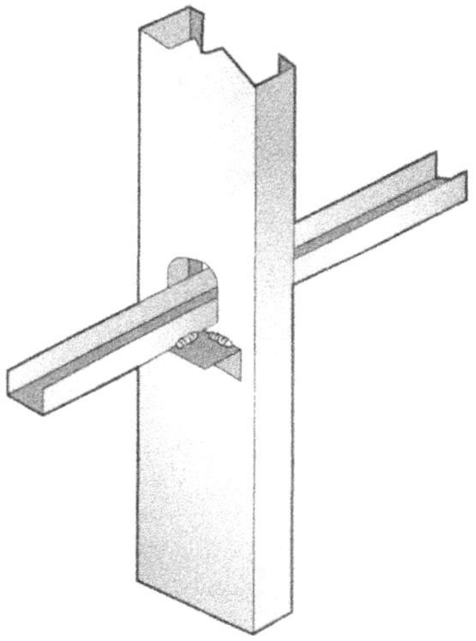

102F65.EPS

Figure 65 Heavy-duty bridging.

Identifying Structural Studs

INSIDE TRACK

Structural studs are marked with a color code for easy identification. The coding is as follows:

Gauge	Color	Mils	Minimum In	Minimum MM
20	White	33	0.0329	0.836
18	Yellow	43	0.0428	1.087
16	Green	54	0.0538	1.367
14	Orange	68	0.0677	1.720
12	Red	97	0.0966	2.454

Note that there are both light and structural gauge studs made of 20-gauge steel. The difference is in the dimensions.

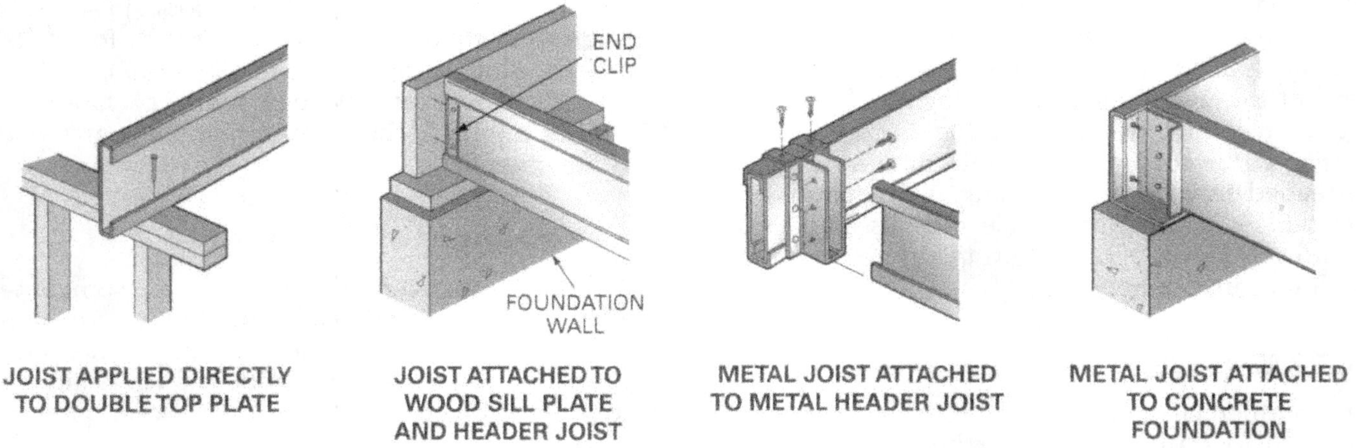

JOIST APPLIED DIRECTLY TO DOUBLE TOP PLATE

JOIST ATTACHED TO WOOD SILL PLATE AND HEADER JOIST

METAL JOIST ATTACHED TO METAL HEADER JOIST

METAL JOIST ATTACHED TO CONCRETE FOUNDATION

102F66.EPS

Figure 66 ◆ Examples of metal joist installations.

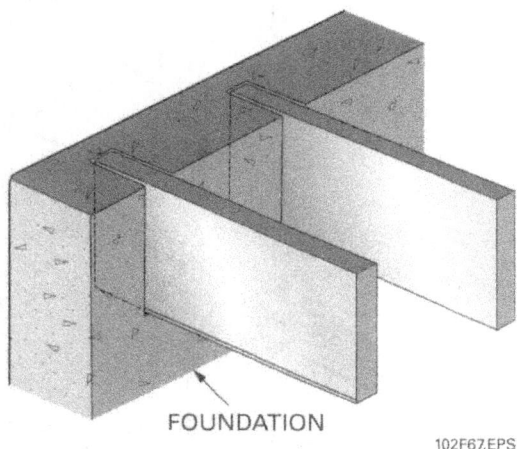

102F67.EPS

Figure 67 ◆ Installing metal floor joists in a slotted foundation wall.

- In many commercial buildings, the area between the suspended ceiling and the floor above acts as the return air plenum for air conditioning and heating, eliminating the need for some of the sheet metal ductwork.
- The use of suspended ceilings eliminates the need for ceiling framing, as well as the need to box in horizontal runs of ductwork and piping.

There are a wide variety of suspended acoustical ceiling systems. They use the same basic materials, but their appearances are completely different. The focus in this module is on the following systems:

- Exposed grid systems
- Metal pan systems

- Direct-hung concealed grid systems
- Integrated ceiling systems
- Luminous ceiling systems
- Suspended drywall furring ceiling systems
- Special ceiling systems

4.4.1 Exposed Grid Systems

For an exposed grid suspended ceiling, also called a direct-hung system, a light metal grid is hung by wire from the original ceiling or the deck above. Ceiling panels, which are usually 2' × 2' or 2' × 4', are then placed in the frames of the metal grid. Exposed grid systems are constructed using the components and materials shown in *Figure 68* and described here:

- *Main runners* – These are the primary support members of the grid system for all types of suspended ceiling systems. They are 12' in length and are usually made in the form of an inverted T. When it is necessary to lengthen the main runners, they are usually spliced together using extension inserts. However, the method of splicing the main runners will vary with the type of system being used.
- *Cross runners (cross ties or cross tees)* – These supports are inserted into the main runners at right angles and spaced an equal distance from each other, forming a complete grid system. They are held in place by either clips or automatic locking devices. Typically, they are either 4' or 2' in length and are usually constructed in the form of an inverted T. Note that 2' cross runners are only required for use with 2' × 2' ceiling panels.

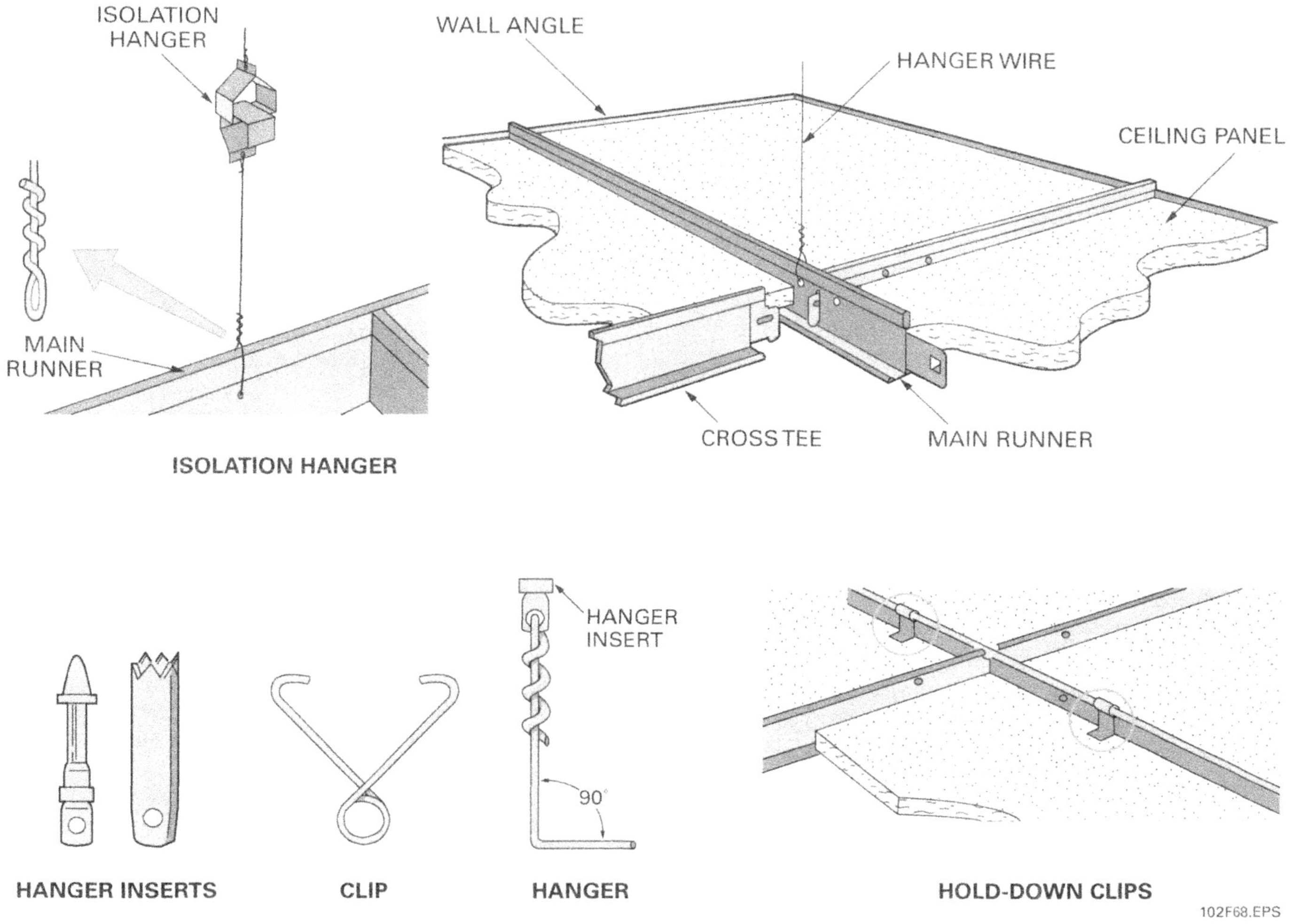

Figure 68 Typical exposed grid system components.

- *Wall angle* – These supports are installed on the walls to support the exposed grid system at the outer edges.
- *Ceiling panels* – These panels are laid in place between the main runners and cross ties to provide an acoustical treatment. Acoustical panels used in suspended ceilings stop sound reflection and reverberation by absorbing sound waves. These panels are typically designed with numerous tiny sound traps consisting of drilled or punched holes or fissures, or a combination of both. A wide variety of ceiling panel designs, patterns, colors, facings, and sizes are available, allowing most environmental and appearance demands to be met. Panels are typically made of glass or mineral fiber. Generally, glass panels have a higher sound absorbency than mineral fiber panels. Panel facings are typically made of embossed vinyl and are available in a variety of patterns, such as fissured,

pebbled, and striated. The specific ceiling panels used must be compatible with the ceiling suspension system, however, because there are variations among manufacturers' standards, and not all panels fit all systems.

NOTE

The terms *ceiling panel* and *ceiling tile* have specific meanings. Ceiling panels are typically any lay-in acoustical board that is designed for use with an exposed mounting system. They do not have finished edges or precise dimensional tolerances because the exposed grid system support members provide the trimout. Ceiling tiles are acoustical ceiling boards, usually 12" × 12" or 12" × 24", which are nailed, cemented, or suspended by a concealed grid system. The edges are often kerfed and cut back.

- *Hanger inserts and clips* – There are many types of fastening devices used to attach the grid system hangers or wires to the building's horizontal structure above the suspended ceiling. Screw eyes and star anchors are commonly used, and require an electric hammer for installation. Hanger clips are also commonly used to fasten into reinforced concrete with a powder-actuated fastening tool. Clips are used where beams are available and are typically installed over the beam flanges, then the hanger wires are inserted through the loops in the clips and secured. These devices must be adequate to handle the load.
- *Hangers* – These are attached to the hanger inserts, pins, and clips to support the suspended ceiling's main runners. The hangers can be made of No. 12 wire or heavier rod stock. Ceiling isolation hangers are also available that isolate ceilings from noise traveling through the building structure.
- *Hold-down clips* – These clips are used in some systems to hold the ceiling panels in place.
- *Nails, screws, tappets* – These fasteners are used to secure the wall angle to the wall. The specific item used depends on the wall construction and material.

4.4.2 Metal Pan Systems

The metal pan system is similar to the conventional suspended acoustical ceiling system, but metal tiles or pans are used in place of the conventional acoustical panel (*Figure 69*).

The pans are made of steel or aluminum and are generally painted white; however, other colors are available by special order. Pans are also available in a variety of surface patterns. Tests have indicated that metal pan ceiling systems are effective for sound absorption. They are durable and easily cleaned and disinfected. In addition, the finished ceiling has little or no tendency to have sagging joint lines or drooping corners. The metal pans are die-stamped and have crimped edges that snap into the spring-locking main runner and provide a flush ceiling.

Take care in handling the pans if you have to remove them. Use white gloves or rub your hands with cornstarch to keep any perspiration marks from the surface of the pans. If care is not taken, fingerprints will be plainly visible when the units are reinstalled.

When installing pans, grasp the pan at its edge and force its crimp into the tee bar slots. Use the palms of your hands to seat the pan. After installing several of the pans as noted above, slide them along the tee bars into position. Use the side of your closed fist to bump the pan into level position if it does not seat readily. If metal pan insulating pads are required, slip them into position over the pans as they are installed. The purpose of the pad is to reduce the travel of sound through the ceiling into the room.

If a metal pan must be removed, a pan pulling device is available (*Figure 70*). To pull out a pan, insert the free ends of the device into two of the perforations at one corner of the pan and pull down sharply. Repeat this at each corner of the pan. By following this removal procedure, there is no danger of bending the pan out of shape. A metal pan ceiling is shown in *Figure 71*.

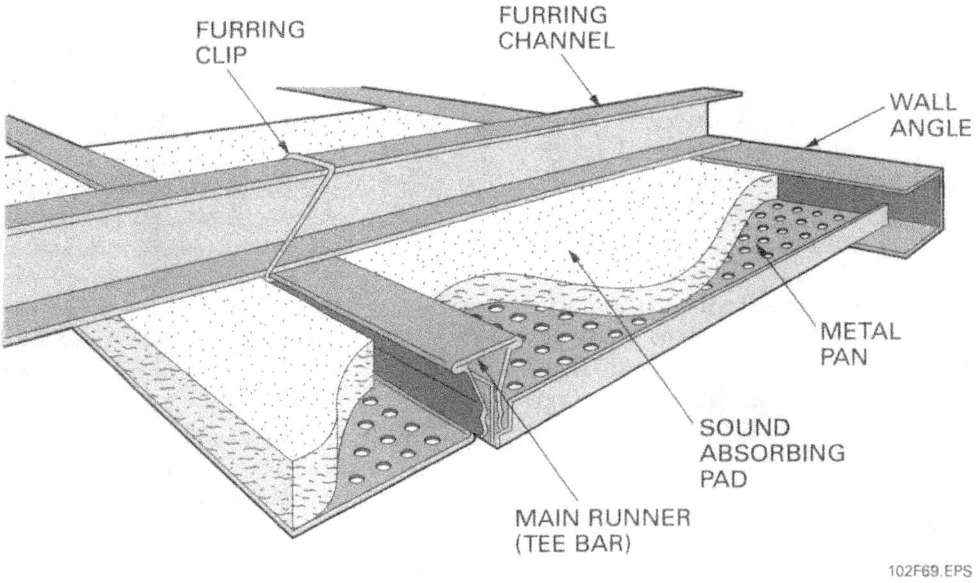

Figure 69 ◆ Typical metal pan ceiling components.

102F69.EPS

Figure 70   Pan removal tool.

Figure 71   Metal pan ceiling.

4.4.3 Direct-Hung Concealed Grid Systems

In this type of suspended acoustical ceiling system, the support runners are hidden from view, resulting in a patterned ceiling that is not broken by the pattern of the runners (*Figure 72*).

The tiles used for this system are similar in composition to conventional acoustical tile, but are manufactured with a kerf on all four edges. Kerfed and rabbeted 12" × 12" or 12" × 24" tiles are used with this system. Tiles of various colors and finishes are available. Refer to *Figure 73* for a diagram of a concealed grid system installation.

If regular access is needed to the area above the ceiling, special access systems can be incorporated into the ceiling (*Figure 74*).

Figure 72   Concealed grid system.

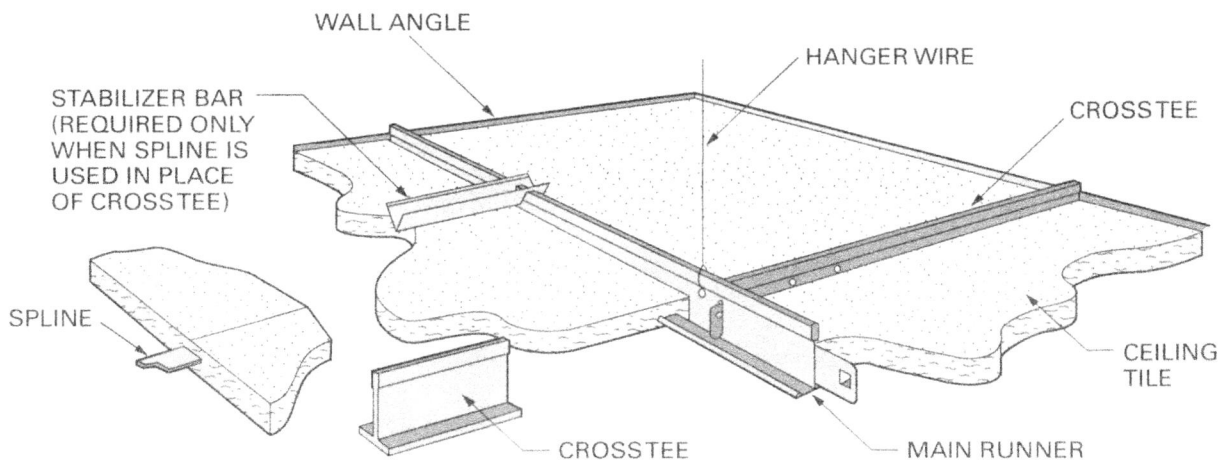

Figure 73   Direct-hung concealed grid system components.

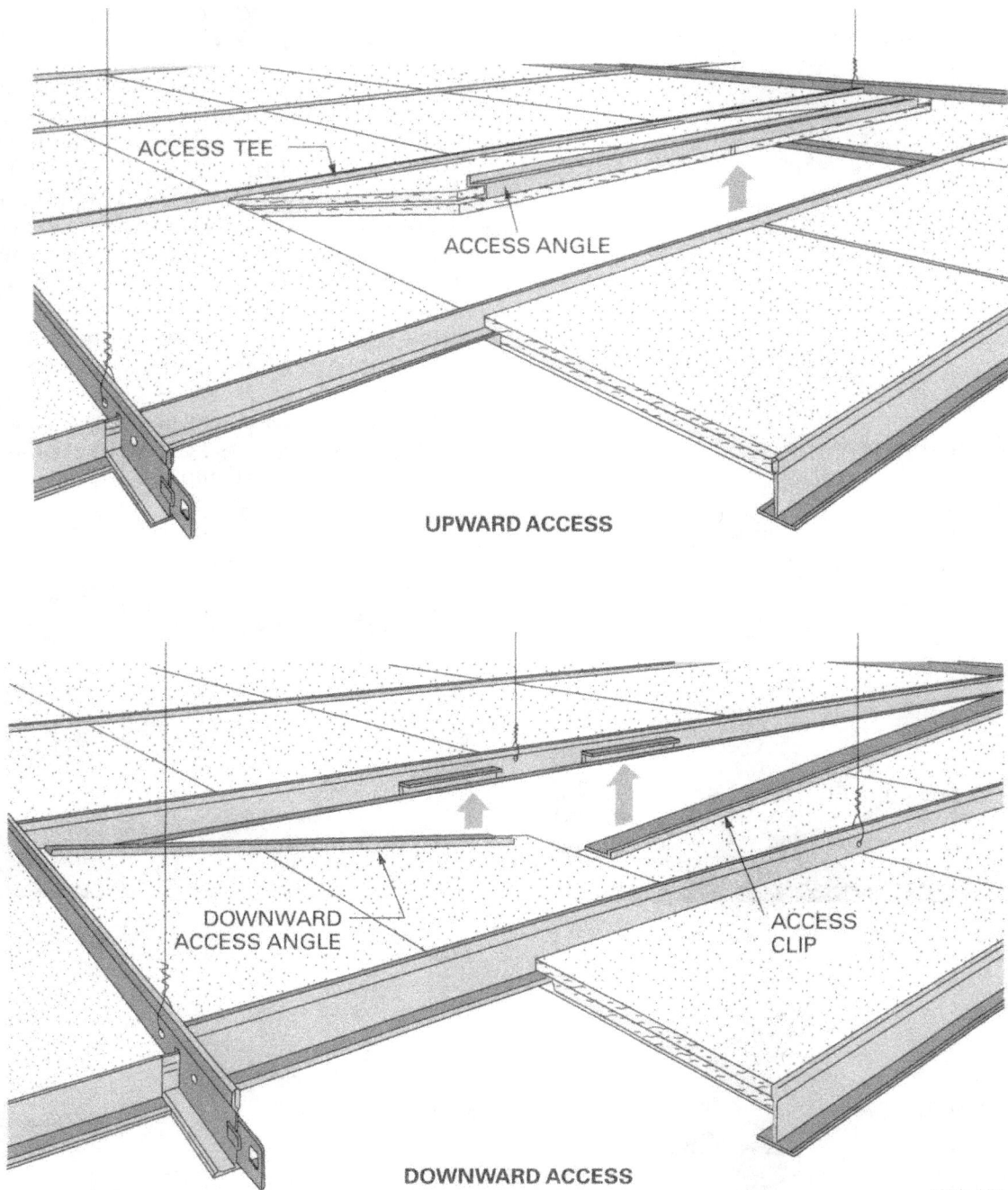

ACCESS TEE

ACCESS ANGLE

UPWARD ACCESS

DOWNWARD
ACCESS ANGLE

ACCESS
CLIP

DOWNWARD ACCESS

102F74.EPS

Figure 74 ✦ Typical access for concealed grid ceilings.

4.4.4 Integrated Ceiling Systems

As indicated by its name, the integrated ceiling system incorporates the lighting and/or air supply diffusers as part of the overall ceiling system (*Figure 75*).

This system is available in units called modules. The common sizes are 30" × 60" and 60" × 60". The dimensions refer to the spacing of the main runners and cross tees.

4.4.5 Luminous Ceiling Systems

Luminous ceiling systems are available in many styles, such as exposed-grid systems with drop-in plastic light diffusers and aluminum or wood framework with translucent acrylic light diffusers (*Figure 76*).

Fluorescent fixtures are generally installed above the translucent diffusers. Standard modules of 2' × 2' up to sizes of 5' × 5' are available,

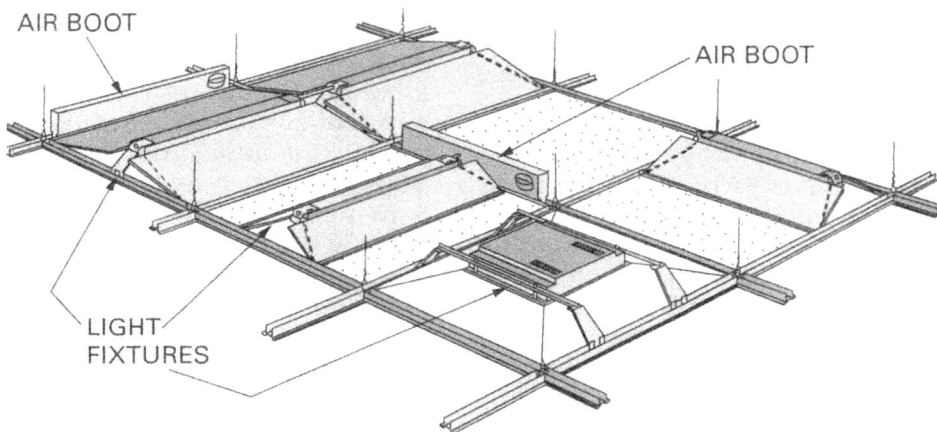

Figure 75 Typical integrated grid system.

102F75.EPS

Hold-Down Clips

Some ceiling panels require hold-down clips to secure the ceiling panels to the grid. For example, clips are used with lightweight panels to prevent them from reacting to drafts. One manufacturer specifies that clips be used if the panels weigh less than one pound. Hold-down clips are not necessarily required for ceilings used in fire-rated applications.

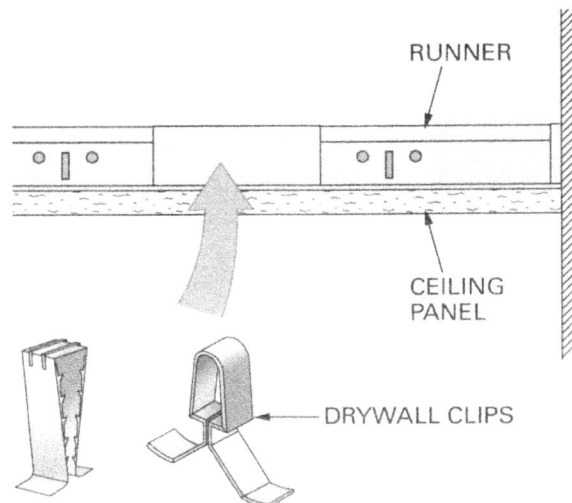

RUNNER

CEILING PANEL

DRYWALL CLIPS

102SA07.EPS

Figure 76 ❖ Typical integrated luminous ceiling system.

as are custom sizes for special applications. There are two types of luminous ceilings—standard and non-standard. Standard systems are, as their name indicates, those that are available in a series of standard sizes and patterns. Non-standard systems deviate from the normal spacing of main supports and/or have unusual sizes, shapes, and configurations of diffusing panels.

All surfaces in the luminous space, including pipes, ductwork, ceilings, and walls, are painted with a 75 to 90 percent reflectance matte white finish. Any surfaces in this area that might tend to flake, such as fireproofing and insulation, should receive an approved hard surface coating prior to painting to prevent flaking onto the ceiling below.

The installation of a standard luminous system is the same as for the exposed grid suspended system, with the exception of the border cuts. Luminous ceilings are placed into the grid members in full modules. Any remaining modules are filled in with acoustical material that has been cut to size.

With a 2' × 2' or 2' × 4' standard exposed grid system, luminous panels are used to provide the light diffusing element in the system. These panels are laid in between the runners. A variety of sizes and shapes of panels are available.

4.4.6 Suspended Drywall Furring Ceiling Systems

The suspended drywall furring system is used when it is desirable or specified to use a drywall finish or a drywall backing for an acoustical tile ceiling. Most suspended ceiling system manufacturers have their own proprietary direct-hung drywall suspension system (*Figure 77*). A job-built drywall suspension grid can be fabricated from the 1½" C-channel and furring channel members commonly used in steel framing. *Figure 78* shows how such a suspension system would look.

After the furring channels are in place, the drywall sheets are installed with drywall screws driven into the furring channel (*Figure 79*).

In some cases, the furring channels are attached directly to structural members such as steel beams or wooden joists, instead of to suspended carrying channels. In other cases, ceiling tiles are attached to the drywall.

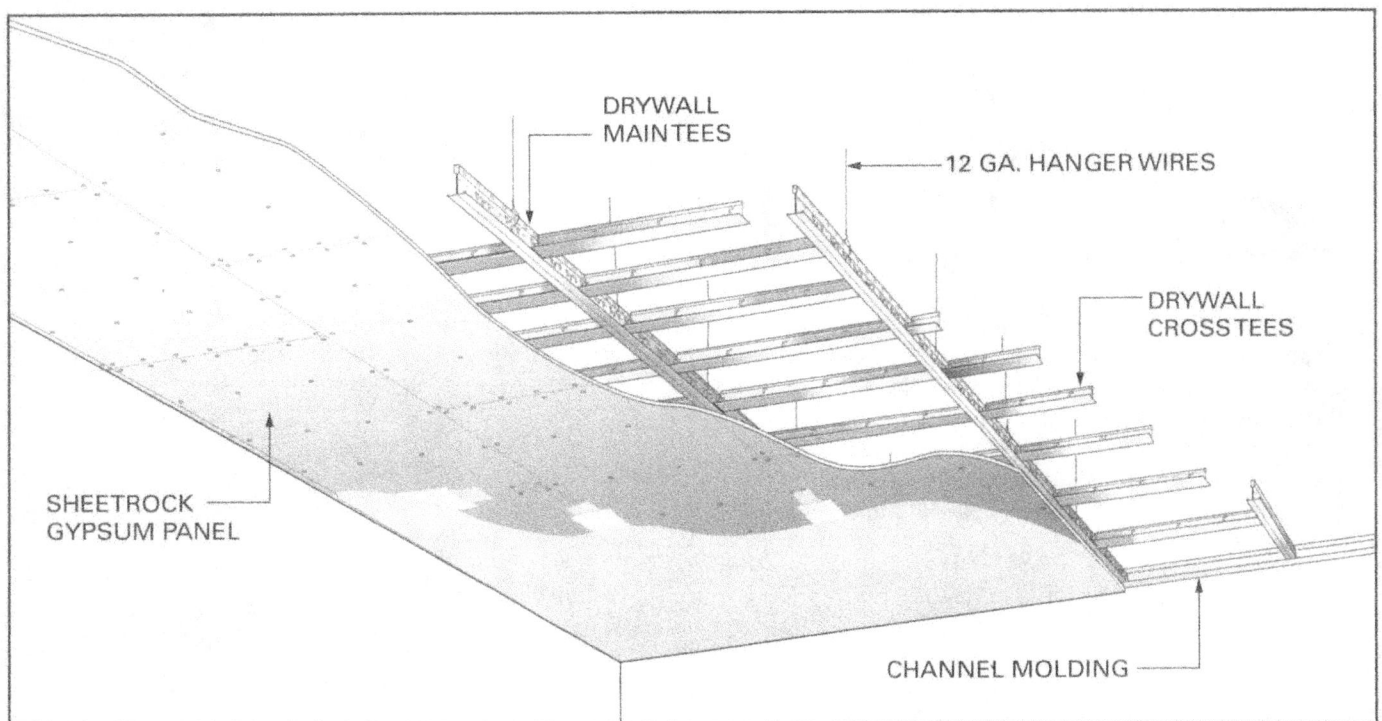

Figure 77 ❖ Proprietary drywall suspension system.

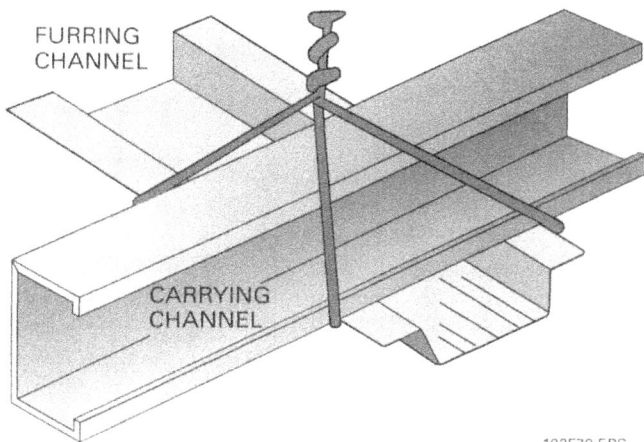

Figure 78　Direct-hung drywall suspension system.

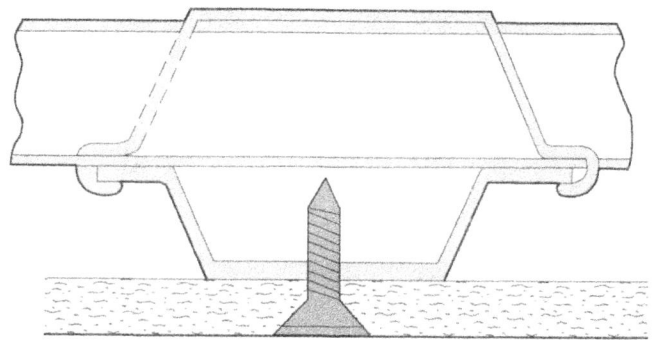

Figure 79　Drywall secured to furring channel.

INSIDE TRACK　　Plenum Ceilings

The systems that provide heating and cooling for most commercial buildings are forced-air systems. Blower fans are used to circulate the air. The blower draws air from the space to be conditioned and then forces the air over a heat exchanger, which cools or heats the air. In a cooling system, for example, the air is forced over an evaporator coil that has very cold refrigerant flowing through it. The heat in the air is transferred to the refrigerant, so the air that comes out the other side of the evaporator coil is cold. In homes, the air is delivered to the conditioned space and returned to the air conditioning/heating system through ductwork that is usually made of sheet metal. In commercial buildings with suspended ceilings, the space between the ceiling and the overhead decking is often used as the return air plenum. (A plenum is a sealed chamber at the inlet or outlet of an air handler.) This approach saves money by eliminating about half the cost of materials and labor associated with ductwork.

Anything in the plenum space (electrical or telecommunications cable, for example) must be specifically rated for plenum use in order to meet fire ratings. Plastic sheathing used on standard cables gives off toxic fumes when burned. Plenum-rated cable uses non-toxic sheathing.

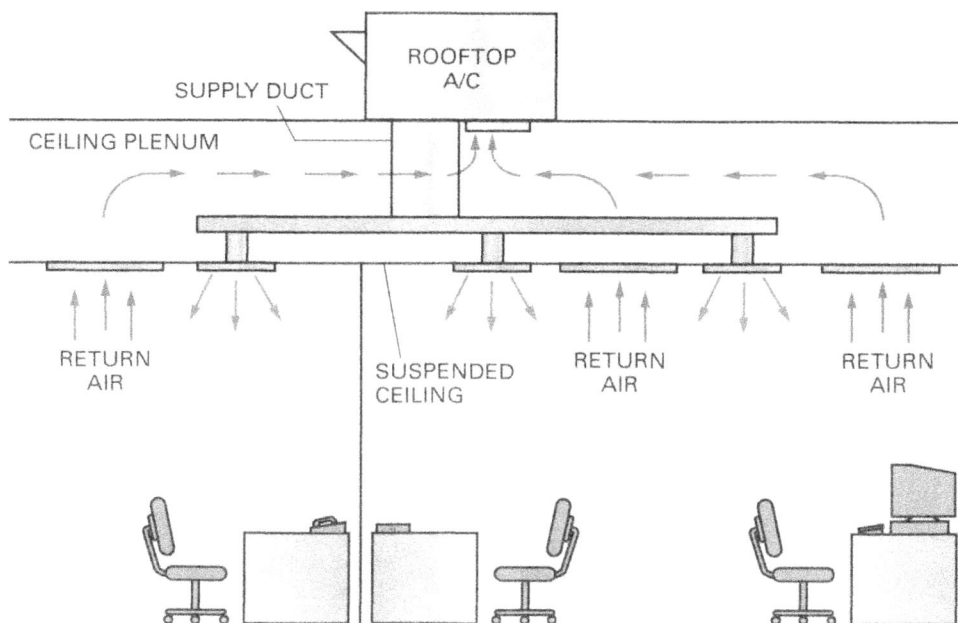

4.4.7 Special Ceiling Systems

There are numerous special ceiling systems that differ from those covered in this module. Some of these are the special metallic system, special pan system, planar system, mirrored (reflective) system, and translucent panel system (*Figures 80* through *84*).

Figure 80 ◆ Special metallic system.

Figure 81 ◆ Special pan system.

Figure 82 ◆ Planar system.

4.4.8 General Guidelines for Accessing Suspended Ceilings

The following are some general guidelines for working with suspended ceilings:

- Contact building maintenance personnel to find out how the ceiling is constructed and how to obtain spare panels in case some of the existing panels get damaged. They should also have the special tools you will need to get access to some types of ceilings. One type of concealed grid system, for example, has a special hook that is used to reach under the panel and release it from the cross member. As previously discussed, pan ceilings require special procedures for removing and installing pans.

- Do not force ceiling panels. Some panels are clipped to the gridwork. If that is the case, you will need to find the panel that is not clipped and start there. A special tool may be needed to release the clips.

Figure 83 ◆ Reflective ceiling.

Figure 84 ◆ Translucent panels.

- As discussed earlier, some ceilings have hinged panels that can be raised or lowered to provide access to the area above.
- Keep your hands clean to avoid staining the ceiling panels. If a panel gets dirty, try cleaning it with a damp sponge or an art gum eraser. Vinyl-faced fiberglass and mylar-faced ceilings can be cleaned with mild detergents or germicidal cleaners.
- Pan ceiling panels require special handling. Wear gloves or rub cornstarch on your hands to prevent the transfer of fingerprints to the panels.

5.0.0 ◆ FIRE-RATED AND SOUND-RATED CONSTRUCTION

Every wall, floor, and ceiling in a building is rated for its fire resistance, as established by building codes. The fire rating is stated in terms of hours, such as one-hour wall or two-hour wall. The rating denotes the length of time an assembly can withstand fire and provide safe evacuation of occupants, as determined under laboratory conditions (*Figure 85*). The greater the fire rating, the thicker the wall is likely to be.

In multi-family residential construction, such as apartments and townhouses, the walls and ceilings dividing the occupancies must meet special fire and soundproofing requirements. The code requirements will vary from one location to another and may even vary within areas of a jurisdiction. For example, dwellings in high-risk areas may have stricter standards than those in other areas of the same city or county.

In some cases, the code may require a masonry wall between occupancies. This masonry wall may even be required to penetrate the roof of the building so that if a fire occurs, it is contained within the unit in which it started because it is unable to travel through the walls or across the attic space.

There are many different construction methods for so-called party walls. Each is designed to meet different fire and soundproofing standards.

QUICK SELECTOR FOR FIRE-RATED ASSEMBLIES
PARTITIONS/WOOD FRAMING (LOAD BEARING)

SINGLE LAYER	REF.	DESIGN NO.	DESCRIPTION	STC	TEST NO.
45 MIN. 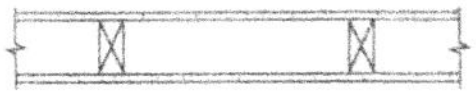	UL FM	U317 WI-45 MIN	½" FIRE-SHIELD GYPSUM WALLBOARD NAILED ON BOTH SIDES 2 × 4 STUDS, 16" OC.	34	NGC 2161
1 HR.	UL FM	U305 WI6A-1HR WP 3605	⅝" FIRE-SHIELD GYPSUM WALLBOARD OR ⅝" FIRE-SHIELD MR BOARD NAILED ON BOTH SIDES 2 × 4 WOOD STUDS, 16" OC.	35	NGC 2403
1 HR. 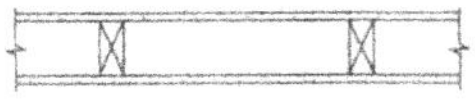	UL FM	U309 WI6B-1HR WP 3605	⅝" FIRE-SHIELD GYPSUM WALLBOARD OR ⅝" FIRE-SHIELD MR BOARD NAILED ON BOTH SIDES 2 × 4 WOOD STUDS, 24" OC.	38	NGC 2404
1 HR. 	FM GA	WIA-1HR (WP) 45 WP-1200	⅝" FIRE-SHIELD GYPSUM WALLBOARD OR ⅝" FIRE-SHIELD MR BOARD SCREW ATTACHED HORIZONTALLY TO BOTH SIDES 3⅝" SCREW STUDS, 24" OC. ALL WALLBOARD JOINTS STAGGERED.	42	NGC 2385
	OSU	T-1770	⅝" FIRE-SHIELD GYPSUM WALLBOARD SCREW ATTACHED VERTICALLY TO BOTH SIDES 3⅝" SCREW STUDS, 24" OC. ALL WALLBOARD JOINTS STAGGERED.		
DOUBLE LAYER					
2 HR. 	UL FM	U301 BASED ON WP 4135	2 LAYERS ⅝" FIRE-SHIELD GYPSUM WALLBOARD NAIL APPLIED TO 2 × 4 WOOD STUDS SPACED 16" OC. BOARDS MAY BE APPLIED HORIZONTALLY OR VERTICALLY WITH ALL JOINTS STAGGERED.	40	NGC 2363
2 HR. 	OSU GA	T-1771 BASED ON WP 1711	FIRST LAYER ⅝" FIRE-SHIELD GYPSUM WALLBOARD SCREW ATTACHED VERTICALLY TO BOTH SIDES 3⅝" SCREW STUDS, 24" OC. SECOND LAYER LAMINATED VERTICALLY ON BOTH SIDES. VERTICAL JOINTS STAGGERED.	48	NGC 2282

102F85.EPS

Figure 85 ◆ Specifications for typical fire-rated walls.

The wall is likely to be more than 3" thick and contain several layers of gypsum wallboard and insulation. A fire-rated wall may abut a non-rated partition or wall. In this case, the rated wall must be carried through to maintain the fire rating (*Figure 86*).

5.1.0 Firestopping

Firestopping means cutting off the air supply so that fire and smoke cannot readily move from one location to another. You will hear the term firestop used in two different ways.

In frame construction, a firestop is a piece of wood or fire-resistant material inserted into an opening such as the space between studs. This firestop acts as a barrier to block airflow that would allow the space to act as a chimney, carrying fire rapidly to upper floors. It does not put out the fire, but it slows the fire's progress.

In commercial construction and some residential applications, firestopping material is used to close wall penetrations such as those created to run conduit, piping, and air conditioning ducts. If such openings are not sealed, fire will travel through the openings in its search for oxygen.

In order to meet the fire rating standards established by the building and fire codes, the openings must be sealed. The firestopping methods are classified as mechanical and nonmechanical.

Mechanical firestops are devices that mechanically seal the opening.

Nonmechanical firestops are fire-resistant materials, such as caulks and putties, that are used to fill the space around the conduit or piping. You may be required to install various nonmechanical firestopping materials when working with fire-rated walls and floors. Holes or gaps affect the fire rating of a floor or wall. Properly filling these penetrations with firestopping materials maintains the rating.

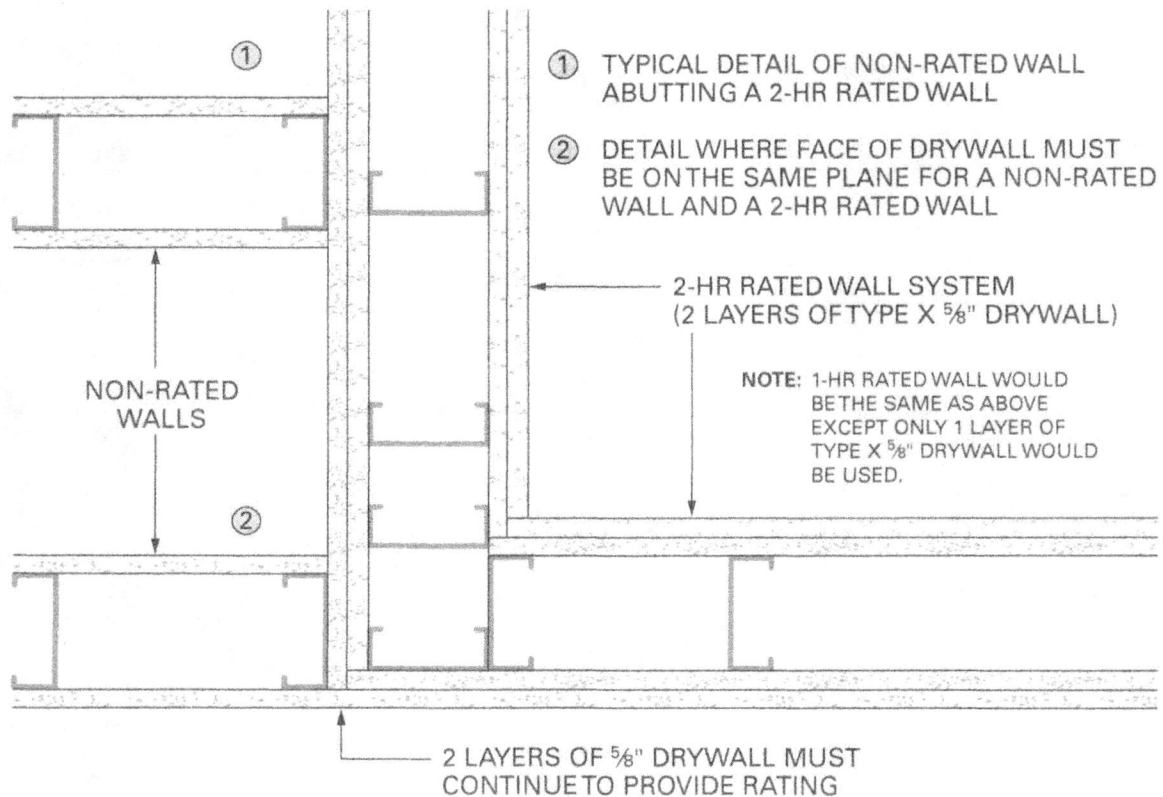

① TYPICAL DETAIL OF NON-RATED WALL ABUTTING A 2-HR RATED WALL

② DETAIL WHERE FACE OF DRYWALL MUST BE ON THE SAME PLANE FOR A NON-RATED WALL AND A 2-HR RATED WALL

2-HR RATED WALL SYSTEM (2 LAYERS OF TYPE X ⅝" DRYWALL)

NOTE: 1-HR RATED WALL WOULD BE THE SAME AS ABOVE EXCEPT ONLY 1 LAYER OF TYPE X ⅝" DRYWALL WOULD BE USED.

NON-RATED WALLS

2 LAYERS OF ⅝" DRYWALL MUST CONTINUE TO PROVIDE RATING

102F86.EPS

Figure 86 ● Example of a fire-rated wall abutting a non-rated wall.

5.2.0 Sound Isolation Construction

Walls and ceilings must also be built for sound containment. To effectively isolate sound, all air leaks and flanking paths must be closed off. Ensure that areas where noise can easily travel around, such as around walls, through windows and doors, air ducts, and crawl spaces, are properly closed off. Buildings must meet an STC rating, which is a numeric rating representing how effective the structure is at isolating airborne sound transmission. The higher the STC rating, the better the sound containment. STC ratings can be negatively impacted by hairline cracks or other openings. You will learn more about STC ratings in the module on *Drywall Installation*.

6.0.0 PROJECT SCHEDULES

A construction project requires a lot of planning and scheduling because different trades, equipment, and materials are needed at different times in the process. Drywall is normally installed when the building has been dried-in (*i.e.*, the exterior siding and roofing are applied so the building remains dry, but the framing is exposed in the interior of the building).

Project planning and scheduling will be covered in more detail later in your training. For now, *Figures 87* and *88* provide an overview of where each trade fits into the construction process for residential and commercial projects, respectively.

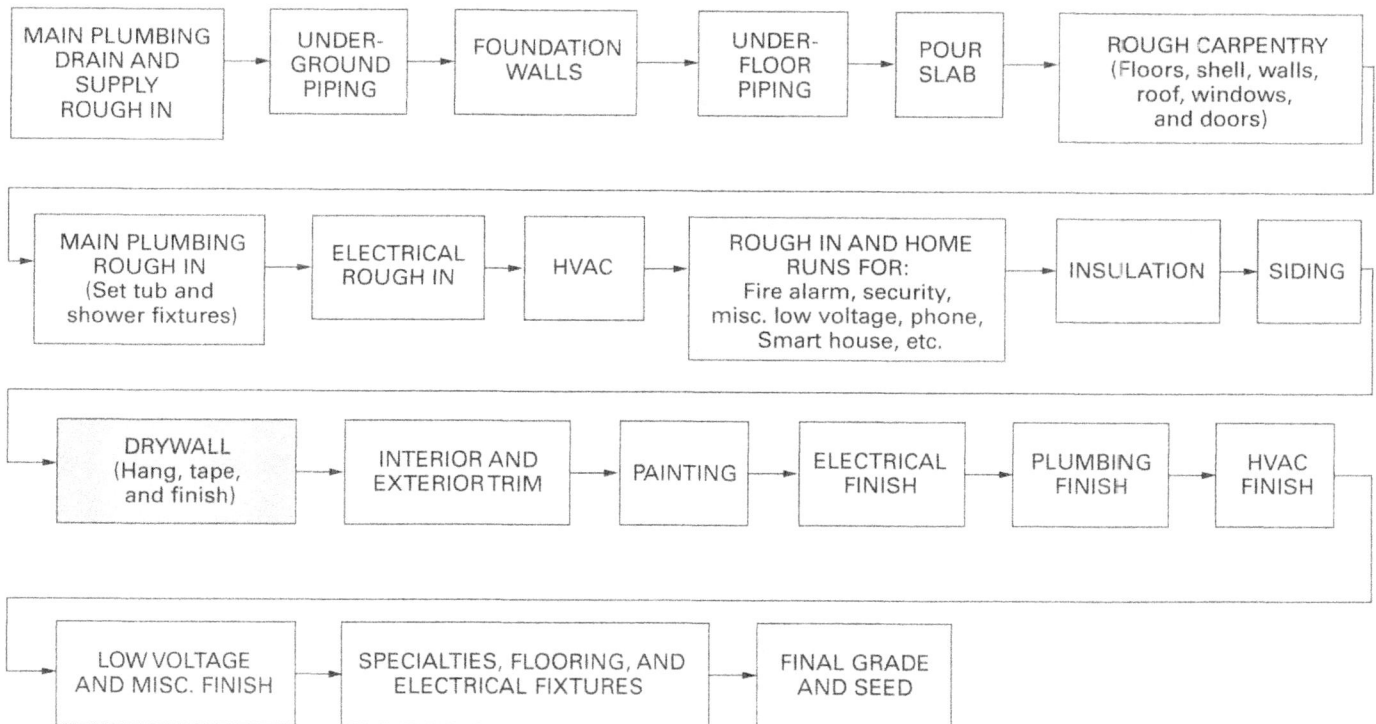

MAIN PLUMBING DRAIN AND SUPPLY ROUGH IN → UNDER-GROUND PIPING → FOUNDATION WALLS → UNDER-FLOOR PIPING → POUR SLAB → ROUGH CARPENTRY (Floors, shell, walls, roof, windows, and doors)

MAIN PLUMBING ROUGH IN (Set tub and shower fixtures) → ELECTRICAL ROUGH IN → HVAC → ROUGH IN AND HOME RUNS FOR: Fire alarm, security, misc. low voltage, phone, Smart house, etc. → INSULATION → SIDING

DRYWALL (Hang, tape, and finish) → INTERIOR AND EXTERIOR TRIM → PAINTING → ELECTRICAL FINISH → PLUMBING FINISH → HVAC FINISH

LOW VOLTAGE AND MISC. FINISH → SPECIALTIES, FLOORING, AND ELECTRICAL FIXTURES → FINAL GRADE AND SEED

102F87.EPS

Figure 87 Typical residential construction schedule.

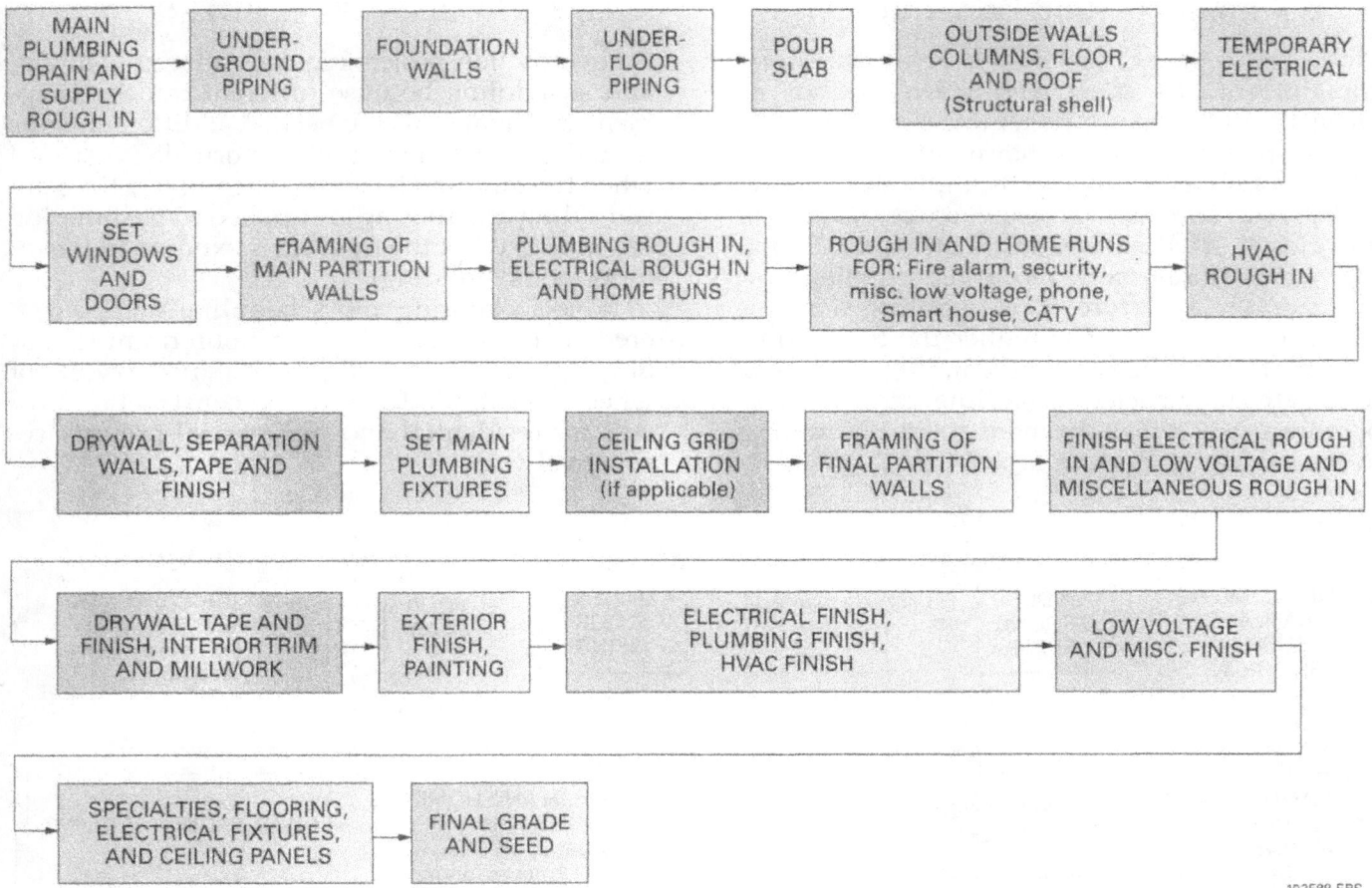

Figure 88 ◆ Typical commercial construction schedule.

102F88.EPS

1. Which of the following materials must be handled with special care because it is treated with hazardous chemicals?
   a. Mold-resistant gypsum drywall
   b. Metal pans
   c. Brick
   d. Plywood

2. Gypsum wallboard with foil backing is used as a _____.
   a. backer for tile
   b. decorator panel
   c. vapor barrier
   d. liner in elevator shafts

3. The typical concrete block used in loadbearing construction is about _____ deep.
   a. 2"
   b. 4"
   c. 6"
   d. 8"

4. Which of the following is an *incorrect* statement regarding trusses?
   a. Ductwork and wiring are easily run through a truss's webbing.
   b. Trusses are generally faster and easier to erect.
   c. Trusses place more restrictions on building design than dimension lumber.
   d. Trusses are stronger than comparable lengths of dimension lumber.

5. The main horizontal framing member of a floor is known as a _____.
   a. girder
   b. joist
   c. lally column
   d. rafter

6. Gypsum board substrate is used primarily for _____.
   a. tile backing
   b. shaft liners
   c. fire resistance
   d. floor and roof assemblies

7. The purpose of a strongback is to _____.
   a. provide a foundation for the ribband
   b. attach the joist to the sill plate
   c. support ceiling joists
   d. brace studs

8. In plank-and-beam construction, framing members are typically installed _____ OC.
   a. 12"
   b. 16"
   c. 24"
   d. 48"

9. Floors in large office buildings are typically made of _____.
   a. concrete
   b. metal joists and wood sheathing
   c. steel beams
   d. galvanized aluminum

10. Concrete tilt-up panels are typically _____ thick.
    a. 1" to 2"
    b. 2" to 3"
    c. 3" to 4"
    d. 5" to 8"

11. The wall shown in *Figure 1* is most likely located between _____.
    a. a doctor's office and the waiting room
    b. the garage and kitchen of a home
    c. a manufacturing area and an office
    d. a bedroom and a bathroom

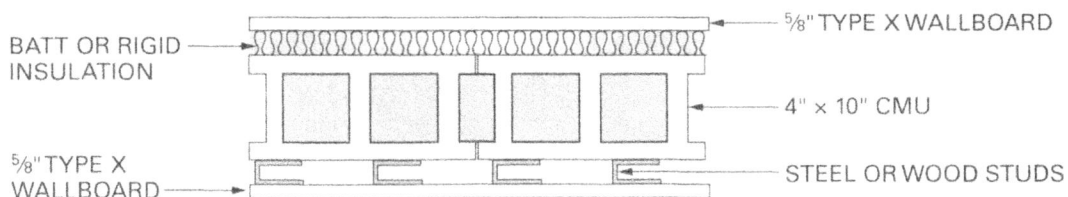

BATT OR RIGID INSULATION

⅝" TYPE X WALLBOARD

⅝" TYPE X WALLBOARD

4" × 10" CMU

STEEL OR WOOD STUDS

102RQ01.EPS

Figure 1

12. Gypsum core board is manufactured in what thickness?
    a. ⅜ inch
    b. ½ inch
    c. ⅝ inch
    d. 1 inch

13. When drywall is used in large commercial construction projects, the drywall sheets are typically attached to _____.
    a. the underside of the concrete floor above using powder-actuated fasteners
    b. the wood joists
    c. suspended furring channels
    d. steel trusses

14. Fire resistance ratings are measured in _____.
    a. hours
    b. inches (of wall thickness)
    c. minutes
    d. millimeters (of wall thickness)

15. In a commercial construction schedule, drywall installation first occurs _____.
    a. after exterior finishing
    b. after plumbing, electrical, and HVAC rough-ins
    c. before partition wall framing
    d. before underground piping is set

Summary

When you are called upon to install drywall in any kind of environment, whether it is residential or commercial, new construction or modification, it will be important for you to know how the building is put together. With that knowledge, you can anticipate potential problems and select the proper board for an application.

In addition to construction materials and schedule, you will need to have a basic understanding of fire and sound ratings, and how drywall complies with them.

Notes

Clay Kubicek

Clay Kubicek is the education director for Crossland Construction, a midwestern construction company that was recently ranked among the top 400 construction companies in the United States by the *Engineering News-Record*.

Tell us a little bit about what you do for Crossland and its employees.

The Crossland family asked me to join their team to coordinate and build upon their existing education program. As Crossland Construction's education director I work with all in-house training and I assist area public schools in the development of a 5th through 12th grade construction technology program. In these two areas my job is to pull together people and resources to build the education vision of Crossland Construction.

We develop and maintain training programs for all employees. This is done through the Crossland Academy. The Academy currently offers over 30 classes for Crossland employees. These classes range from a one-day policies and procedures class, to a four-year Kansas Department of Labor approved carpentry apprenticeship program. In this four-year program Crossland trainees have the opportunity to earn a journeyman carpentry card. In addition to the journeyman carpentry course, all our safety programs use NCCER curriculum, so our employees also have a transcript at the National Registry. From computer training to project management seminars, the Crossland Academy is actively trying to meet the educational needs of a growing company. In the coming year we will be working to develop a four-year ironworker program in addition to our other programs.

With area public schools my job is to help them develop their construction education programs. Crossland has developed a partnership plan with some area schools in which we provide curriculum, resources, and materials to provide construction career paths for high school students. The Crossland school program is arranged to expose 5th and 6th grade students to commercial construction through field trips; give 7th and 8th graders a chance to explore construction through field trips and one-week modules using Pitsco learning systems, and culminate with 9th through 12th graders using NCCER's Construction Technology curriculum. The students become Core certified in their 9th grade year. In grades 10 and 11 the Construction Technology curriculum has been developed with the Kansas State Department of Education to give students an overview and background in masonry, concrete, carpentry, HVAC, plumbing, and electrical work. The 12th grade option is a dual-credit course in project management.

Interested students are able to earn college credit using this curriculum program. Field trips, guest speakers, subject matter experts are offered to the schools at every grade level as the students move through the program.

The students have the option of taking their NCCER certifications and entering the labor force on their way to becoming a skilled craftsperson, as well as entering a two- or four-year college program in a construction-related field of study. This "Crossland Connection" with area public schools is Crossland's way of showing the students all the opportunities they can find in the construction industry.

The Crossland family is committed to education, and my job is to assist in developing a culture of education. Schools interested in examining this program can find an outline at www.crossland construction.com and use the education link.

What advice do you have for trainees as they embark on their careers? What should they look for in an employer?

Keeping a positive attitude is one bit of advice I would give to anyone pursuing a goal. I know it is someone else's thunder, but there is power in positive thinking. Have a can-do attitude. Complete your training with superb attendance and participation. Technology is and will continue to change the way training and education take place in the workplace. Stay current with technology. Where the term "life-long learner" was a catchphrase, it is now a necessity for advancement in any field. Another thing is to be careful not to just focus on today's pay. Look long-term and have vision. When looking for an employer in the construction industry, I believe a trainee would benefit in the long run by looking for a company that wants to invest in the trainee's future, not just their own. Find a company with plans to expand and grow. Look for a company that offers comprehensive training opportunities, a retirement plan, insurance, and a history of looking after their employees. Make sure you are working with people of integrity.

What should trainees do to ensure their progress toward their career goals? What do employers look for in potential employees?

There are multiple things a trainee can do to ensure they reach their career goals. Ask and seek the answers to their questions. If they are curious about how something is done, take the time to find the answer. We are in the information age, and to reach any career goal it is imperative we know how to find needed and relevant information. Keep current on materials and construction trends. Once again technology has created many changes in the construction industry. Stay current. It is also important to plan. They need to have a plan to reach their goals, and follow that plan. Most quality companies reach their goals by having integrity. Trainees should follow that lead. Find a role model and mentor. Find a person you would like to emulate. Learn from success. Finally, develop your craft skills as well as your human relationship skills.

Crossland looks for people who have a can-do attitude, have a sense of urgency in their performance style, and are willing to learn and grow. Crossland is looking for team players who can cooperate and work within a diverse and growing company. Any person with a strong work ethic could find many opportunities with Crossland Construction.

This book is designed for first-year drywall trainees. In a few years, most of them will have moved into their careers and will begin progress toward their goals. What should craftworkers do to keep up with industry developments?

One thing is to take advantage of relevant training opportunities available within their company. If their company does not offer the training they desire, seek out another source. Craftworkers should inquire of their superintendents about their company's future needs and training options. It is important to stay open-minded and flexible. Once again, we should all keep our eyes on the growing role of technology in the construction industry. As with any profession, staying current with industry trends and practices requires reading and research. Subscribing to a couple of industry-driven magazines and journals helps in keeping up-to-date on industry developments. For tools and techniques there are a number of journals available. For a larger perspective, the *Engineering News-Record* provides quality industry insight. To keep abreast with the business side of the industry *Construction Executive: the Magazine for the Business of Construction* provides good insight. There are numerous sources available on the Internet also. If trainees are interested in pursuing a professional career in construction, they should be on a constant search for knowledge and information concerning the industry. A craftworker must work with many aspects of a construction job. Exemplary craftworkers have a knowledge base of all aspects of the construction process.

Trade Terms Introduced in This Module

Admixture: Any material that is added to a concrete mixture to obtain additional properties.

APA-rated: Building material that has been rated by the American Plywood Association for a specific use.

Blocking: A wood block used as a filler piece and support member between framing members.

Bridging: Wood or metal pieces placed diagonally between joists to provide added support.

Cantilever: A beam, truss, or floor that extends beyond the last point of support.

Corrugated: Material formed with parallel ridges or grooves.

Cripple stud: In wall framing, a short framing stud that fills the space between the header and the top plate or between the sill and the soleplate.

Dimension lumber: Any lumber within a range of 2" to 5" thick and up to 12" wide.

Dormer: A framed structure that projects out from a sloped roof.

Double top plate: A length of lumber laid horizontally over the top plate of a wall to add strength to the wall.

Fire rating: A classification indicating in time (hours) the ability of a structure or component to withstand fire conditions.

Firestop: A piece of lumber or fire-resistant material installed in an opening to prevent the passage of fire.

Firestopping: A material or mechanical device used to block openings in walls, ceilings, and floors to prevent the passage of fire and smoke.

Footing: The foundation for a column or the enlargement placed at the bottom of a foundation wall to distribute the weight of the structure.

Furring strips: Strips of wood or metal applied to a wall or other surface to make it level, form an air space, and/or provide a fastening surface for finish covering.

Gable: The triangular wall enclosed by the sloping ends of a ridged roof.

Girder: The main steel or wood supporting beam for a structure.

Green concrete: Concrete that has hardened, but has not yet gained its full structural strength.

Gypsum: A chalky type of rock that serves as the basic ingredient of plaster and gypsum wallboard.

Gypsum wallboard: A generic term for gypsum core panels covered with paper on both sides. It is commonly used to finish walls.

Header: A horizontal member that supports the load over an opening such as a door or window. Also known as a lintel.

Kerf: A groove or notch made by a saw.

Millwork: Various types of manufactured wood products such as doors, windows, and moldings.

Oriented strand board (OSB): Panels made from layers of wood strands bonded together.

Plastic concrete: Concrete in a liquid or semi-liquid workable state.

Plenum: A sealed chamber for moving air under slight pressure at the inlet or outlet of an air conditioning system. In some commercial buildings, the space above a suspended ceiling often acts as a return air plenum.

Post-tensioned concrete: Concrete placed around steel reinforcement such as rods or cables that are isolated from the concrete. After the concrete has cured, tension is applied to the rods or cables to provide greater structural strength.

Pre-stressed concrete: Concrete that is placed around pre-stressed reinforcing steel in a casting bed. This type of concrete cannot be cut without first consulting a structural engineer.

Rabbeted: A board or panel with a groove cut into one or more of its edges.

Rafter: A sloping structural member of a roof frame to which sheathing is attached.

Reinforced concrete: Concrete that has been placed around some type of reinforcing material, usually steel.

Ribband: A 1 × 4 nailed to ceiling joists to prevent twisting and bowing of the joists.

Shakes: A handsplit wood shingle.

Sheathing: The sheet material or boards used to close in walls and roofs.

Shiplap: Lumber with edges that are shaped to overlap adjoining pieces.

Sill plate: A horizontal timber that supports the framework of a building. It forms the transition between the foundation and the frame.

Soleplate: The lowest horizontal member of a wall or partition. It rests directly on the rough floor.

Striated: A surface design that has the appearance of fine parallel grooves.

Stringer: The support member at the sides of a staircase; also, a timber used to support formwork for a concrete floor.

Strongback: An L-shaped arrangement of lumber used to support ceiling joists and keep them in alignment. In concrete work, it represents the upright support for a form.

Stucco: A type of plaster used to coat exterior walls.

Sub floor: Panels or boards fastened to the tops of floor joists.

Substrate: The underlying material to which a finish is applied.

Top plate: The upper horizontal member of a wall or partition frame.

Trimmer joist: A full-length horizontal member that forms the sides of a rough opening in a floor. It provides stiffening for the frame.

Trimmer stud: The vertical framing member that forms the sides of a rough opening for a door or window. It provides stiffening for the frame and supports the weight of the header.

Truss: An engineered assembly made of wood or metal that is used in place of individual structural members such as the joists and rafters used to support floors and roofs.

Underlayment: A material such as plywood or particleboard that is installed on top of a subfloor to provide a smooth surface for the finish flooring.

Vaulted ceiling: A high, open ceiling that generally follows the roof pitch.

Veneer: The covering layer of material for a wall or the facing materials applied to a substrate.

Additional Resources and References

Additional Resources

This module is intended to be a thorough resource for task training. The following reference work is suggested for further study. This is optional material for continued education rather than for task training.

Gypsum Construction Handbook. Chicago, IL: United States Gypsum Company, 2000.

Figure Credits

NCCER CURRICULA — USER UPDATE

NCCER makes every effort to keep its textbooks up-to-date and free of technical errors. We appreciate your help in this process. If you find an error, a typographical mistake, or an inaccuracy in NCCER's curricula, please fill out this form (or a photocopy), or complete the online form at **www.nccer.org/olf**. Be sure to include the exact module ID number, page number, a detailed description, and your recommended correction. Your input will be brought to the attention of the Authoring Team. Thank you for your assistance.

Instructors – If you have an idea for improving this textbook, or have found that additional materials were necessary to teach this module effectively, please let us know so that we may present your suggestions to the Authoring Team.

NCCER Product Development and Revision
13614 Progress Blvd., Alachua, FL 32615

Email: curriculum@nccer.org
Online: www.nccer.org/olf

❏ Trainee Guide   ❏ Lesson Plans   ❏ Exam   ❏ PowerPoints   Other _____

Craft / Level: _____   Copyright Date: _____

Module ID Number / Title: _____

Section Number(s): _____

Description: _____

_____

_____

_____

_____

Recommended Correction: _____

_____

_____

_____

_____

Your Name: _____

Address: _____

_____

Email: _____   Phone: _____

45103-07

Thermal and
Moisture Protection

45103-07
Thermal and Moisture Protection

Topics to be presented in this module include:

Overview

A properly insulated building will be comfortable to live or work in and will be economical to heat and cool. Without insulation, warm air will escape the building in cold weather, causing the heating system to operate constantly. This results in an increased use of energy. In hot weather, the air conditioning system will have to work harder, with the same results. Proper insulation in walls, floors, and roof decks will minimize this problem. Vapor barriers are also important. They must be used to prevent moisture from penetrating the building. Moisture can cause a variety of serious problems, including wood decay and mold growth. A skilled craftsman will know how to select and install insulating materials and vapor barriers.

Objectives

When you have completed this module, you will be able to do the following:

1. Describe the requirements for insulation.
2. Describe the characteristics of various types of insulation material.
3. Calculate the required amounts of insulation for a structure.
4. Install selected insulation materials.
5. Describe the requirements for moisture control and ventilation.
6. Install selected vapor barriers.
7. Describe various methods of waterproofing.
8. Describe air infiltration control requirements.
9. Install selected building wraps.

Trade Terms

Condensation	Permeability
Convection	Permeable
Dew point	Permeance
Diffusion	Vapor barrier
Exterior insulation finish	Water stop
system (EIFS)	Water vapor
Perm	

Required Trainee Materials

1. Pencil and paper
2. Appropriate personal protective equipment

Prerequisites

Before you begin this module, it is recommended that you successfully complete *Core Curriculum*; and *Drywall Level One,* Modules 45101-07 and 45102-07.

This course map shows all of the modules in the first level of the *Drywall* curriculum. The suggested training order begins at the bottom and proceeds up. Skill levels increase as you advance on the course map. The local Training Program Sponsor may adjust the training order.

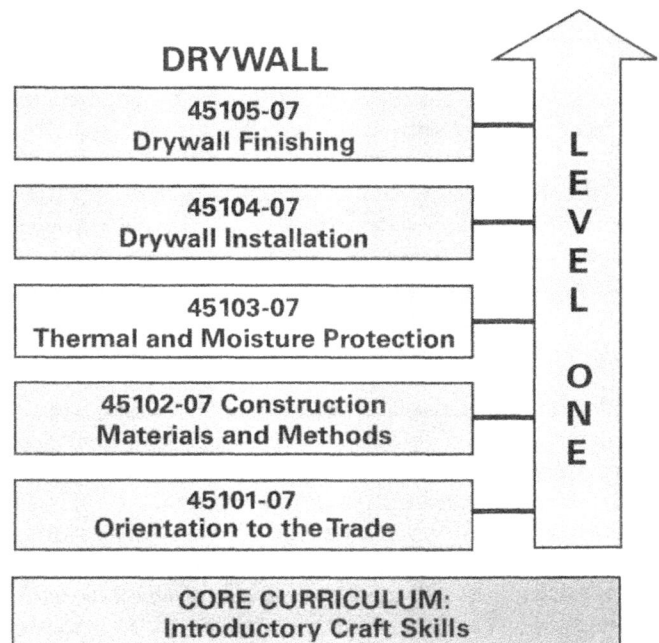

DRYWALL

45105-07 **Drywall Finishing**	L
45104-07 **Drywall Installation**	E
45103-07 **Thermal and Moisture Protection**	V E L
45102-07 Construction **Materials and Methods**	O
45101-07 **Orientation to the Trade**	N E

CORE CURRICULUM:
 Introductory Craft Skills

103CMAP.EPS

1.0.0 ◆ INTRODUCTION

Four important considerations for the construction of any building include the following: thermal insulation, moisture control and ventilation, waterproofing, and air infiltration control. This module covers these areas and presents materials and procedures that can be applied to ensure effective installations.

Vapor retarders, also called vapor barriers, are an important part of moisture control. A vapor barrier is any material that prevents the passage of water. A properly installed vapor barrier will protect ceilings, walls, and floors from moisture originating within a heated space.

Some vapor barrier materials, such as kraft paper, are attached to blanket or batt insulation. They are installed when insulation is installed. Others, such as aluminum foil, may be applied to the back of gypsum drywall during its installation. Polyethylene film used as a vapor barrier is applied over studs and ceiling joists after insulation is installed. Vapor barriers are also installed under slabs, between the gravel cushion and the poured concrete.

2.0.0 ◆ THERMAL INSULATION

Most materials used in construction have some insulating value. Air is an excellent insulator if it is confined to very small spaces and is kept very still. Manufactured insulation material is based on trapping a large amount of air in a large number of very small spaces to provide resistance to the transfer of heat and sound. Double-pane and triple-pane windows use this method to reduce heat loss.

The amount of insulation in a building directly affects heating and cooling costs. It also affects the value of the building. Some jurisdictions require that a permanent certificate be placed on the building's electrical box stating the insulative properties of material used in the structure. When required, this certificate is completed by the builders or designer.

2.1.0 Thermal Resistance Value of Materials

A law of physics is that heat will always flow (or conduct) through any material or gas from a higher temperature area to a lower temperature area.

The term *R-value* refers to the resistance to conductive heat flow through a material or gas. R-value is expressed as:

$$R = \frac{1}{k} \text{ or } \frac{1}{C}$$

There are many different types of insulation. Each type has specific applications for which it is best suited.

103SA01.EPS

Where

k = amount of heat in British thermal units (Btu) transferred in one hour through 1 sq ft of a material that is 1" thick and has a temperature difference between its surfaces of 1°F; also called the coefficient of thermal conductivity

C = conductance of a material, regardless of its thickness; the amount of heat in Btus that will flow through a material in one hour per sq ft of surface with 1°F of temperature difference

R = thermal resistance; the reciprocal (opposite) of conductivity or conductance

The higher the R-value, the lower the conductive heat transfer. *Table 1* shows the R-values of a number of common building materials including some common insulating materials.

The total heat transmission through a wall, roof, or floor of a structure in Btu per sq ft per hour with a 1°F temperature difference is called the total heat transmission or U-value. It is expressed as follows:

$$U = \frac{1}{R_1 + R_2 + \ldots R_n}$$

Where

$R_1 + R_2 + \ldots R_n$ represents the sum of the individual R-values for the materials that make up the thickness of the wall, roof, or floor

Table 1 R-Values of Common Materials

Material	Thickness	R-Value° F/sq ft/hr in Btus
Air film and spaces		
Air space bound by ordinary materials	¾"	0.96
Air space bound by ordinary materials	¾" to 4"	0.94
Exterior surface resistance	--	0.17
Interior surface resistance	--	0.68
Masonry units		
Sand and gravel concrete block	4"	0.71
Sand and gravel concrete block	8"	1.11
Sand and gravel concrete block	12"	1.28
Lightweight concrete block	4"	1.50
Lightweight concrete block	8"	2.00
Lightweight concrete block	12"	2.27
Face brick	4"	0.44
Common brick	4"	0.80
Masonry materials		
Concrete, oven-dried sand, and gravel aggregate	1"	0.11
Concrete. undried sand, and gravel aggregate	1"	0.08
Stucco	1"	0.20
General building materials		
Wood sheathing or subfloor	¾"	0.94
Fiberboard sheathing (regular density)	½"	1.32
Fiberboard sheathing (intermediate density)	½"	1.14
Fiberboard sheathing (nailbase)	½"	1.14
Plywood	⅜"	0.47
Plywood	½"	0.62
Plywood	¾"	0.93
Bevel-lapped wood siding	½" × 8"	0.81
Bevel-lapped wood siding	¾" × 10"	1.05
Vertical tongue-and-groove (cedar or redwood)	¾"	1.00
Gypsum board	⅜"	0.32
Gypsum board	½"	0.45
Interior plywood paneling	¼"	0.31
Building paper (permeable felt)	--	0.06
Plastic film	--	0.00
Insulating materials		
Fibrous batts (from rock slag or glass)	2" to 2¾"	7.00
Fibrous batts (from rock slag or glass)	3" to 3½"	11.00
Fibrous batts (from rock slag or glass)	5" to 5½"	19.00
SM brand Styrofoam® plastic foam	1"	5.41
TG brand Styrofoam® plastic foam	1"	5.41
IB brand Styrofoam® plastic foam	1"	4.35
Molded polystyrene beadboard	1"	4.17
Polyurenthane foam	1"	5.88
Woods		
Fir, pine, and similar softwoods	¾"	0.94
Fir, pine, and similar softwoods	1½"	1.89
Fir, pine, and similar softwoods	2½"	3.12
Fir, pine, and similar softwoods	3½"	4.35
Maple, oak, and similar hardwoods	1"	0.91

103T01.EPS

Air Versus Inert Gas

The air trapped between the panes of double-pane and triple-pane windows makes a good insulator, but there are better alternatives. Inert gases that block more heat than air create a stronger barrier. That's why some window manufacturers fill the space between the panes with argon—an inert, nontoxic, nonflammable gas.

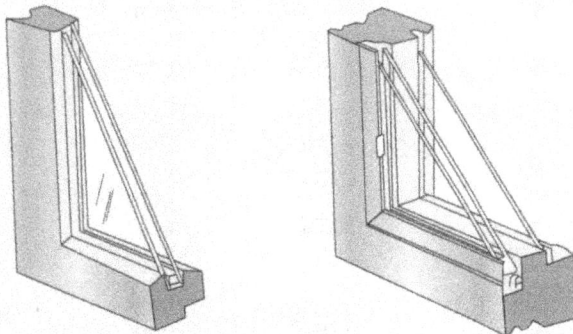

103SA02.EPS

The lower the U-value, the lower the heat transmission.

While the R-values provide a convenient measure to compare heat loss or gain, the total U-value for a structure is used in the calculations for sizing the structure's heating and cooling equipment (*Figure 1*).

By doubling the R-value of a wall or roof, the conductive heat loss or gain can theoretically be reduced by half. However, it is important to note that as insulation thicknesses are increased, the heat transmission (U-value) is decreased, but not in a direct relationship. Increases of insulation will continue to decrease heat loss, but at lower and lower percentages. At some point, it becomes economically useless to add more insulation. The same is true for double-, triple-, and

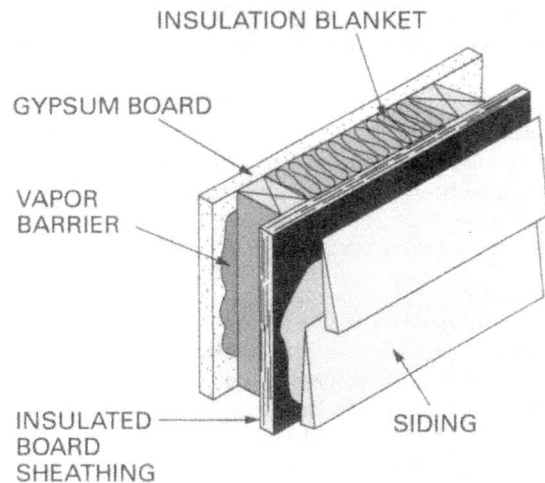

INSULATION BLANKET

GYPSUM BOARD

VAPOR BARRIER

INSULATED BOARD SHEATHING

SIDING

2 × 4 STUD WALL WITH RIGID BOARD:

TYPE	R-VALUE
AIR FILMS*	0.9
¾" WOOD EXTERIOR SIDING	1.0
1" POLYSTYRENE RIGID BOARD	5.0
3½" BATT OR BLANKET INSULATION	11.0
VAPOR BARRIER	0.0
½" GYPSUM BOARD	0.5
	18.4 TOTAL R

2 × 6 INSULATED STUD WALL:

TYPE	R-VALUE
AIR FILMS*	0.9
¾" WOOD EXTERIOR SIDING	1.0
¾" INSULATION BOARD	2.0
5½" INSULATING BLANKET	19.0
VAPOR BARRIER	0.0
½" GYPSUM BOARD	0.5
	23.4 TOTAL R

*Stagnant air film that forms on any surface

103F01.EPS

Figure 1 ◆ R-values of typical wall construction.

quadruple-pane windows. It must also be noted that conductive heat loss or gain does not include heat gains or losses due to air leaks or radiation through windows or other openings.

Over-Insulating

Installing excess insulation wastes money and may cause other problems. If a building is over-insulated and lacks sufficient ventilation and water barrier protection, moisture can collect inside. This promotes the growth of mold and fungus. It is even possible that cancer-causing radon gas could be trapped in the building, accumulating over time to dangerous levels.

2.2.0 Insulation Requirements

Increasing energy costs and mandated government energy conservation have resulted in much higher R-value requirements for new construction. While building code and design standards for insulation have been traditionally based on average low-temperature zones and charts based on the range of low temperatures expected (see *Appendix*), the requirements are constantly changing. The International Code Council now recommends insulation values be based on climate zones, which are determined by local temperature and humidity levels (*Figure 2* and *Table 2*).

In warm climates of the country, many codes now require almost the same amount of insulation for air conditioning as the cold climates require for heating. In some cases, codes are using comfort standards similar to those shown in *Table 3*. The all-weather standard requires that the insulation provided must be adequate to maintain a desired interior temperature during periods of extreme high and low outside temperatures.

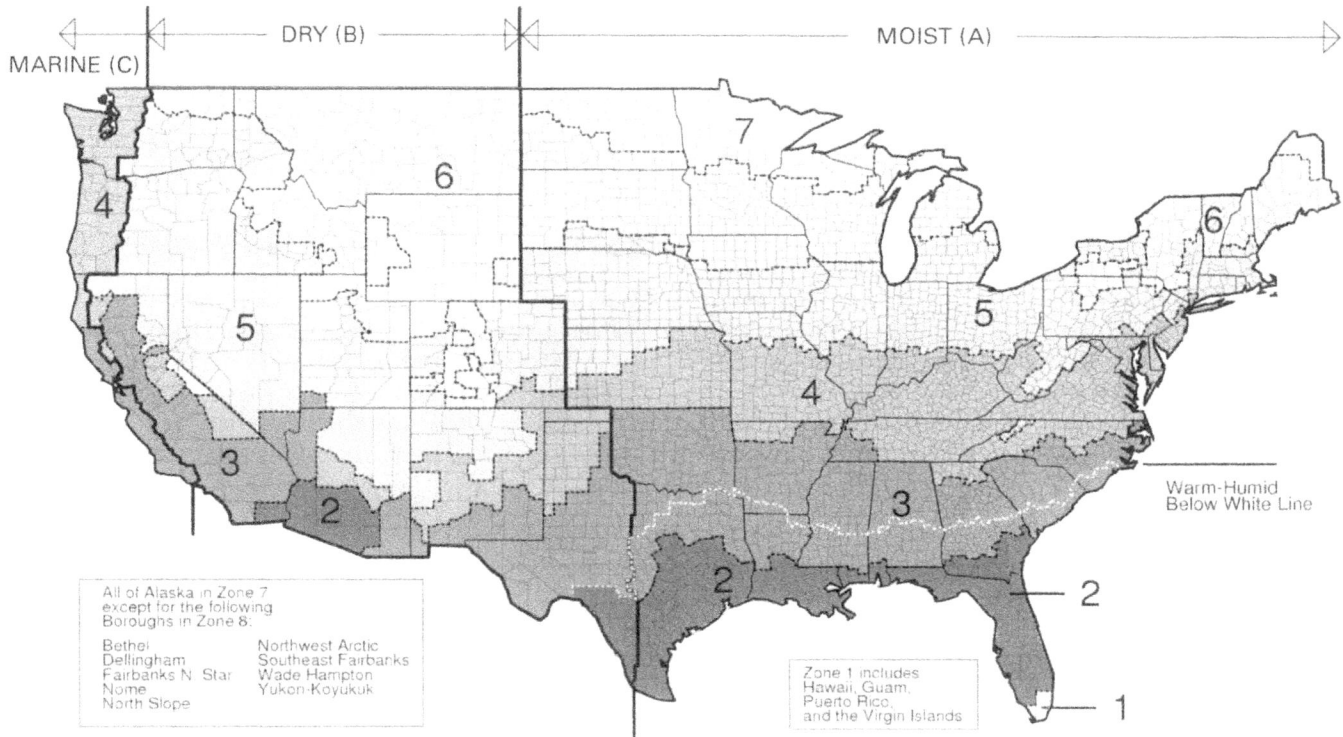

103F02.EPS

Figure 2 Climate zones in the United States.

Table 2 Recommended R-Values of Insulation

Climate Zone	Floors	Walls[1]	Ceilings
1	13	13	30
2	13	13	30
3	19	13	30
4 except Marine	19	13	38
5 and Marine 4	30^2	19^2	38
6	30^2	19^2	49
7 and 8	30^2	21	49

[1]Wood frame walls
[2]Alternative options exist

103T02.EPS

Table 3 Typical Comfort Standards

Comfort Standard	Insulation Location	Insulation R-Value
All-weather	Walls	R-19
	Ceilings	R-30 to R-38
	Floors	R-19
Moderate	Walls	R-13
	Ceilings	R-26
	Floors	R-13
Minimum	Walls	R-11
	Ceilings	R-19
	Floors	R-11

103T03.EPS

To meet the moderate standard, the insulation provided must be adequate to maintain a desired temperature during periods of average outside temperature extremes.

In general, insulation must be installed where any exterior surface of a structure is exposed to a thermal difference relative to its internal surface. These areas are:

- Roofs
- Above ceilings
- In exterior walls
- Beneath floors over crawl spaces
- Around the perimeter of concrete floors and around foundations

As you study the information in *Tables 1* and *2*, you will notice that the ceiling insulation has the greatest R-value.

2.3.0 Insulation Materials and Types

Insulation materials can be divided into four general classifications (*Table 4*). These materials are used in the manufacture of five basic categories of insulation: flexible, loose-fill, rigid or semi-rigid, reflective, and miscellaneous.

The R-value of the insulation is marked on the insulation itself or its packaging (*Figure 3*).

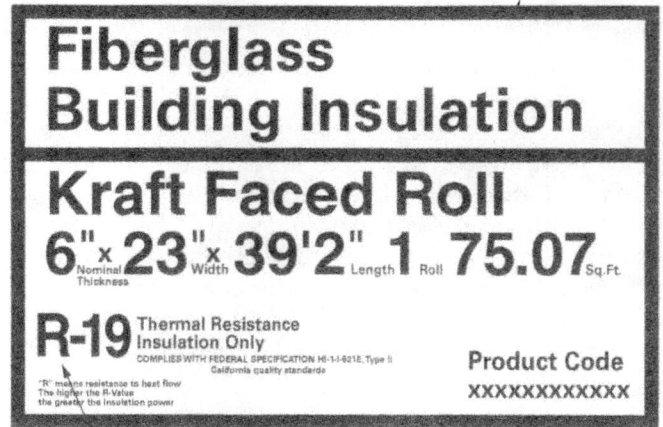

Figure 3 ◆ Typical R-value identification.

2.3.1 Flexible Insulation

Flexible insulation is usually manufactured from fiberglass in blanket form (*Figure 4*) and fiberglass or mineral wool in batt form (*Figure 5*). In some cases, it is manufactured from wood fiber or cotton and treated for resistance to fire, decay, insects, and rodents. The blankets are available in 16" or 24" widths and the batts in 15" or 23" widths. Both are furnished in thicknesses ranging from

Table 4 Insulation Materials

Classification	Material	Comments
Mineral	Rock Slag Glass Vermiculite Perlite	Rock and slag are used to produce wool by grinding and melting the materials and blowing them into a fine mass.
Vegetable (natural)	Wood Sugar cane Corn stalks Cotton Cork Redwood bark Sawdust or shavings	Many vegetable products are processed and formed into various shapes, including blankets and rigid boards.
Plastic	Polystyrene Polyurethane Polyisocyanurate Phenolic	
Metal	Foil Tin plate Copper Aluminum	Metallic insulating materials are generally applied to rigid boards or papers and used primarily for their reflective value.

103T04.EPS

1" to 12". The batts are packaged in flat bundles in lengths of 24", 48", or 93" and may be unfaced or faced with asphalt-laminated kraft paper or fire-resistant foil scrim (FSK) with or without nailing flanges. The blankets are furnished in rolls that may be encased in an asphalt-laminat-ed kraft paper or plastic film. In most cases, they have a facing with nailing flanges. Some blankets are available with an FSK facing. Blankets and batts with kraft or film casing and/or facings are combustible and must not be left exposed in attics, walls, or floors.

Fiberglass batt insulation at a 3½" thickness (standard rating or high rating) may be used on exterior walls between the studs. This has been the normal insulation thickness in the past, because of the use of 3½"-wide studs spaced 16" on center (OC). However, in the northern parts of the country, some builders have been using 2 × 6 studs spaced 16" or 24" OC, which has allowed for an increase in the wall insulation to 5½". This thickness is ample insulation for all parts of the U.S.

103F04.EPS

Figure 4 Flexible blanket insulation.

103F05.EPS

Figure 5 Flexible batt insulation.

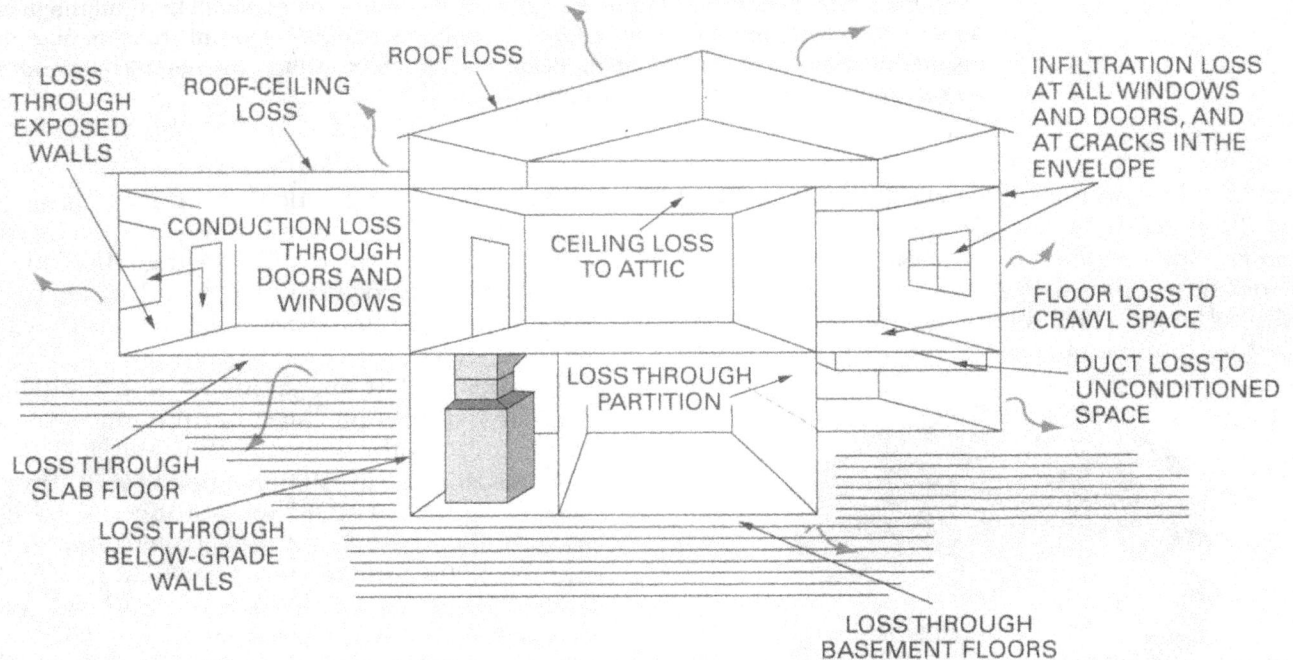
2.3.2 Loose-Fill Insulation

Loose-fill insulation is supplied in bulk form packaged in bags or bales (*Figure 6*). In new construction, it is usually blown or poured and spread over the ceiling joists in unheated attics. In existing construction that was not insulated when it was built, the material can be blown into the walls as well as the attic.

The materials used in loose-fill insulation include rock or glass wool, wood fiber, shredded redwood bark, cork, wood pulp products such as shredded newspaper (cellulose insulation), and vermiculite. All wood products, including paper, must be treated for resistance to fire, decay, insects, and rodents.

Shredded paper absorbs water easily and loses considerable R-value when damp. In addition to wall surface and/or ceiling vapor barriers, it is essential to install a waterproof membrane along the eaves to prevent water leakage.

The R-value of loose-fill insulation depends on proper application of the product. The manufacturer's instructions must be followed to obtain the correct weight per square foot of material as well

103F06.EPS

Figure 6 ◆ Loose-fill insulation.

as the minimum thickness. Before loose insulation is installed, the area of the space to be insulated is calculated (minus adjustments for framing members). Then, the required number of bags or pounds of insulation is determined from the bag label charts for the desired R-value.

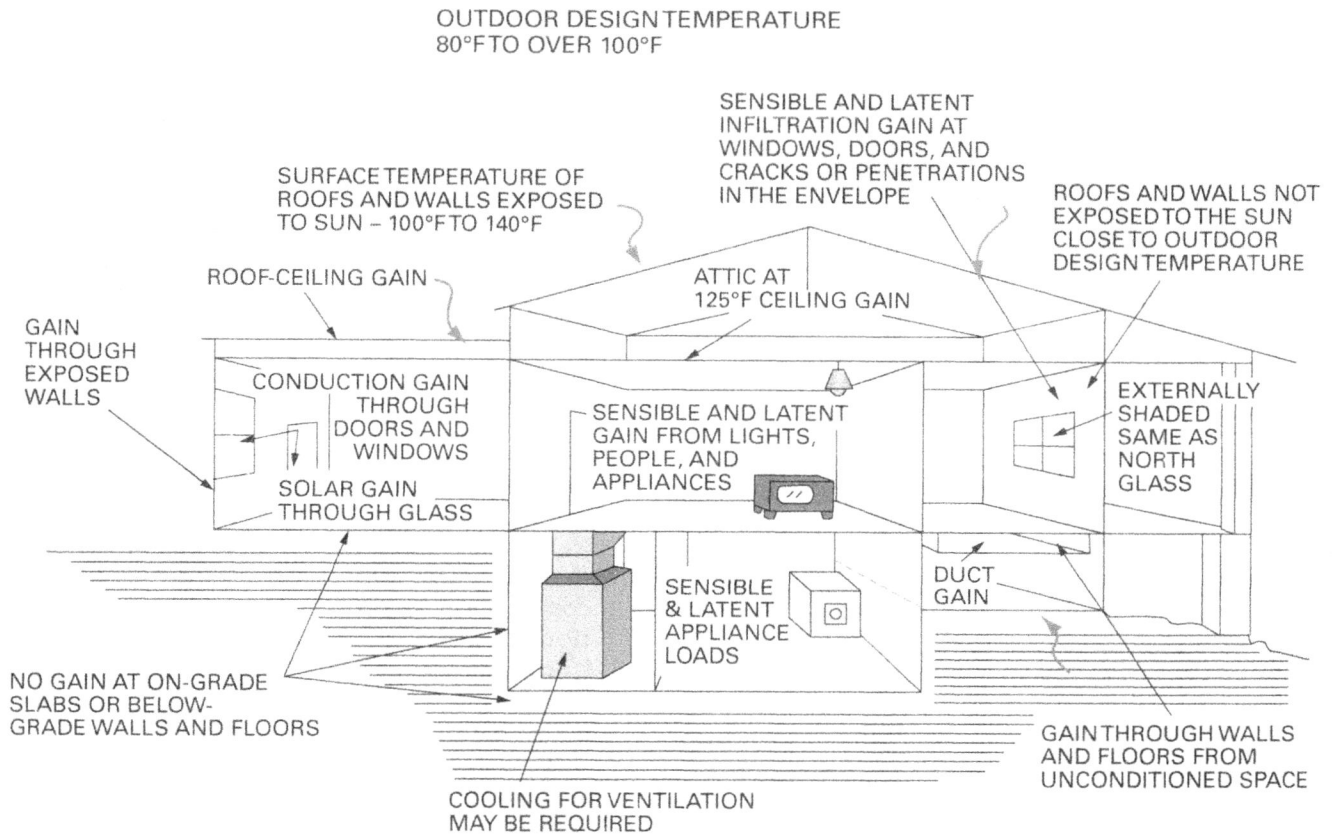

OUTDOOR DESIGN TEMPERATURE
80°F TO OVER 100°F

SENSIBLE AND LATENT INFILTRATION GAIN AT WINDOWS, DOORS, AND CRACKS OR PENETRATIONS IN THE ENVELOPE

SURFACE TEMPERATURE OF ROOFS AND WALLS EXPOSED TO SUN – 100°F TO 140°F

ROOFS AND WALLS NOT EXPOSED TO THE SUN CLOSE TO OUTDOOR DESIGN TEMPERATURE

ROOF-CEILING GAIN

ATTIC AT 125°F CEILING GAIN

GAIN THROUGH EXPOSED WALLS

CONDUCTION GAIN THROUGH DOORS AND WINDOWS

SENSIBLE AND LATENT GAIN FROM LIGHTS, PEOPLE, AND APPLIANCES

EXTERNALLY SHADED SAME AS NORTH GLASS

SOLAR GAIN THROUGH GLASS

DUCT GAIN

SENSIBLE & LATENT APPLIANCE LOADS

NO GAIN AT ON-GRADE SLABS OR BELOW-GRADE WALLS AND FLOORS

GAIN THROUGH WALLS AND FLOORS FROM UNCONDITIONED SPACE

COOLING FOR VENTILATION MAY BE REQUIRED

203SA04.EPS

2.3.3 Rigid or Semi-Rigid Insulation

Rigid or semi-rigid insulation is available in sheet or board form and is generally divided into two groups: structural and non-structural. It is available in widths up to 4' and lengths up to 12'.

Structural insulating boards come in densities ranging from 15 to 31 pounds per square foot.

They are used as sheathing, roof decking, and wallboard. Their primary purpose is structural, while their secondary purpose is insulation. The structural types are usually made of processed wood, cane, or other fibrous vegetable materials.

Non-structural rigid foam board (*Figure 7*) or semi-rigid fiberglass insulation is usually a lightweight sheet or board made of fiberglass

Figure 7 ◆ Rigid foam board.

103F07.EPS

or foamed plastic such as polystyrene, polyurethane, polyisocyanurate, and expanded perlite. Most of these products are waterproof and can be used on the exteriors or interiors of foundations, under the perimeters of concrete slabs, over wall sheathing, and on top of roof decks. When used above grade with proper flashing, a protective and decorative coating is sometimes applied directly to the panel. This is known as an exterior insulation finish system (EIFS).

In other cases, it is sealed with an air infiltration film, and normal siding is applied. The foam boards generally range in thickness from 1" to 4" with R-values up to R-30. Because all foam insulation is flammable, it cannot be left exposed. It must be covered with at least ½" of fireproof material.

Some manufacturers also provide rigid foam cores that can be inserted in concrete blocks or used with masonry products to provide additional insulation in concrete block or masonry walls.

The most common rigid and semi-rigid insulation types include the following:

- *Rigid expanded polystyrene* – This material has an R-value of R-4 per inch of thickness. Water significantly reduces this value because rigid expanded polystyrene is not water-resistant. It is not recommended for below-grade insulation. It is the lowest in cost. This material is also called beadboard.
- *Rigid extruded polystyrene* – The R-value of this material is about R-5 per inch of thickness. It is water-resistant and can be used below grade. When used in above-grade applications, it is subject to damage by ultraviolet light and must be coated or covered.
- *Rigid polyurethane and polyisocyanurate* – Initially, these boards have R-values up to R-8 per inch. However, over time the R-value drops to between R-6 and R-7 due to escaping gases. This is referred to as aged R-value. These products are subject to damage by ultraviolet light and must be coated or covered.
- *Semi-rigid fiberglass* – The R-value of this material is about R-4 per inch. Boards of this kind are used on below-grade slabs and walls to provide water drainage as well as insulation. They are also used under membrane roofs as insulation. On below-grade applications, the walls or floors must be waterproofed with a coating or membrane between the wall or floor and the insulation to block water penetration.

2.3.4 Reflective Insulation

This type of insulation (*Figure 8*) usually consists of multiple outer layers of aluminum foil bonded to inner layers of various materials for strength. The number of reflecting surfaces (not the thickness of the material) determines its insulating value. To be effective, the metal foil must face an open air space that is ¾" or more in depth. In some cases, reflective material is bonded to flexible insulation as the inside surface for both insulation and vapor seal purposes.

INSIDE TRACK

Insulation Weight

Insulation materials have weight. This weight must be considered when designing a building. One of the advantages of using lightweight insulation board, such as rigid polyurethane and polyisocyanurate, is that it permits greater freedom of design. The lighter the insulation, the less weight loadbearing members of the structure must support.

Reflective Insulation

Reflective insulation, by itself, can only block radiated heat. At relatively hot temperatures, it helps keep buildings cooler by deflecting heat from the sun. At relatively cold temperatures, it can do little to prevent heat from escaping the building.

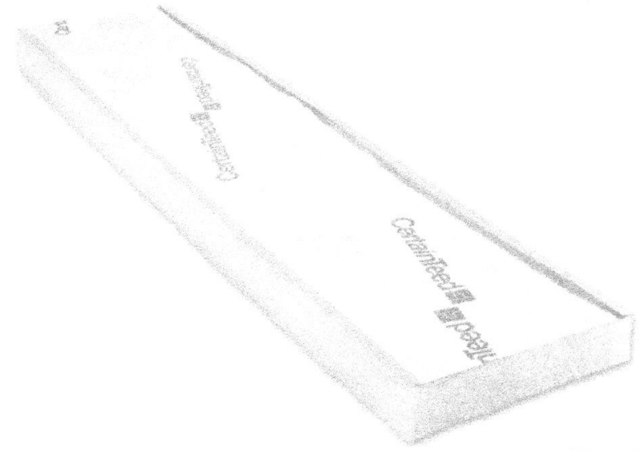

103F08.EPS

Figure 8 Reflective foil-faced batt insulation.

103F09.EPS

Figure 9 Sprayed-in-place insulation.

2.3.5 Miscellaneous Types of Insulation

There are other types of insulation that do not fit the previous four categories. These types are as follows:

- *Foamed-in-place insulation* – This type of insulation can be applied to new or existing construction using special spray equipment. It can be injected between brick veneer and masonry walls; between open studs or joists; and inside concrete blocks, exterior wall cavities, party walls, and piping cavities. The material must be applied by trained and certified contractors.
- *Sprayed-in-place insulation* – Usually, these types of insulation (*Figure 9*) consist of confetti-like or fibrous inorganic material either mixed with an adhesive or sprayed against a wall with an

adhesive coating. They are often left exposed for acoustical as well as insulating properties. Like foamed-in-place products, sprayed-in-place insulation should be applied by trained contractors.

- *Lightweight aggregates* – Insulation material consisting of perlite, vermiculite, blast furnace slag, sintered clay products, or cinders is often added to concrete, concrete blocks, or plaster to improve their insulation quality and reduce heat transmission.

In the 1970s, urea formaldehyde foamed-in-place insulation was injected into many homes. However, due to improper installation, the foam shrank and gave off formaldehyde fumes. As a result, its use was banned in the United States and

Sprayed- and Foamed-in-Place Insulation

Sprayed- and foamed-in-place insulation materials are well suited for irregular surfaces. These include walls and ceilings that are curved or that have beams, pipes, or other equipment protruding from them. Foams and sprays can be built up in layers to the desired insulation thickness.

Canada. Later, it was allowed back on the market in certain areas of the United States. A urethane foam that expands on contact can also be used. It does not have a formaldehyde problem, but it does emit cyanide gas when burned. As a result, it requires fire protection and, like urea formaldehyde, it may also be banned in some areas of the country.

Another foamed-in-place product is a phenol-based synthetic polymer (Tripolymer®–C.P. Chemical Co.) that is fire-resistant and does not drip or create smoke when exposed to high heat. This material does not expand once it leaves the delivery hose of the proprietary application equipment.

3.0.0 ◆ INSULATION INSTALLATION GUIDELINES

Before installation, building plans and codes must be checked to determine the R-values and the types of insulation required or permitted for the structure being insulated. Then, the required amount of insulation for the structure must be calculated. Any specific instructions provided by the selected manufacturer must be followed when installing the insulation.

3.1.0 Estimating Typical Insulation Requirements

The following is a method of estimating the amount of insulation for the walls, ceilings, and floors of a single-story structure. If no plans are available and the codes specify only a minimum R-value for a structure, refer to the comfort level standards discussed previously and select a desired comfort level based on the occupancy of the structure. Then, perform the following steps to calculate the amount of required insulation material:

Step 1   Determine the square footage of exterior walls to be insulated:

- Determine from the plans or measure the perimeter length of each exterior wall in feet.

- Add the lengths of all exterior walls to find the total perimeter of the structure.

- Multiply the total perimeter by the ceiling height to find the total square footage of the walls:

Exterior perimeter (ft) × ceiling
height (ft) = sq ft of walls

- Determine from the plans or measure the square footage of each opening in the perimeter walls:

Height (in) × width (in) =
sq in ÷ 144 = sq ft of opening

- Add the square footage of all openings to determine the total square footage.

- Subtract the total square footage of all openings from the total square footage of the walls to find the square footage of insulation required for the walls:

Total wall area (sq ft) – total opening
area (sq ft) = sq ft of wall insulation

Step 2   Calculate the square footage of ceiling/floor to be insulated. The square footage of a floor will be the same as the ceiling. The square footage of either one can be calculated, and the result can be used for both.

- Divide the ceiling or floor plan into rectangular or square areas and determine from the plans or measure the length and width of each area (*Figure 10*).

- For each area, multiply the length by the width to determine the square footage:

Length (ft) × width (ft) =
sq ft of each area

- Add the square footage of all areas to find the total square footage of insulation required for the ceiling or floor.

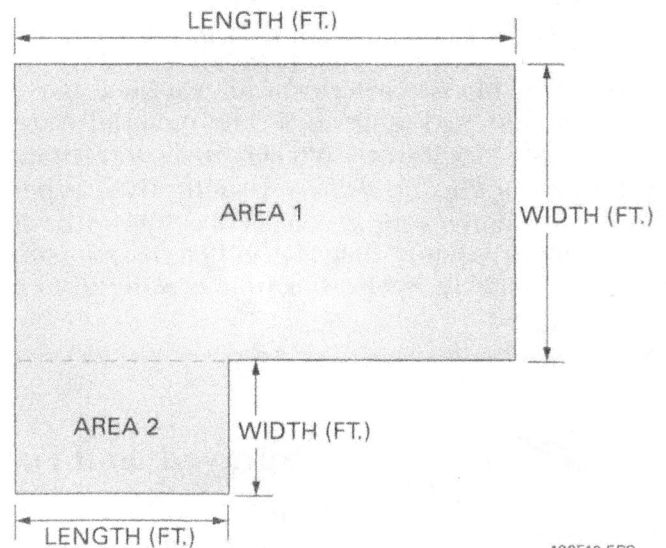

Figure 10 ◆ Dividing a plan view into rectangular or square areas.

Step 3 Add the total insulation square footage required for the walls, the ceiling, and, if required, the floor to determine the total square footage of insulation required for the structure.

Step 4 Divide the square footage of the structure by the coverage per package of insulation for the R-value required. The coverage information will be given in the manufacturer's information for the insulation to be used. See *Table 5* for several examples of package coverage for various R-values and package sizes.

NOTE

The walls, ceiling, and floor of a structure may require different insulation R-values. Check your local code to determine what the requirements are for each installation.

3.2.0 Typical Flexible Insulation Installation

WARNING!

Wear proper eye protection, respiratory equipment, and gloves when handling and installing insulation.

Use the following procedure when installing typical flexible insulation:

Step 1 For walls, measure the inside cavity height and add 3". From the wall, lay the distance out on the floor and mark it. Unroll blanket insulation or lay batts on the floor. Use two layers or more. At the cut mark, compress the insulation with a board and cut it with a utility knife. On blanket or faced insulation, remove about 1" of insulation from the ends to provide a stapling flange at the top and bottom.

Step 2 If a separate interior vapor seal will be installed, install blanket or faced insulation so that the stapling flange is fastened to the inside surfaces of the wall studs (*Figure 11*).

- If the facing of the blanket or batt is the vapor seal, install the stapling flange on the face of the studs and overlap them by at least 1" (*Figure 12*). For faced or blanket insulation, use a power, hand, or hammer stapler to first staple the top flange to the plate.

- Align and staple down the sides.

- Staple the bottom flange to the sole plate. Pull the flanges tight and keep them flat when stapling. Space staples about 12" apart if stapling to the face of the studs; on the sides, space them about 6" apart.

- For unfaced batt insulation, install the batt at the top and bottom first and push it tight against the plates.

- Evenly push the rest of the batt into the cavity (*Figure 13*). For narrow spaces around windows and doors, stuff the spaces with pieces of insulation and cover it with a plastic or tape vapor seal.

WARNING!

Exercise caution when installing insulation around electrical outlet boxes and other wall openings or devices. Failure to do so may result in electrocution.

Table 5 Typical Insulation Coverage for Various Types of Packaging and R-Values

R-Value	Thickness	Width × Length	Square Feet per Package	Pieces per Package
11	3½"	15" × 94"	88	9
13	3⅝"	15" × 94"	88	9
19	6½"	15" × 94"	49	5
30	9½"	16" × 48"	37	7
38	12"	24" × 48"	48	6

103T05.EPS

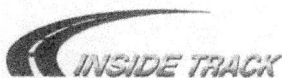

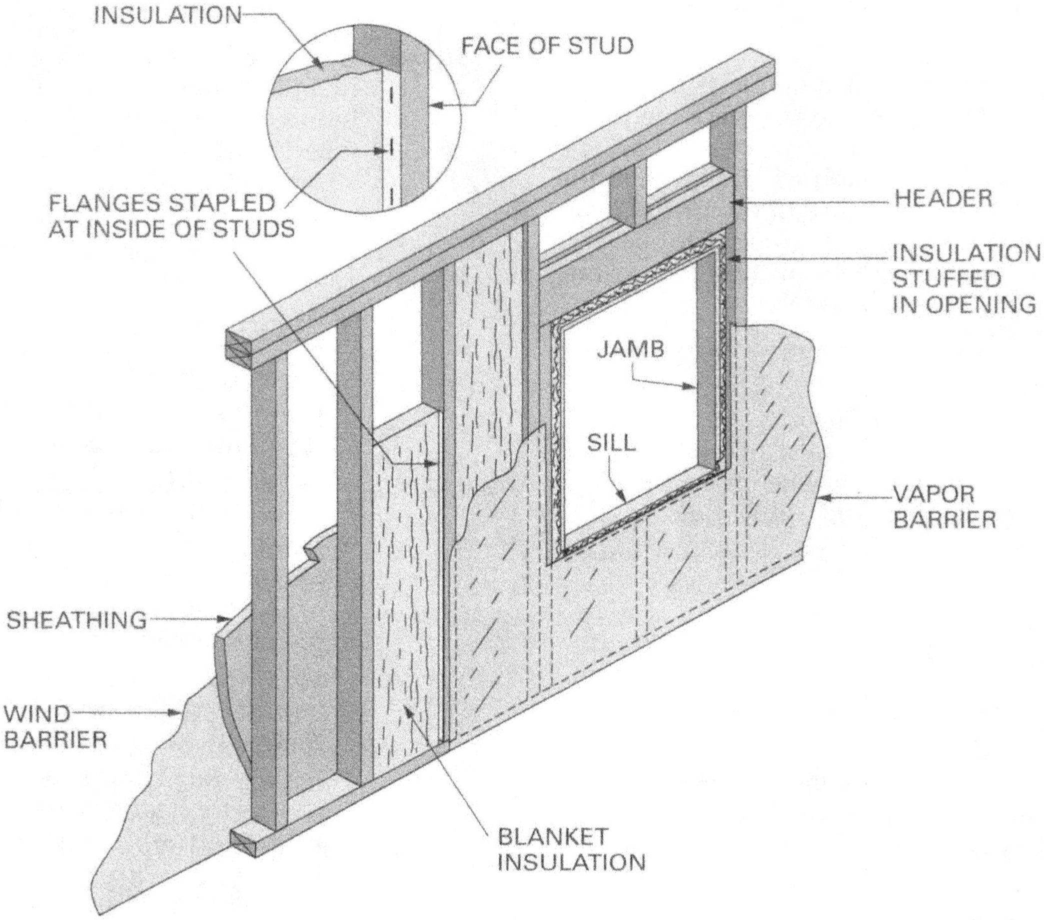
INSULATION
FACE OF STUD
FLANGES STAPLED
AT INSIDE OF STUDS
HEADER
INSULATION
STUFFED
IN OPENING
JAMB
SILL
VAPOR
BARRIER
SHEATHING
WIND
BARRIER
BLANKET
INSULATION
103F11.EPS

Figure 11 ◆ Blanket installation without integral vapor seal.

THINK ABOUT IT

Fiberglass Insulation

1. Are the flanges on faced fiberglass insulation always stapled to the inside of the stud?
2. Can you increase the effectiveness of fiberglass insulation by squeezing more into a smaller space?

INSIDE TRACK

Cathedral Ceilings

If a cathedral ceiling incorporates gypsum drywall attached to the bottom of the rafters, airflow must be maintained from the soffit to the ridge. Proper ventilation must also be maintained above and below skylights to prevent buildup of heat and moisture. Check your local code for the appropriate methods to use when working with cathedral ceilings.

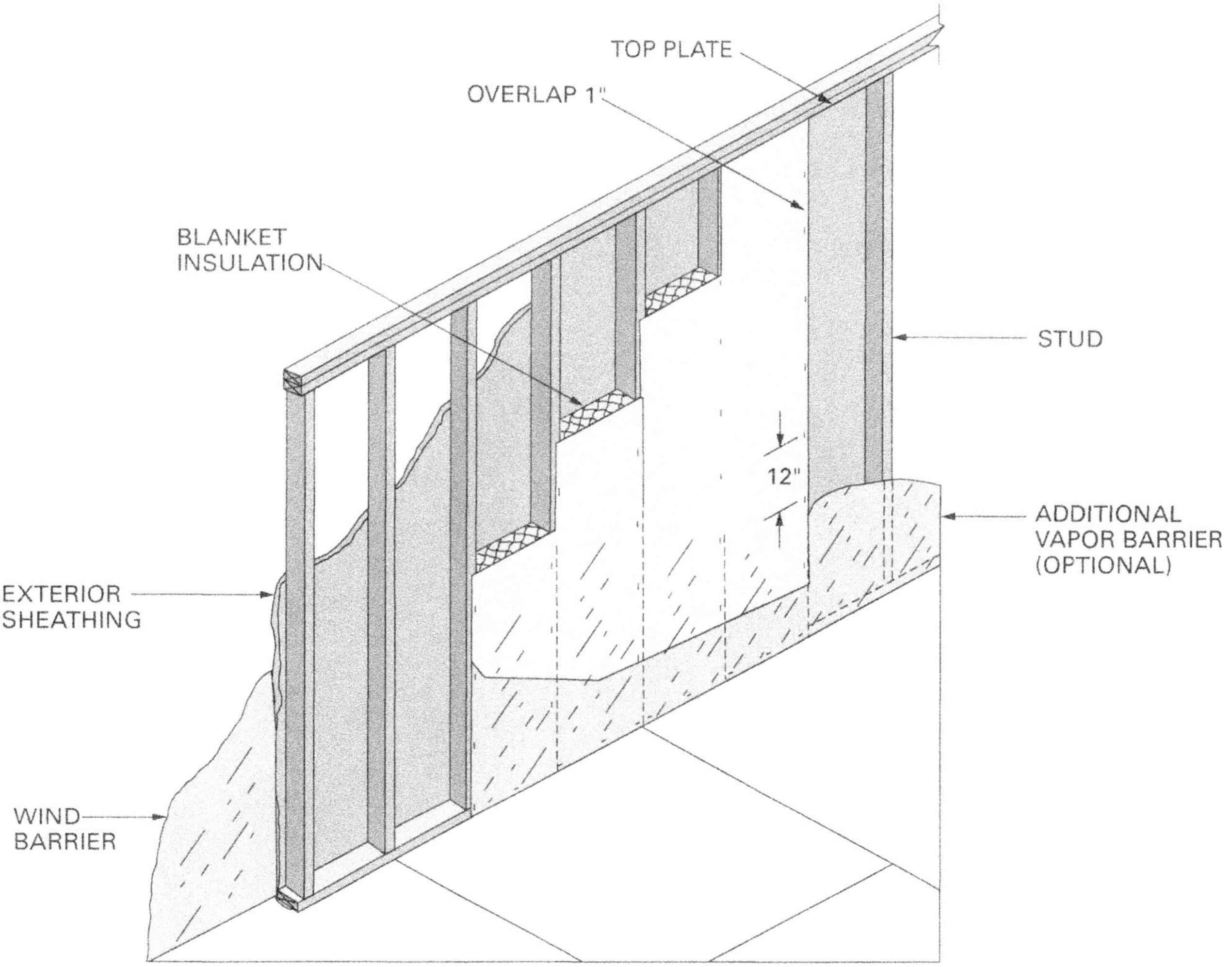

TOP PLATE

OVERLAP 1"

BLANKET
INSULATION

STUD

12"

ADDITIONAL
VAPOR BARRIER
(OPTIONAL)

EXTERIOR
SHEATHING

WIND
BARRIER

103F12.EPS

Figure 12   Blanket installation with integral vapor seal.

Step 3   Faced or blanket insulation for ceilings or floors is usually installed from the bottom in the same manner as the walls. Unfaced batts can be installed from either the top or the bottom.

    – Make sure that ceiling insulation extends over the wall into the soffit area (*Figure 14*). Also make sure soffit baffles (*Figure 15*) are inserted over and cover the ceiling insulation. The baffles should be fastened to the roof deck to hold them in place so that they do not slide down into the soffit and block ventilation.

    – For floors, ensure that the insulation is installed around the perimeter of the floor against the header (*Figure 16*). Floor insulation over a basement is installed with the vapor barrier facing down.

    – Over a crawl space, the vapor barrier faces up. In either case, the insulation can be supported below by a wire mesh (chicken wire), if desired.

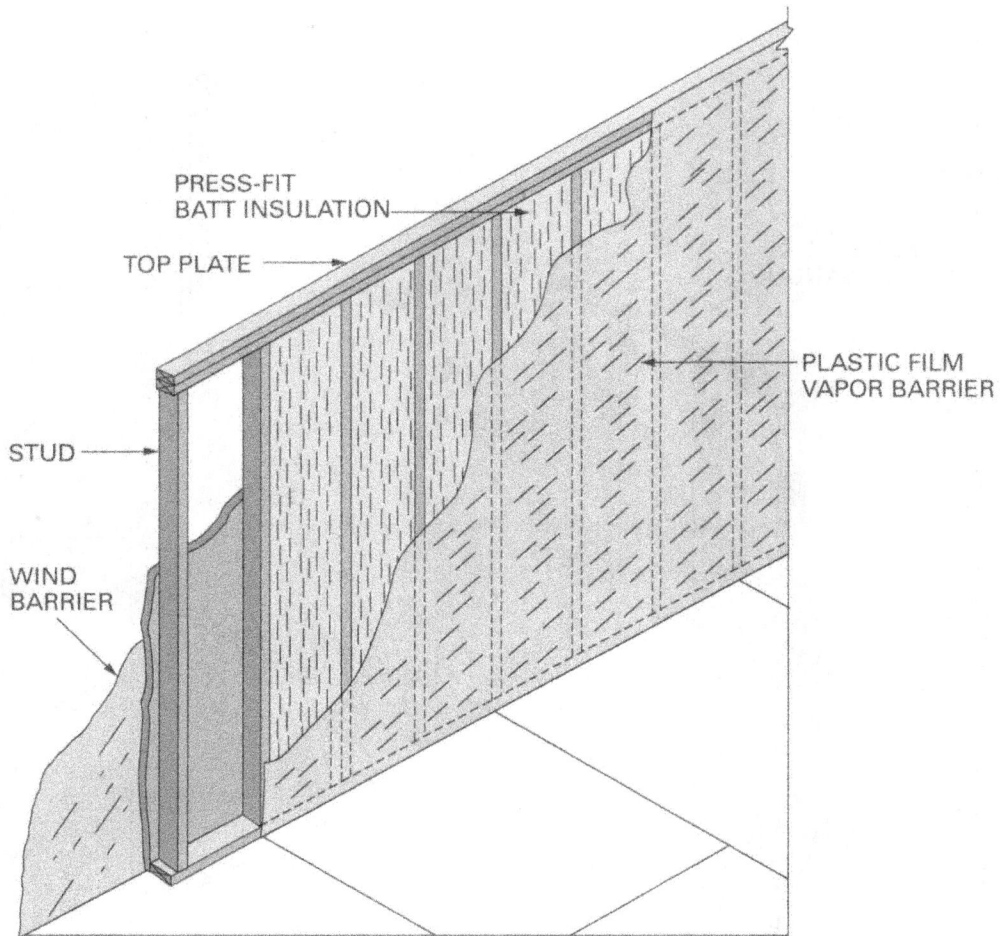

Figure 13 ◆ Batt insulation with separate vapor barrier.

PRESS-FIT
BATT INSULATION

TOP PLATE

STUD

WIND
BARRIER

PLASTIC FILM
VAPOR BARRIER

103F13.EPS

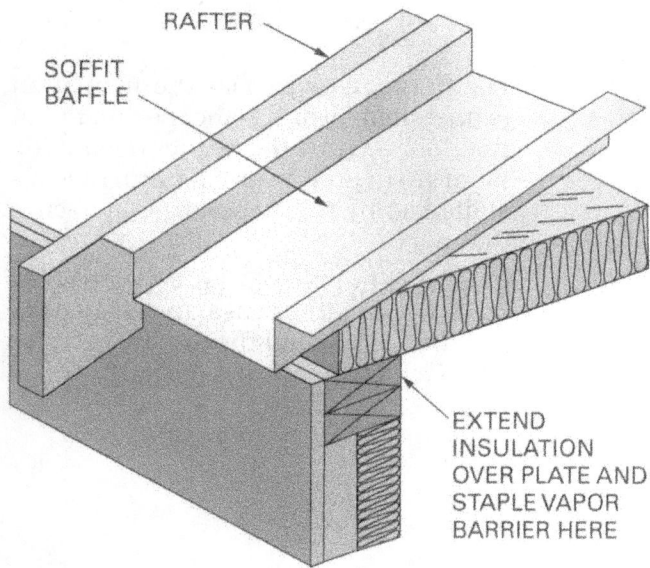

RAFTER

SOFFIT
BAFFLE

EXTEND
INSULATION
OVER PLATE AND
STAPLE VAPOR
BARRIER HERE

103F14.EPS

Figure 14 ◆ Ceiling insulation at wall and soffit.

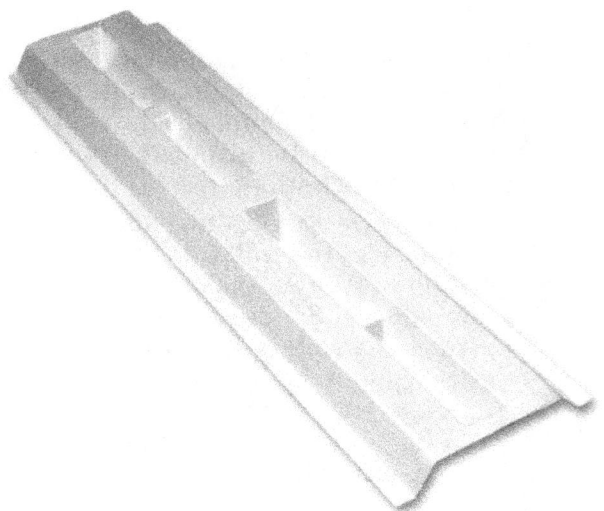

103F15.EPS

Figure 15 ◆ Typical plastic soffit baffle (shown upside down).

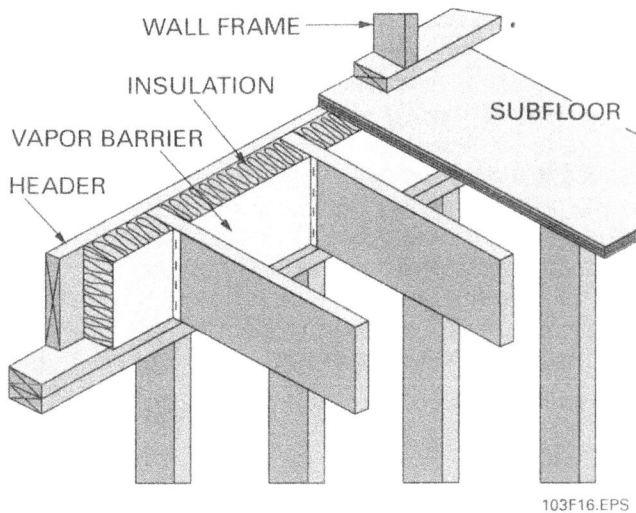

WALL FRAME
INSULATION
VAPOR BARRIER
HEADER
SUBFLOOR

103F16.EPS

Figure 16 Perimeter floor insulation.

3.3.0 Typical Loose-Fill Insulation Installation

For new construction, loose-fill insulation is used primarily for attic insulation. On older construction, it can also be blown into wall cavities through holes drilled at the center and tops of exterior walls. The following steps only cover attic or ceiling installation.

WARNING!

Wear proper eye protection, respiratory equipment, and gloves when handling and installing insulation.

Step 1 Make sure that the finished ceiling below has been installed. Also, ensure that a separate vapor barrier has been installed to prevent moisture penetration of the insulation and to prevent the fine dust from the insulation from penetrating the ceiling in the event of future cracks (*Figure 17*). Make sure that soffit baffles and blocking have been installed to prevent the material from spilling into the soffits.

Step 2 If the final insulation depth will be higher than the ceiling joists, permanently install strike-off boards (*Figure 18*). Pour the insulation from bags or blow the insulation over the ceiling joists using special equipment. Using a straightedge, tamp the insulation and then level it to the required depth for the R-value desired.

INSIDE TRACK

Polystyrene Forms

Structural forms made of polystyrene are sometimes used for residential and light commercial construction. Concrete is poured into the forms, which are left in place to provide insulation for the walls. The forms usually provide sufficient insulation by themselves, but check the local code for these requirements.

103SA05.EPS

3.4.0 Typical Rigid Insulation Installation

Rigid insulation panels can be fastened like sheathing over the studs or wood sheathing of a structure. Nails with large heads/washers or screws with washers are used to prevent crushing the insulation.

Rigid insulation panels may be installed on the exterior of a foundation. Typically, the exterior of the foundation is waterproofed first. Then, the panels are applied over special mastic and secured with concrete nails to hold them in place until the mastic sets. For existing construction, the panels may be installed on the interior of the foundation if the walls are adequately waterproofed.

Figure 19 shows typical methods of installing rigid insulation under surface slabs. Usually, the insulation is only applied around the perimeter of the slab, anywhere from 24" to 36" from the edge of the slab and/or down the inside of the slab footings to below the frost line (*Figure 20*). A vapor barrier should be applied under the slab and over any insulation under the slab.

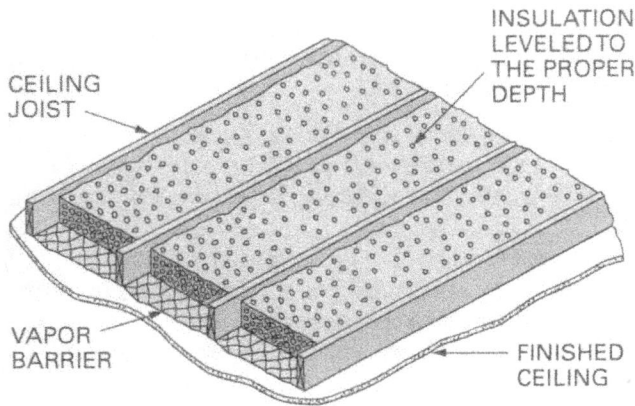

203F17.EPS

Figure 17 ◆ Loose-fill insulation.

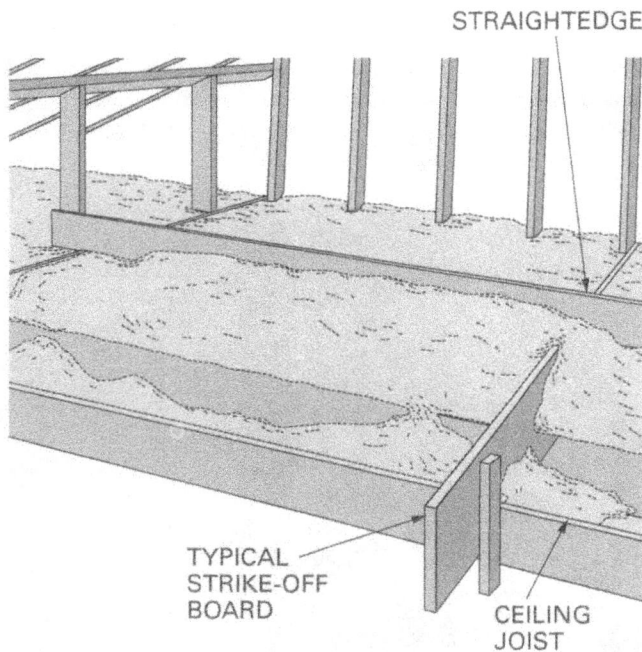

103F18.EPS

Figure 18 ◆ Leveling loose-fill insulation.

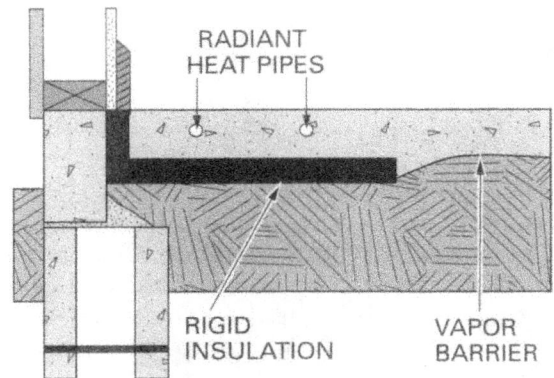

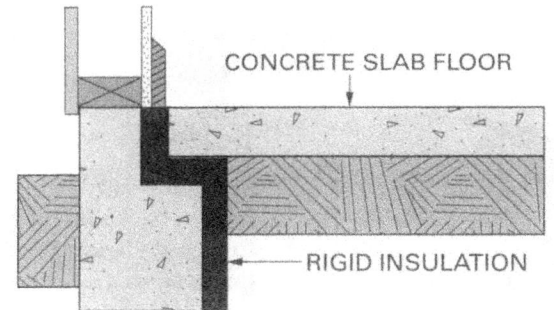

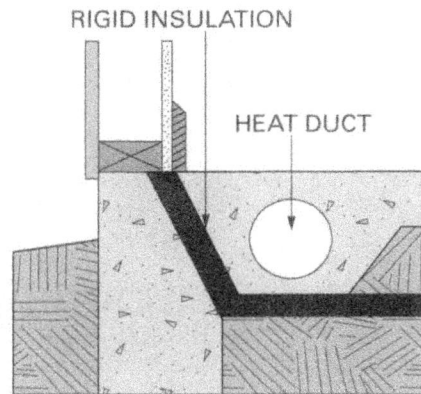

103F19.EPS

Figure 19 ◆ Rigid insulation installed under a concrete slab.

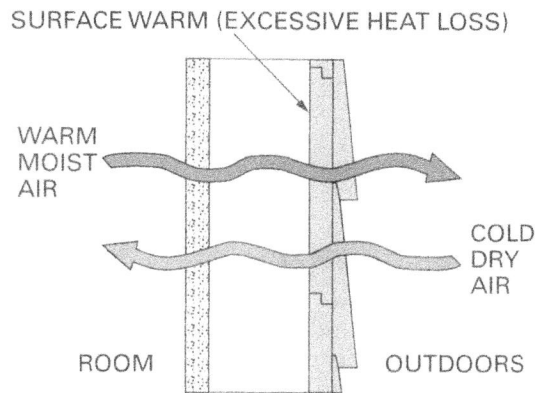

Figure 20 ◆ Rigid insulation installed under a slab-and-down footing.

4.0.0 ◆ MOISTURE CONTROL

Water vapor contained in air can readily pass through most building materials used for wall construction. This vapor caused no problem when walls were porous because it could pass from the warm wall to the outside of the building before it could condense into liquid water (*Figure 21*).

When buildings were first constructed with insulation in the walls to cut down on heat loss, moisture in the air passed through the insulation until it reached a point cold enough to cause it to condense. The condensed moisture froze in very cold weather and reduced the efficiency of the insulation. The ice contained within the wall thawed as the weather warmed, and the resulting water in the wall caused studs and sills to decay over time.

For these reasons, it is important to keep cellars, basements, crawl spaces, exterior walls, and attics dry. Moisture in crawl spaces, basements, and attics also encourages woodchewing insects such as termites, as well as the growth of mold. In the case of crawl spaces, moisture often rises from the ground into the crawl space during periods of heavy rain. To prevent the concentration of this damaging moisture, some precautions must be taken in the original design of the structure:

• The earth must slope down and away about 20' from the structure, carrying surface water away.

• The crawl space should be protected from moisture by a vapor barrier on the ground.

• The foundation walls should be penetrated with vents so that moisture will not be trapped in the crawl space.

• A vapor barrier should be installed between the insulation and the subfloor.

Basements usually have the most trouble with condensation in summer during humid weather. The earth under the concrete basement floor is

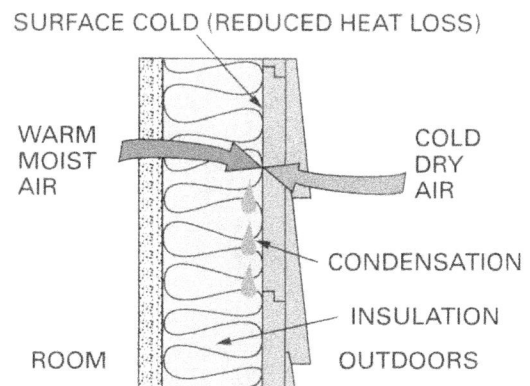

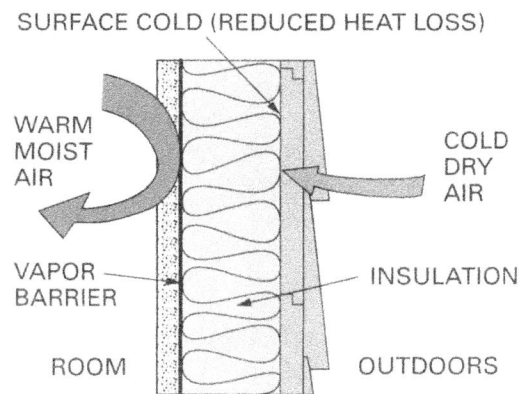

Figure 21 ◆ Effects of insulation and vapor barrier.

comparatively cool, causing the floor of the basement to be a cold surface. The hot air is saturated with moisture and condenses when it comes in contact with the cooler surfaces of the floor and walls. This problem is difficult to control. If the surface of the concrete is rough and porous, the moisture will sink in and not cause a wetness problem. If, however, the floor is dense and smoothly finished, the tightly knit grains of concrete form a vapor barrier of sorts, and the water collects on the slab. This problem can usually be solved by the use of dehumidification devices during the summer months.

Moisture weeping through the concrete floor is a different problem. In new construction, this is controlled by installing perimeter drainage and a vapor barrier under the concrete slab. When installing polyethylene film as an underslab vapor barrier, be careful not to tear, puncture, or damage the film in any way. Any passageways for moisture will defeat the purpose of the vapor barrier. Prior to pouring the concrete slab, make sure the polyethylene film is placed properly and is free of punctures. Keep all construction debris away from the vapor barrier.

To keep moisture from rising up into the basement, 6" of coarse gravel should be placed over the compacted earth to provide drainage to the perimeter drain before the slab is poured. A polyethylene film should be placed on top of the gravel to keep the concrete from penetrating into the gravel and possibly weakening the slab. In very wet areas or areas with a high water table, floor drainage, in addition to a gravel bed, may also be required.

4.1.0 Interior Ventilation

One of the best ways to reduce or eliminate the chances of moisture damage in attics or in the space between the rafters and the finished roof is through proper ventilation. Ventilation provides a stream of outside air to remove trapped moisture before it is allowed to do any damage. In insulated attics, baffles (blocking strips) are used to keep the insulation material from getting into the vented areas. With the increased use of blown-in insulation in attics, baffles are being required by code in some areas.

The amount of ventilation required varies by climate and building codes. Attics and gable and hip roofs may be ventilated with a variety of louvers and vents. Flat roofs are ventilated with a combination of eave vents and roof stacks (*Figure 22*).

Ice dams can usually be avoided by installing plenty of insulation and providing ample ventilation in the attic.

Properly designed subroof ventilation is the best weapon for preventing water vapor infiltration into a steeply sloped roof, but is less effective

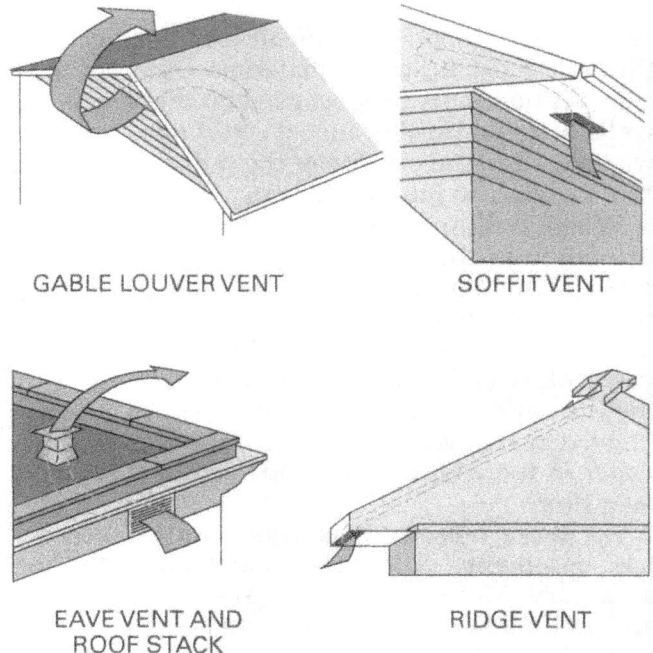

GABLE LOUVER VENT   SOFFIT VENT

EAVE VENT AND   RIDGE VENT
ROOF STACK

103F22.EPS

Figure 22 ⬧ Various methods of roof ventilation.

on roofs with low slopes because natural convection decreases with diminishing roof height. Moisture dissipation occurs through diffusion and wind-induced ventilation.

Normally, the ventilation requirement for a gable roof is 1 sq ft of free air ventilation for every 300 sq ft of ceiling area if a vapor barrier exists under the ceiling. If no vapor barrier is present, the requirement is 1 sq ft for every 150 sq ft of ceiling area. The total requirement must be split evenly between the inlet vents and the outlet vents.

Free air ventilation is the rating of the ventilation devices, taking into account any restrictions caused by screening, louvers, and other devices.

Ice Dams

In colder climates, ice dams can be a problem. Ice dams are formed along the edge of a sloping roof when a building's attic is not properly insulated and ventilated. Heat escaping through the roof melts accumulated snow, forming icicles along the edge of the roof. Over time, water collects under the outer layer of snow and is trapped by the ice. This water backs up under the shingles and penetrates the roof, causing water damage and other problems.

Some homeowners use special snow rakes to remove snow from the roof. This helps prevent ice dams from forming, but it is labor intensive. If used improperly, the rake may actually damage the roof. The best way to prevent ice dams is to properly insulate and ventilate the building's attic.

4.2.0 Vapor Retarders

Vapor retarders, also known as vapor barriers or vapor diffusion retarders (VDR), are any material or substance that will not permit the passage of water vapor or will do so only at an extremely slow rate. The permeability of a substance is a measure of its capacity to allow the passage of liquids or gases. Water vapor permeability is the property of a substance to permit the passage of water vapor and is equal to the permeance of a substance that is 1" thick. The measure of water vapor permeability is the perm. This equals the number of grains of water vapor passing through a 1 sq ft piece of material per hour, per inch of mercury difference in vapor pressure. All you really have to remember is that any material that has a perm rating of 1.0 or less is considered a vapor retarder and will not allow the passage of any appreciable or harmful amounts of water vapor. Any material with a rating higher than 1.0 is a breathable material that will permit the passage of water vapor in whatever degree its perm rating indicates. The higher the perm number, the greater the amount of water vapor that will pass through the material in a given time; 0.0 is totally impermeable (*Table 6*).

A properly installed vapor barrier will protect ceilings, walls, and floors from moisture originating within a heated space (*Figure 23*).

An insulated wall will divide two temperature gradients. The area on the inside of the structure will normally be warmer than the air on the out-

Table 6 Perm Ratings of Various Vapor Retarder Materials

Material	Permeance
Aluminum foil (1 mil)	0.0*
Aluminum foil (0.35 mil)	0.05*
Polyethylene (4 mil)	0.08*
Polyethylene (6 mil)	0.06*
Polyester (1 mil)	0.07*
Saturated and coated roll roofing	0.05**
Reinforced kraft and asphalt-laminated paper	0.3**
Asphalt-saturated and coated vapor barrier paper	0.2 to 0.3*
15 lb tarred felt	4.0**
15 lb asphalt felt	1.0**
12.5 lb asphalt	0.5**
22 lb asphalt	0.1**
Built-up membrane (hot-mopped)	0.0**

*   Per *ASTM E96-66, Water Vapor Transmission of Materials in Sheet Form*
** Per *ASTM C355-64, Water Vapor Transmission of Thick Material*

103T06.EPS

WALL WITH NO VAPOR BARRIER

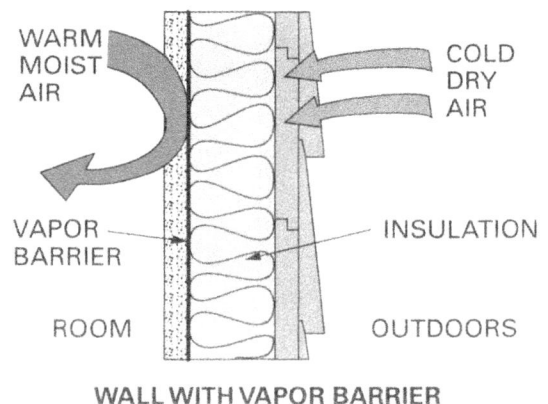

WALL WITH VAPOR BARRIER

103F23.EPS

Figure 23 Vapor barrier installation.

side. The vapor barrier is usually located on the warm side to prevent moisture from moving through the insulation to the cool side and condensing.

4.2.1 Materials

Common vapor barrier materials include asphalted kraft paper, aluminum foil, and polyethylene film.

Asphalted kraft paper is usually incorporated with blanket or batt insulation. It serves as a means for attaching the insulation to the building framework and as a reasonably good vapor barrier when installed on the warm side of the wall or ceiling (*Figure 24*).

Aluminum foil may be incorporated with blanket or batt insulation in the same manner as kraft paper. It is also applied to the back of gypsum lath and gypsum wallboard where it works as a relatively effective vapor barrier.

Polyethylene film is applied over the studs and ceiling joists after the insulation is installed. When wallboard with polyethylene film or foil backing is used, the insulation will normally be plain batts or blankets that do not have an integral vapor barrier. As a vapor barrier, polyethylene film is stapled over the studs and also covers the window frames. This helps to keep the window frames and sashes clean during application and finishing of the gypsum wallboard. The film should be overlapped 2" to 4" and sealed with special mastic or tape.

4.2.2 Installation in Crawl Spaces

The ground under a ventilated crawl space should be covered with a vapor barrier ground cover to protect the underside of the house from condensation (see *Figure 25*). A vapor barrier should also be installed over the subfloor above the crawl space.

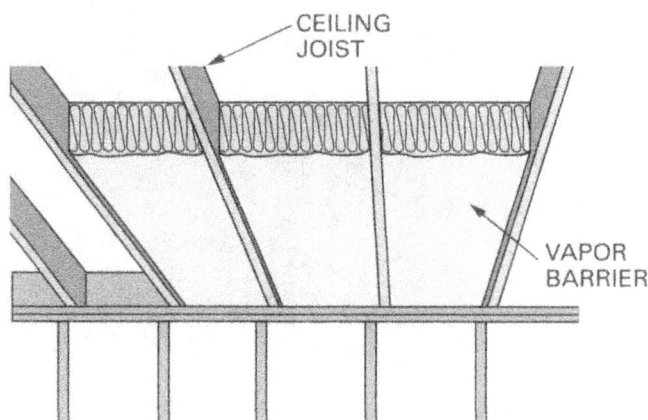

Figure 24 ✦ Installing insulation batts between ceiling joists with vapor barrier down.

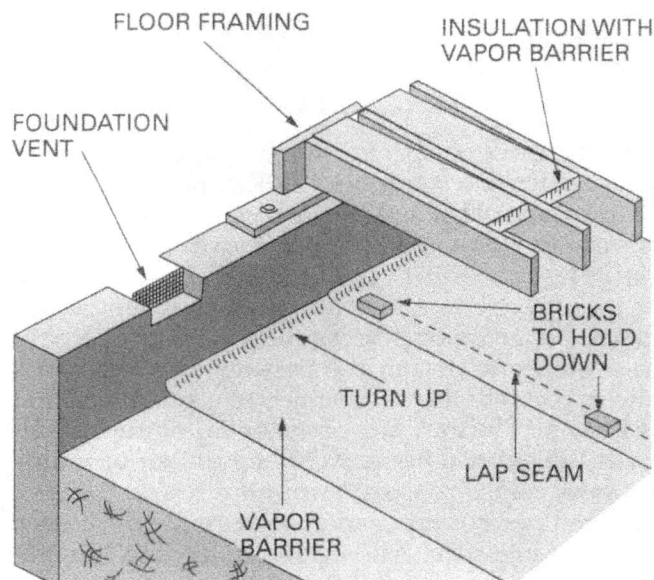

Figure 25 ✦ Vapor barrier installation for crawl spaces.

INSIDE TRACK

Vapor Barrier Backing

Many of the insulation materials produced today have a vapor barrier applied to the inside surface. Many interior wall surface materials are also backed with vapor barriers. When these materials are properly installed, they usually provide satisfactory resistance to moisture penetration. If the insulating materials do not include a vapor barrier, then one should be installed as a separate element.

Besides installing vapor barriers, crawl spaces should be properly vented to permit the escape of moisture. Usually, this is accomplished by the use of a proper number of screened foundation vents installed in the above-grade foundation surrounding the crawl space. The normal requirement is 1 sq ft of free air ventilation for every 150 sq ft of crawl space area when a vapor barrier ground cover is used.

4.2.3 Installation in Slabs

When allowed to proceed unchecked, moisture will migrate from the ground upward through concrete and into the building, where it can cause moisture problems, damage, and higher energy costs. Even though the water table may be several feet below the slab, moisture vapor will migrate up to and through concrete slabs.

Up to 80 percent of the moisture entering a structure does so by migrating from the ground beneath the structure. Moisture vapor passes through concrete more readily than liquid moisture.

Moisture in a building can cause deterioration of interior finishes, especially floors and equipment. Moisture can also add to energy costs by raising humidity and taxing cooling systems that require dehumidification.

Vapor barriers should be continuous under the slab. Great care must be taken not to tear or puncture the barrier. Keep all construction debris away from the barrier location. Vapor barrier installation must be done by qualified contractors.

When used in thickened-edge slab construction, as shown in *Figure 26*, a vapor barrier is placed between the gravel cushion and the poured concrete. The same arrangement is used for other types of slab-on-grade construction.

Figure 27 shows a method of constructing a finished floor over a concrete slab, which affords double protection against moisture. The sealer or waterproofer is placed on the slab itself, and a vapor barrier is suspended above the slab.

4.2.4 Installation in Walls

A polyethylene sheet vapor barrier is easy to apply to frame walls where no integral barrier is provided or where a supplementary barrier is preferred. A flap would normally overlap both floor and ceiling barriers to seal the interior off completely. Adjacent sheets of the film are overlapped 2" to 4" and are sealed with a special mastic or tape.

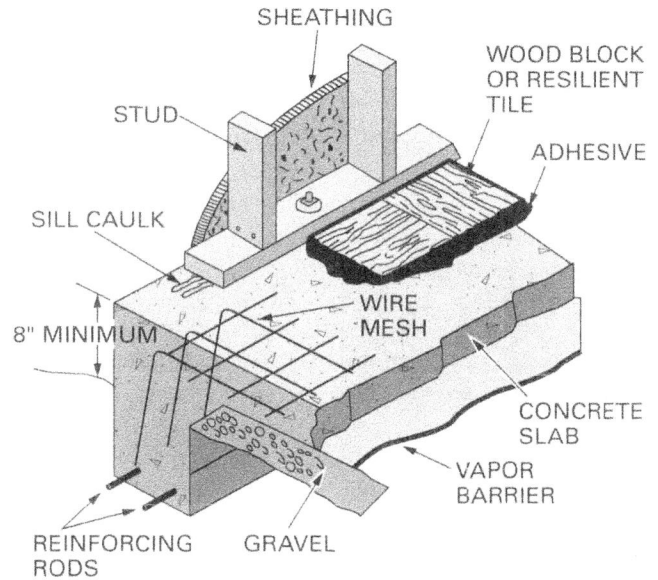

Figure 26 Thickened-edge slab vapor barrier installation.

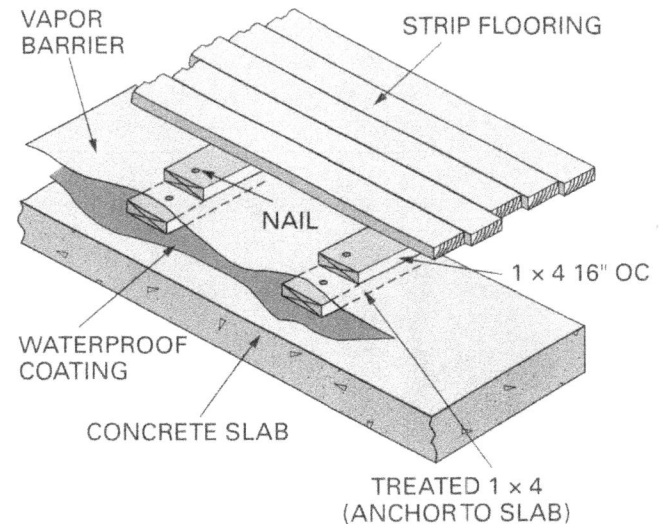

Figure 27 Surface-mounted vapor barrier on a slab.

When vapor barriers are applied to walls, particular attention should be paid to fitting the material around electrical outlet boxes, exhaust fans, light fixtures, registers, and plumbing. Considerable water vapor can escape through the cracks around the equipment, travel from the warm side of the wall to the cold side, and condense on the sheathing or siding. This is especially true if the insulation is poorly fitted at the top and bottom.

4.2.5 Installation in Roofs

A major cause of failure in built-up roofs is condensation of moisture vapor, which rises from
inside the building and penetrates the roof deck
insulation. When this vapor reaches its dew
point, which can occur inside the insulation or at
the cool outer surface, it condenses. This results in
a reduction or total loss of the thermal efficiency
of the insulation, as well as dripping and damage.
To prevent this, select a vapor barrier that is both
easy to apply and resistant to job site abuse. Install
it on the warm side of the roof deck insulation.

5.0.0 ◆ WATERPROOFING

The single most critical area for waterproofing
construction is the below-grade foundation wall.
Rising water tables, hydrostatic heads, structural
movement, and ground water all require a special
type of protection. A liquid waterproofing system
applied by spray methods ensures the high buildup of film thickness needed to cope with these
problems (*Figure 28*).

For all below-grade applications of waterproofing, be sure to fill all cracks, crevices, and grooves.
Ensure that the coating is continuous and free
from breaks and pinholes.

Carry the coating over the exposed tops and
outside edges of the footing (*Figure 29*), forming a
cove at the junction of the wall and footing.

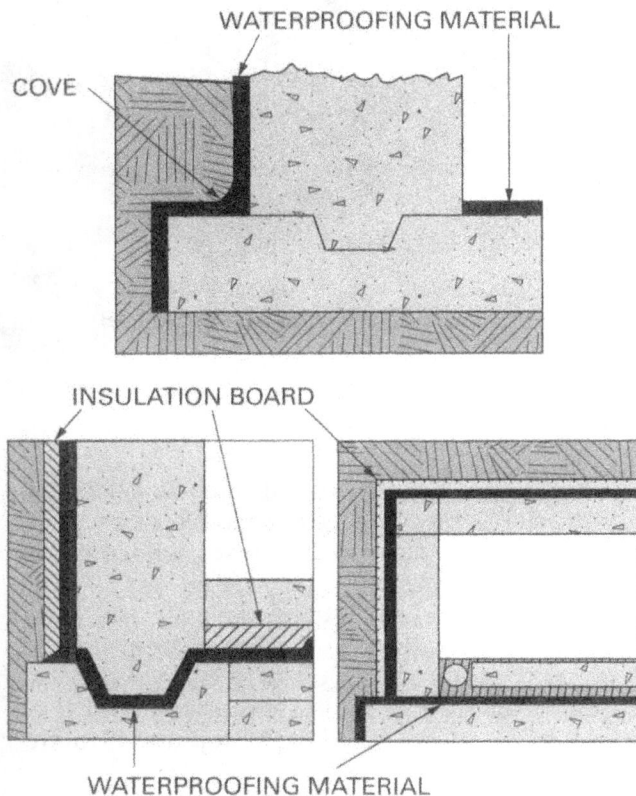

103F28.EPS

Figure 28 ◆ Applying waterproofing material.

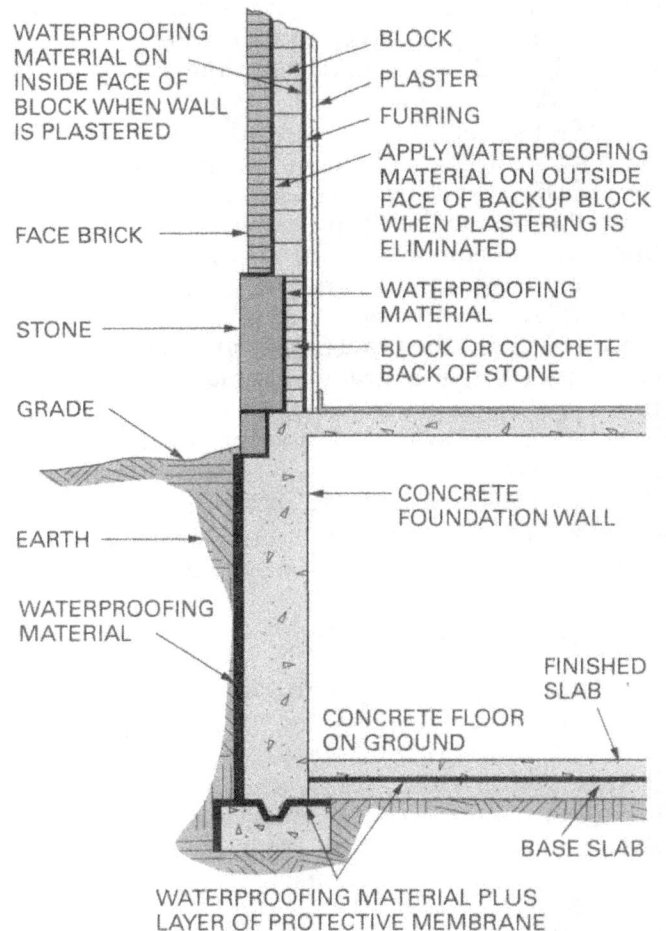

103F29.EPS

Figure 29 ◆ Below-grade waterproofing application.

Spread the coating around all joints, grooves, and slots and into all chases, corners, reveals, and soffits. Bring the coating up to the finished grade.

Do not place backfill for 24 to 48 hours after application. Where possible, backfill should be placed within approximately seven days to avoid any unnecessary damage due to construction activities. Take care to place the backfill in a manner that will not rupture or damage the film or cause the coating or membranes to be displaced on the coated surface.

5.1.0 Water Stops

Water stops are thin sheets of rubber, plastic (PVC), or other material inserted in a construction joint to obstruct the seepage of water through the joint. PVC water stops may be used for installations in underpasses, tunnels, tanks, locks, walls, swimming pools, siphons, sewage disposal plants, reservoirs, culverts, sewage treatment plants, channels, drums, filtration plants, foundations, bridges, basements, abutments and decks, mineshafts, aqueducts, retaining walls, and roofs. Refer to *Figure 30* for various applications of water stops.

5.2.0 Joint Treatment

Joints in structures are critical. They must maintain integrity during movement, yet remain permanently waterproof and airtight. Therefore, it is important to select the proper joint treatment system to avoid problems with moisture penetration at the construction joints.

5.3.0 Vapor Barrier for Cold Storage and Low-Temperature Facilities

Cold-storage vapor barriers are designed for use in areas of extremely low temperatures to halt the migration of damaging moisture vapor into and through the insulation. Most cold storage vapor barriers consist of two layers of kraft paper, each extrusion coated with black polyethylene; a layer of aluminum foil; two layers of non-asphaltic adhesive; and two layers of high-tensile-strength reinforcing fibers embedded in the adhesive.

6.0.0 AIR INFILTRATION CONTROL

In addition to insulation, the exterior sheathing of a structure should be covered to prevent wind pressure from causing infiltration of outside air into the structure. To achieve maximum energy

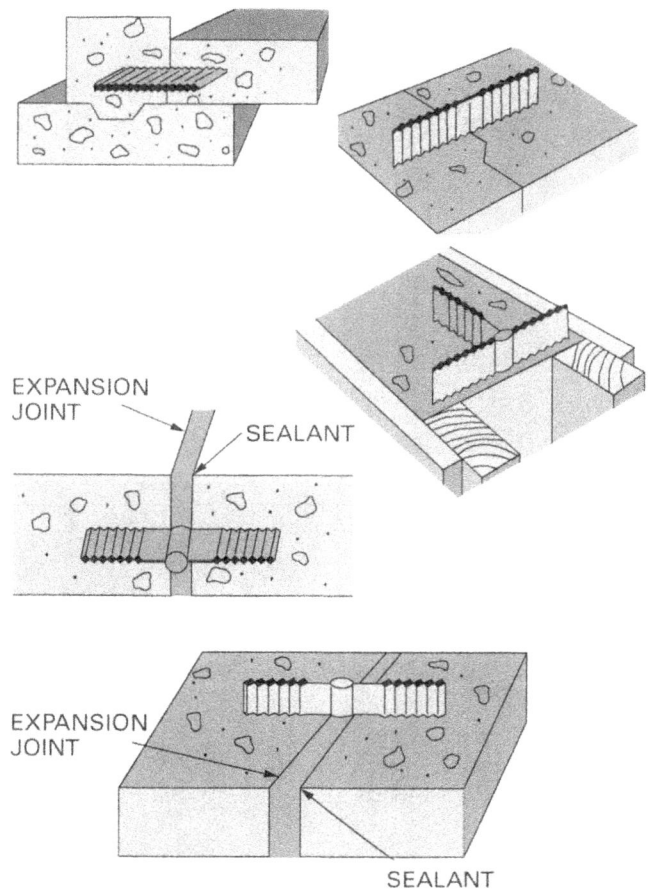

Figure 30 ◇ Water stops used in joints.

103F30.EPS

efficiency in a structure, air infiltration must be strictly controlled.

Traditionally, structures have been covered with water-resistant building paper to help prevent water leakage through the primary barrier (siding) from reaching the structural sheathing or other components of the structure. Additionally, the paper had to be water permeable to allow moisture inside the walls of the structure to pass through and evaporate. To some extent, the paper reduced air infiltration of the structure, especially when board sheathing was used.

For a number of years, products called house wraps or building wraps have been used to replace building paper. These products, under brand names such as Tyvek® or ProWrap®, are easier to apply and perform the same functions as building paper (*Figure 31*). When properly applied and sealed, the wraps provide a nearly airtight structure no matter what sheathing material is used. Most versions of these wraps are an excellent secondary barrier under all siding, including stucco and EIFS.

RESIDENTIAL APPLICATION

COMMERCIAL APPLICATION

103F31.EPS

Figure 31 ❖ Building wrap.

Many of these products are made of spun, high-density, polyethylene fibers randomly bonded into an extremely tough, durable sheet material. They are usually available in several versions and weights for residential and light commercial use. Special versions may be available with ver-

tical water channeling permanently pressed into the material for stucco and EIFS. The material is usually furnished in rolls in various sizes from 18" wide to 10' wide and in lengths from 100' to 200'.

Nails with large heads, nails or screws with plastic washers (*Figure 32*), or 1" wide staples may be used to secure the wrap to wood, plastic, insulating board, or exterior gypsum board. Screws and washers are used for steel construction. Special contractor's tape (*Figure 32*) or sealants compatible with the wrap are used to seal the edges and joints of the wrap.

WARNING!

Some building wraps are slippery and should not be used in any application where they can be walked on. Because the surface will be slippery, use pump jacks or scaffolding for exterior work above the lower floor. If ladders must be used, extra precautions must be taken to prevent the ladders from sliding on the wrap.

Always refer to the manufacturer's instructions for specific installation information. House or building wrap is generally installed as follows:

Step 1   Using two people and beginning at a corner on one side of the structure, leave 6" to 12" of the wrap extended beyond the corner to be used as an overlap on the adjacent side of the structure (*Figure 33*). Align the roll vertically and unroll it for a short distance. Check that the stud marks on the wrap align with the studs of the structure. Also check that the bottom edge of the wrap extends over and runs along the line of the foundation. Secure the wrap to the corner at 12" to 18" intervals.

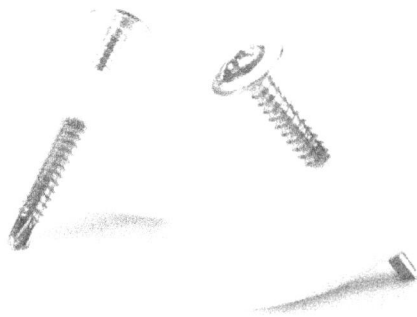

SCREWS WITH PLASTIC WASHERS

CONTRACTOR'S TAPE

103F32.EPS

Figure 32 Building wrap accessories.

Step 2 Unroll the wrap two or three more feet and ensure that it overlaps and runs along the line of the foundation. Secure the wrap vertically at 12" to 18" intervals on each stud using the stud marks as a fastening guide. Continue around the structure, covering all openings. If a new roll is started, overlap the end of the previous roll 6" to 12" to align the stud marks of the new roll with the studs of the structure.

Step 3 If the upper parts of the structure require coverage, repeat Steps 1 and 2, starting above the existing wrap. Make sure that the bottom edge of this layer of wrap aligns along the top edge of the lower wrap and overlaps it by 6" to 12".

Step 4 At the top plate, make sure the wrap covers both the lower and upper (double) top plate (*Figure 34*), but leave the flap loose for the time being.

Step 5 At each opening, use one of the following two methods to cut back the wrap.

NOTE

Always follow the window or door manufacturer's recommendations for flashing windows or doors.

Method 1 – Uninstalled Windows/Doors:

– At the opening, cut the wrap as shown in *Figure 35*. Fold the three flaps around the sides and bottom of the opening and secure every 6". Trim off the excess.

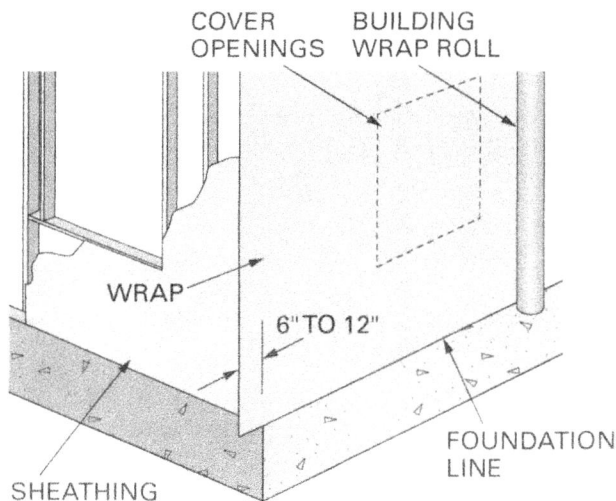

Figure 33 Starting a roll of building wrap.

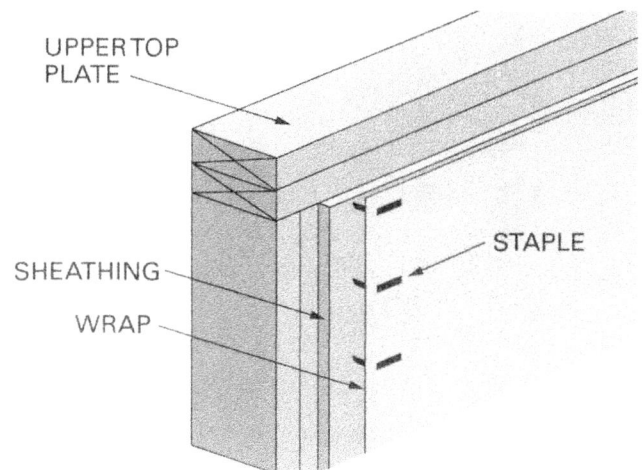

Figure 34 Top plate detail.

103F34.EPS

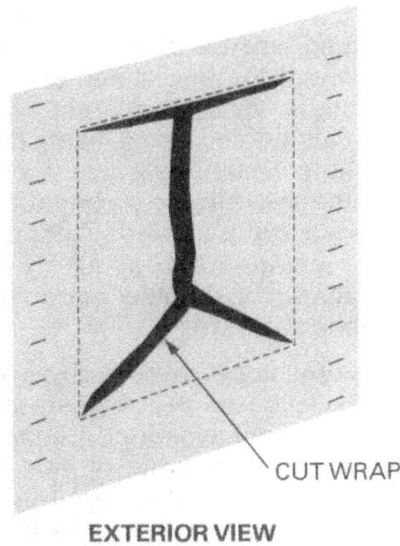

EXTERIOR VIEW

CUT WRAP

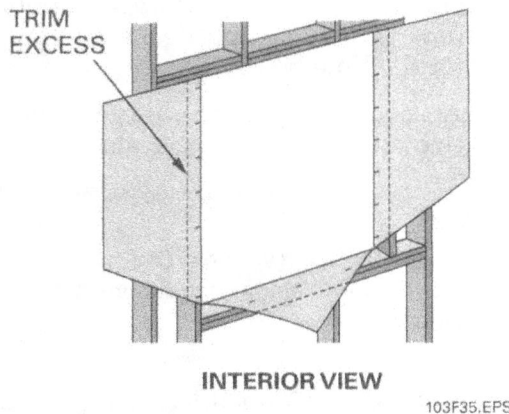

TRIM EXCESS

INTERIOR VIEW

103F35.EPS

Figure 35 ◈ Cutting and folding wrap at an opening.

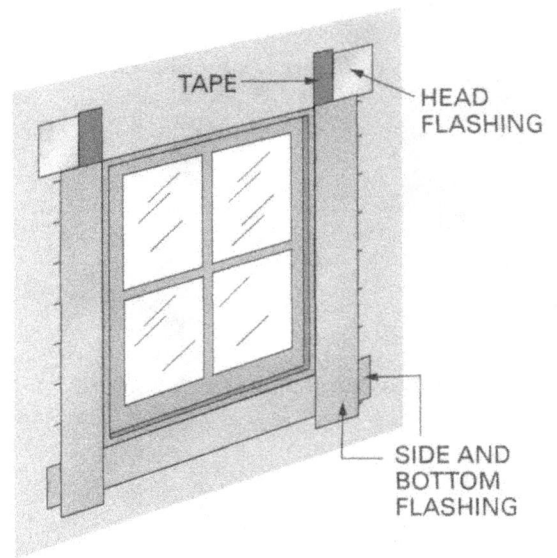

TAPE

HEAD FLASHING

SIDE AND BOTTOM FLASHING

103F36.EPS

Figure 36 ◈ Installing flashing around an opening.

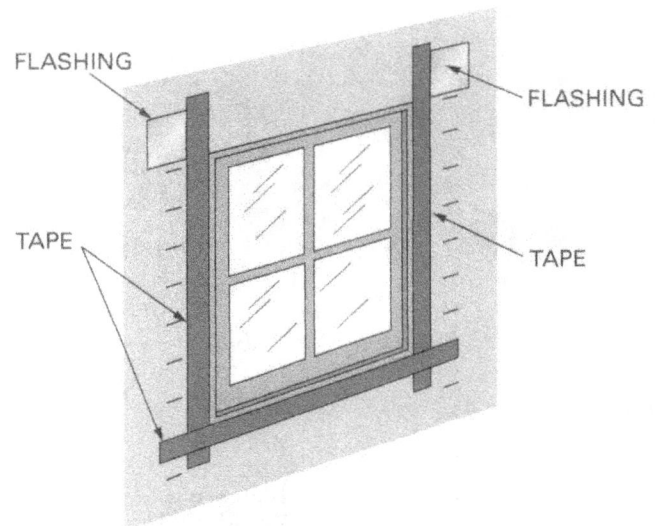

FLASHING

FLASHING

TAPE

TAPE

103F37.EPS

Figure 37 ◈ Installing wrap with a window in place.

- At the outside, install 6" flashing along the bottom of the opening, then up the sides over the top of the wrap.
- Install head flashing at the top of the opening under the wrap and over the side flashing (*Figure 36*). Tape the flap ends to the head flashing using tape approved by the manufacturer.

Method 2 – Windows/Doors with Flanges:

- Create a top flap of the wrap. Insert a head flashing under the flap and over the flange.
- Extend the flashing to the sides about 4" and tape the flap to the head flashing.
- On the remaining sides, trim the wrap to overlap the flange area and tape the edge to the flanges (*Figure 37*).

Step 6   Secure all the bottom edges of the wrap to the foundation with the recommended joint sealer, then fasten the lower edge to the sill. At the top plate, seal the edge to the upper plate with the sealer and fasten the edge to the plate.

Step 7   Seal all vertical and horizontal joints in the wrap with the recommended tape.

Step 8   Before applying the siding, repair any damage or tears in the wrap with tape or sealant.

Window Flashing System

Take special care to prevent moisture from entering the structure around windows and doors. The Dupont® flashing system is designed for this purpose.

103SA06.EPS

1. Insulation is based on trapping _____.
   a. large amounts of air in a few small spaces
   b. large amounts of air in a large number of very small spaces
   c. large amounts of air in a few large spaces
   d. small amounts of air in a few large spaces

2. The R-value is a measure of the ability of a material to _____.
   a. resist the passage of moisture
   b. resist heat conduction
   c. allow cold air to enter a building
   d. convert water vapor into a liquid

3. For thermal transmission control purposes, insulation does not have to be installed _____.
   a. above ceilings
   b. in exterior walls
   c. in interior walls
   d. beneath floors over crawl spaces

4. Which of the following materials is likely to be used in structural insulation?
   a. Polystyrene
   b. Processed wood
   c. Fiberglass
   d. Expanded perlite

5. Strike-off boards are used in the installation of _____ insulation.
   a. loose-fill
   b. rigid
   c. flexible
   d. foamed-in-place

6. Rigid insulation is usually installed around the perimeter of a concrete slab at a distance of _____ from the edge.
   a. 8" to 12"
   b. 12" to 18"
   c. 18" to 24"
   d. 24" to 36"

7. When a vapor barrier is used under the ceiling, proper free air ventilation for a gable roof is defined as 1 sq ft for every _____ sq ft of attic area.
   a. 150
   b. 160
   c. 300
   d. 320

8. Any material with a perm rating of less than _____ is a vapor barrier.
   a. 0.01
   b. 0.05
   c. 0.1
   d. 1.0

9. When waterproofing material is sprayed on exterior below-grade walls, backfilling of the walls should be avoided for _____ days.
   a. one to two
   b. two to three
   c. three to four
   d. five to six

10. The vertical seams of building wrap are usually overlapped _____ at each corner of the building.
    a. 2" to 6"
    b. 3" to 4"
    c. 6" to 12"
    d. 12" to 18"

Summary

This module presented the materials and procedures that can be applied to ensure that effective insulation, moisture control, ventilation, waterproofing, and air infiltration control are achieved.

Some drywall contractors install insulation as well as drywall, so their employees must learn both crafts. In addition, insulation forms an important part of exterior wall systems, so it is important for the drywall installer to be familiar with the various types of insulation and their installation methods.

Notes

Barry Caldwell
Silver Medal Winner
Associated Builders and Contractors National Championships

Barry Caldwell won the silver medal in carpentry at the 2005 ABC Craft Olympics. He remembers the tough competition and the testing areas. In a mock building, contestants had to build a concrete form, construct sub-flooring, a metal wall, and a wooden wall. The competition was timed over six hours, and Barry finished all task areas with just five minutes remaining. The judge's criteria were safety, workmanship, job-site etiquette, cleanliness, efficient use of materials, and of course—accuracy. There was an eighth-of-an-inch tolerance, Barry says.

How did you become interested in carpentry?
I've been working around the trade basically my whole life. I started helping my father (who's also a carpenter) on job sites when I was five years old. After high school, I went to a vocational school for two years, and then I entered an apprenticeship program with Cleveland Construction. I've been there for the past three years.

What are some of the things you do in your job?
I work for the Interior Division of Cleveland Construction, which specializes in interior and exterior framing, and drywall and ceiling tile installation. The Interior Division is where people start out. As they progress, they can move on to other areas in the company, or they can stay with that division. Eventually, I'd like to move into a management position with the company.

What do you like most about carpentry?
I like it because there's a real sense of accomplishment. Fifty years later, you can go back and see what you've done. I love being able to go down the road and say "I did that" or "I helped with that." I love that new technologies, new tools, are always coming out. For example, right now there's a move toward using metal studs in framing instead of wood studs. I also like working outside. And there's always something new—a new project, a new site.

You're still in your apprenticeship now.
What separates a good apprentice from the rest?
Successful apprentices want to stay busy—they want to work. A good apprentice is motivated. The trade is not for everybody. If you don't want to do it, then don't do it. The best apprentices are those people who want to do the work. I love my job. I have fun at work every day.

Condensation: The process by which a vapor is converted to a liquid, such as the conversion of the moisture in air to water.

Convection: The movement of heat that either occurs naturally due to temperature differences or is forced by a fan or pump.

Dew point: The temperature at which air becomes oversaturated with moisture and the moisture condenses.

Diffusion: The movement, often contrary to gravity, of molecules of gas in all directions, causing them to intermingle.

Exterior insulation finish system (EIFS): A protective and decorative coating applied directly to insulation board.

Perm: The measure of water vapor permeability. It equals the number of grains squared of water vapor passing through a one square foot (sq ft) piece of material per hour, per inch of mercury difference in vapor pressure.

Permeability: The measure of a material's capacity to allow the passage of liquids or gases.

Permeable: Porous; having small openings that permit liquids or gases to seep through.

Permeance: The ratio of water vapor flow to the vapor pressure difference between two surfaces.

Vapor barrier: A material used to retard the flow of vapor and moisture into walls and prevent condensation within them. The vapor barrier must be located on the warm side of the wall.

Water stop: Thin sheets of rubber, plastic, or other material inserted in a construction joint to obstruct the seepage of water through the joint.

Water vapor: Water in a vapor (gas) form, especially when below the boiling point and diffused in the atmosphere.

Recommended Regional R-Values

Prior to 2006, the International Code Council recommended insulation R-values based on the region's average low temperature (see *Figure A–1* and *Table A–1*). Today, insulation values are based on the region's average temperature and humidity levels. Interestingly, the United States Department of Energy makes more stringent recommendations based on climate zones. It is important for you to remember that the ICC and other agencies, as well as the federal government may recommend insulation requirements for an area, but the local building code is the final authority.

Table A-1 Recommended R-Values of Insulation

Average Low Temperature	Floors	Walls	Ceilings
+10° F to +40° F	0	11	19
0° F to +10° F	19	17	30
0° F and below	19	17	38

103A01B.EPS

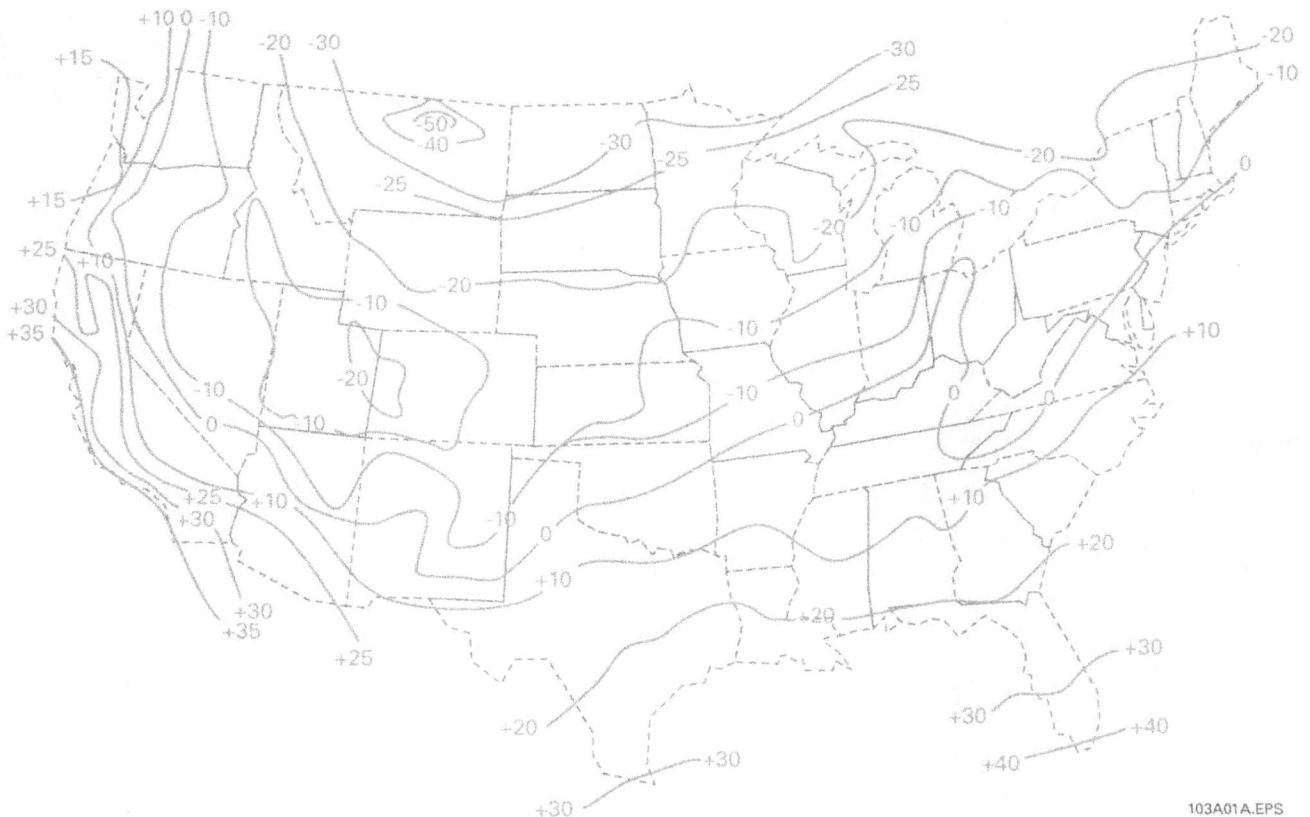

103A01A.EPS

Figure A-1 ◈ Average low temperatures across the United States.

Additional Resources

This module is intended to be a thorough resource for task training. The following reference works are suggested for further study. These are optional materials for continued education rather than for task training.

International Energy Conservation Code®, International Code Council, 2006.

U.S. Department of Energy website, www.eere. energy.gov.

Figure Credits

CertainTeed Corporation, 103SA01, 103F05, 103F08
   Copyright © 2006. Used with permission.

International Code Council, 103F02
   2006 International Energy Conservation Code. Copyright 2006. Falls Church, Virginia: International Code Council. Reproduced with permission. All rights reserved.

Topaz Publications, Inc., 103F04, 103F07, 103SA05

US Greenfiber LLC, 103F06

Johns Manville, 103F09, 103F31 (top)

Owens Corning, 103F15

Dupont™ Building Innovations™, 103F31 (middle, bottom), 103F32, 103SA06

NCCER CURRICULA — USER UPDATE

NCCER makes every effort to keep its textbooks up-to-date and free of technical errors. We appreciate your help in this process. If you find an error, a typographical mistake, or an inaccuracy in NCCER's curricula, please fill out this form (or a photocopy), or complete the online form at **www.nccer.org/olf**. Be sure to include the exact module ID number, page number, a detailed description, and your recommended correction. Your input will be brought to the attention of the Authoring Team. Thank you for your assistance.

Instructors – If you have an idea for improving this textbook, or have found that additional materials were necessary to teach this module effectively, please let us know so that we may present your suggestions to the Authoring Team.

NCCER Product Development and Revision

13614 Progress Blvd., Alachua, FL 32615

Email: curriculum@nccer.org
Online: www.nccer.org/olf

❏ Trainee Guide ❏ Lesson Plans ❏ Exam ❏ PowerPoints Other _____

Craft / Level: _____ Copyright Date: _____

Module ID Number / Title: _____

Section Number(s): _____

Description: _____

_____

_____

_____

_____

Recommended Correction: _____

_____

_____

_____

_____

Your Name: _____

Address: _____

_____

Email: _____ Phone: _____

45104-07

Drywall Installation

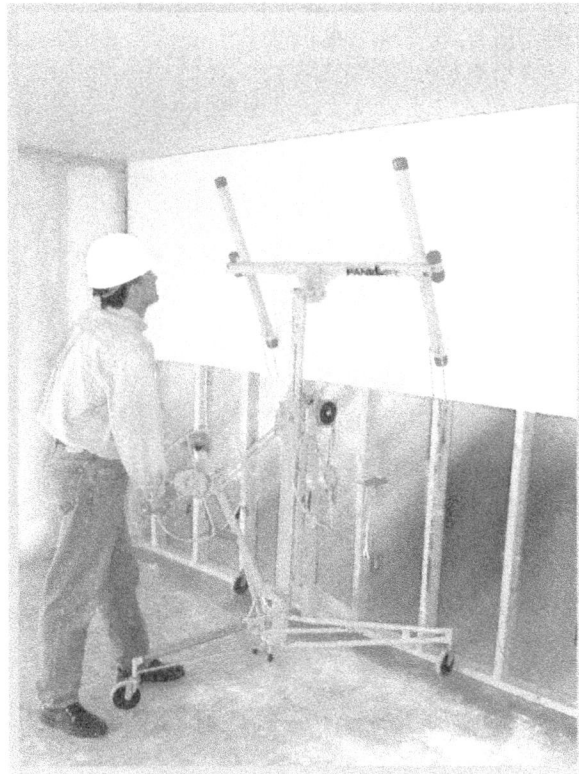

45104-07
Drywall Installation

Overview

Gypsum drywall is the most common wall finish used in residential and commercial construction. There are a variety of drywall materials used for different applications, as well as a number of construction methods used to build walls to meet building codes for fire resistance and sound transmission. There are also many different types of fasteners used in drywall installation. The selection of materials, fasteners, and construction methods is controlled by building codes, and therefore must be carefully considered.

Objectives

When you have completed this module, you will be able to do the following:

1. Identify the different types of drywall and their uses.
2. Select the type and thickness of drywall required for specific installations.
3. Select fasteners for drywall installation.
4. Explain the fastener schedules for different types of drywall installations.
5. Perform single-layer and multi-layer drywall installations using different types of fastening systems, including:
   - Nails
   - Drywall screws
6. Install gypsum drywall on steel studs.
7. Explain how soundproofing is achieved in drywall installations.
8. Estimate material quantities for a drywall installation.

Trade Terms

Corner bead	Joint
Floating interior angle	Nail pop
construction	Slurry
Gypsum board	Substrate

Required Trainee Materials

1. Pencil and paper
2. Appropriate personal protective equipment

Prerequisites

Before you begin this module, it is recommended that you successfully complete *Core Curriculum*; and *Drywall Level One*, Modules 45101-07 through 45103-07.

This course map shows all of the modules in the first level of the *Drywall* curriculum. The suggested training order begins at the bottom and proceeds up. Skill levels increase as you advance on the course map. The local Training Program Sponsor may adjust the training order.

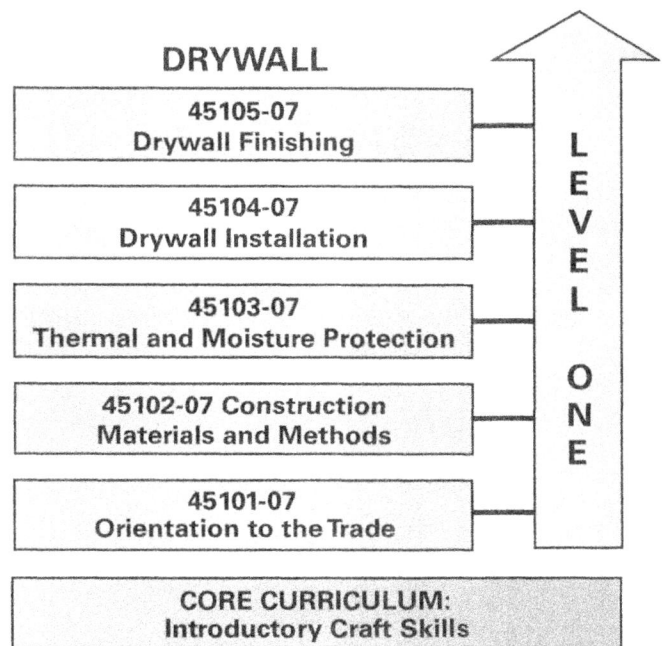

DRYWALL

45105-07
Drywall Finishing

45104-07
Drywall Installation

45103-07
Thermal and Moisture Protection

45102-07 Construction
Materials and Methods

45101-07
Orientation to the Trade

CORE CURRICULUM:
Introductory Craft Skills

LEVEL ONE

104CMAP.EPS

1.0.0 ◆ INTRODUCTION

Gypsum board, also known as gypsum drywall, is one of the most popular and economical methods of finishing the interior walls and ceilings of wood-framed and metal-framed buildings. Gypsum drywall gives a wall or ceiling made from many panels the appearance of being made from one continuous sheet.

The responsibility for drywall installation and finishing varies from job to job and from one locale to another. In some situations, carpenters install the drywall and painters finish it. In others, professional drywall workers do the entire job. The smaller the project, the more likely it is that the carpenter will install and finish the drywall.

2.0.0 ◆ GYPSUM BOARD

As you learned in an earlier module, gypsum board is a generic name for products consisting of a noncombustible core. This product is made primarily of gypsum with a paper covering on the face, back, and long edges. A typical board application is shown in *Figure 1*.

Gypsum board differs from products such as plywood, hardboard, and fiberboard because of its noncombustible core. Gypsum is a mineral found in sedimentary rock formations in a crystalline form known as calcium sulphate dihydrate.

One hundred pounds of gypsum rock contains approximately 21 pounds (10 quarts) of chemically combined water. The gypsum rock is mined and then crushed. The crushed rock is heated to about 350°F, driving out or evaporating three-fourths of the chemically combined water in a process called calcining. The calcined gypsum is then ground into a fine powder used in plaster, wallboard, and other gypsum products.

To produce gypsum board, the calcined gypsum is mixed with water and additives to form a slurry, which is fed between continuous layers of paper on a board machine. As the board automatically moves down a conveyor line, the calcium sulphate recrystallizes or rehydrates, reverting to its original rock state.

The paper becomes chemically and mechanically bonded to the core. The board is then cut to length and conveyed through dryers to remove any free moisture.

Gypsum drywall is generally available in widths of 4' and lengths of 8', 10', 12', and 14'. Other lengths are available by special order. The available board edges are rounded, tapered, beveled, square edge, and tongue and groove (*Figure 2*). Pre-decorated architectural gypsum drywall panels are also available (*Figure 3*).

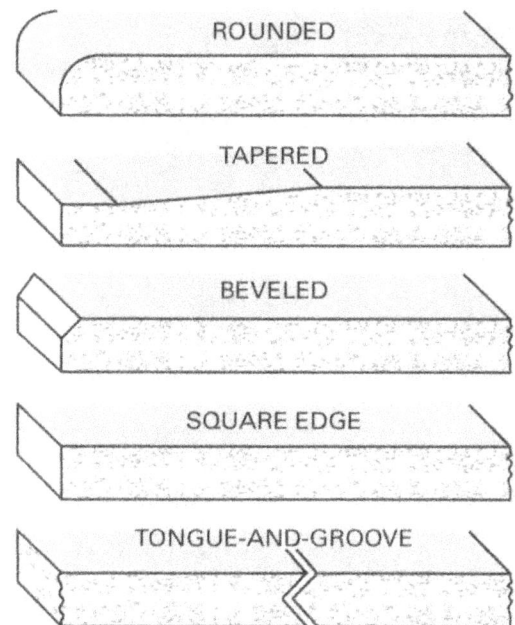

104F01.EPS

Figure 1 ◆ Typical board application.

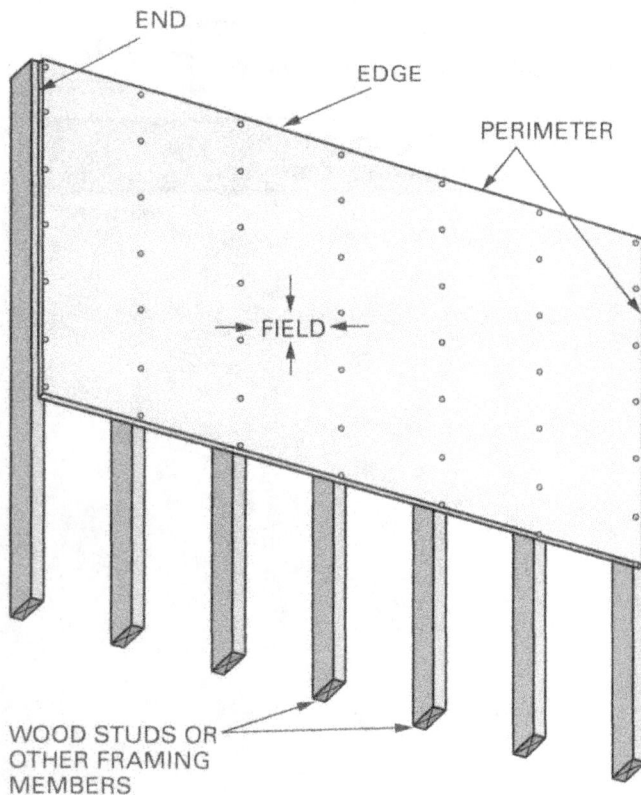

104F02.EPS

Figure 2 ◆ Standard edges of gypsum board.

The Way It Was

Until the 1930s, walls were typically finished by installing thin, narrow strips of wood or metal known as lath between studs, and then coating the lath with wet plaster. Skilled plasterers could produce a very smooth wall finish, but the process was time-consuming and messy. In the early 1930s, paper-bound gypsum board was introduced and soon came into widespread use as a replacement for the tedious lath and plaster process.

104F03.EPS

Figure 3   Architectural gypsum board.

2.1.0 Advantages of Gypsum Board Construction

Gypsum board walls and ceilings have a number of outstanding advantages:

- Fire resistance
- Sound insulation
- Durability
- Economy
- Versatility

2.1.1 Fire Resistance

Gypsum board is an excellent fire-resistive material. It is the most commonly used interior finish where fire resistance classifications are required.

Its noncombustible core contains chemically combined water, which, under high heat, is slowly released as steam, effectively retarding heat transfer. Even after complete calcination, when all of the water has been released, it continues to act as a heat-insulating barrier.

In addition, tests conducted in accordance with the American Society for Testing Materials International, *ASTM Method E84*, show that it has low flame spread and low fuel and smoke contribution factors. When installed in combination with other materials, it serves to effectively protect building elements from fire for prescribed time periods. Type X board is most often used in fire-rated assemblies. Be sure all local codes and regulations are met.

2.1.2 Sound Isolation

Control of unwanted sound that might be transmitted to adjoining rooms is a key consideration in the design of a building. It has been determined that low-density paneling transmits an annoying amount of noise. Sound-absorbing acoustical surfacing materials, while they reduce the reflection of sound within a room, do not greatly reduce transmission of sound into adjoining rooms. Gypsum board wall and ceiling systems effectively help to control sound transmission.

2.1.3 Durability

Gypsum board makes strong, high-quality walls and ceilings with excellent dimensional stability. Their surfaces are easily decorated and refinished.

2.1.4 Economy

Gypsum board products are easy to apply. They are the least expensive of the wall surfacing materials that offer a fire-resistant interior finish. Both regular and architectural wallboard may be installed at relatively low cost. When architectural board is used, further decorative treatment is unnecessary.

2.1.5 Versatility

Gypsum board products satisfy a wide range of architectural requirements for design. Ease of application, performance, availability, ease of repair, and adaptability to all forms of decoration combine to make gypsum board unmatched by any other surfacing product.

Gypsum Drywall—
A Versatile Finish Material

It is common to think of gypsum drywall as a finish for flat walls and ceilings. These photographs show how it can be used in much more complex designs.

IN PROCESS

104SA01.EPS

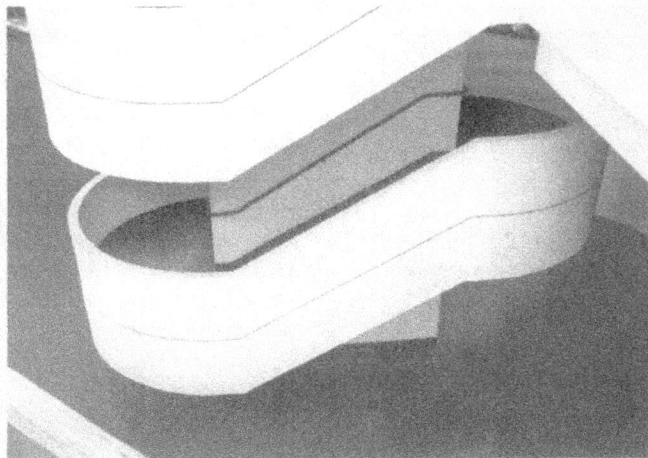

FINISHED

104SA02.EPS

3.0.0 ◆ TOOLS USED FOR GYPSUM BOARD APPLICATION

The following tools are used for gypsum board application:

- *Carbide cutter* – The carbide cutter shown in *Figure 4* is a dual-purpose tool. With the blade positioned as shown at the top of the figure, the tool is used to score drywall, cement board, and other sheet materials. With the blade reversed, the tool can be used to chisel openings in cement board and masonry backing panels.
- *4' T-square* – This tool is indispensable for making accurate cuts across the narrow dimension of gypsum board. See *Figure 5*.
- *Utility knife* – This is the standard knife used for cutting gypsum board. It has replaceable blades stored in the handle. See *Figure 6*.

- *Rasp* – The rasp (*Figure 7*) is used to quickly and efficiently smooth rough-cut edges of gypsum board. The tool has both a file for finishing and a rasp for rough shaping.
- *Circle cutter* – The circle cutter (*Figure 8*) has a calibrated steel shaft that allows accurate cuts up to 16" in diameter. The cutter wheel and center pin are heat-treated.
- *Utility saw* – This saw (also known as a keyhole saw) is used for cutting small openings and making odd-shaped cuts. See *Figure 9*. A power cutout tool (*Figure 10*) can be used for the same purpose.
- *Drywall saw* – This saw has a short blade and coarse teeth. It cuts gypsum quickly and easily. The sharp points of the teeth and the stiffness of the blade allow the saw to be punched through the board for starting the cut. See *Figure 11*.

Figure 4   Carbide cutting tool.

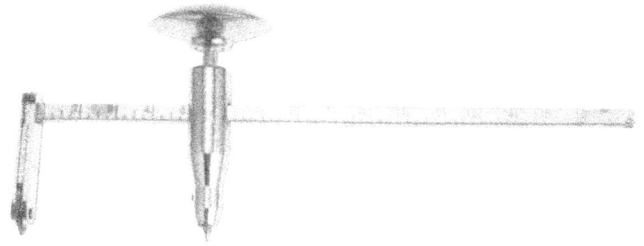

Figure 5   T-square.

Figure 6   Utility knife.

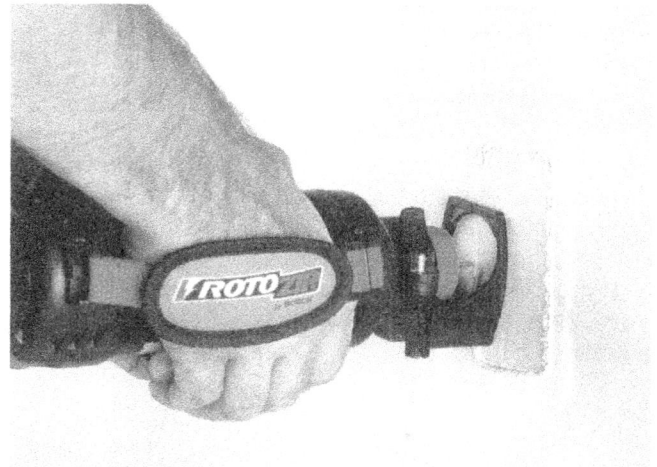

Figure 7   Drywall rasp.

Figure 8   Circle cutter.

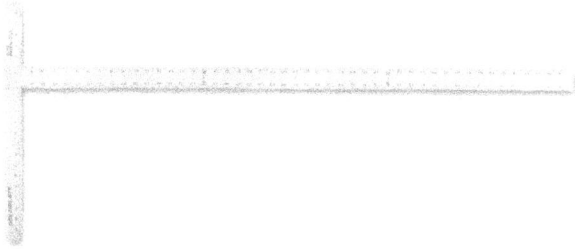

Figure 9   Utility saw.

Figure 10   Power cutout tool.

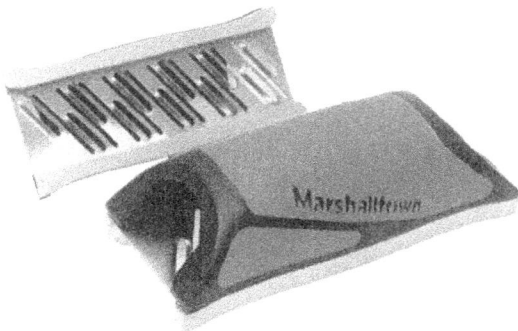

Figure 11   Drywall saw.

 INSIDE TRACK

Measuring and Marking

When measuring drywall, use a soft lead pencil to mark the drywall. A ballpoint pen mark may bleed through the joint compound and paint.

- *Gypsum board lifter* – A gypsum board lifter is used to move the board forward as it is being lifted. The lifter can be used for either parallel or perpendicular board applications. See *Figure 12.*
- *Drywall hammer* – This hammer has a symmetrical convex face designed to compress the gypsum panel face and leave the desired dimple. The blade end is not for cutting. It is dull and is used for wedging and prying. See *Figure 13.*
- *Screw gun* – Electric drywall screw guns are designed to drive steel screws to a precise depth below the gypsum board face; at that point, the drive is disengaged by a clutch mechanism. The depth setting is adjustable. The screws are held in place by a magnetic bit tip. The corded screw gun shown in *Figure 14* is designed to operate at a speed of 0 to 4,000 rpm. Cordless models operate at somewhat lower speeds. *Figure 15* shows a newer type of corded screw gun that operates in the 0 to 6,000 rpm range. Note that this screw gun has a belt clip. Belt clips are a common feature on modern drywall screw guns.
- *Drywall lift* – This special device is designed to raise and support drywall panels during ceiling or high wall installations. See *Figure 16.*
- *T-brace* – A job-built T-brace is used to hold drywall in place against ceiling joists while fasteners are being installed or while adhesive is setting. See *Figure 17.*

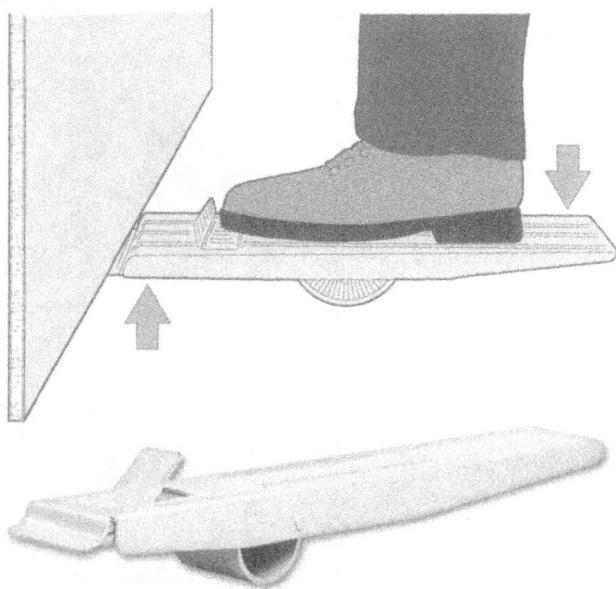

104F13.EPS

Figure 13 ◈ Drywall hammer.

104F14.EPS

Figure 14 ◈ Screw guns.

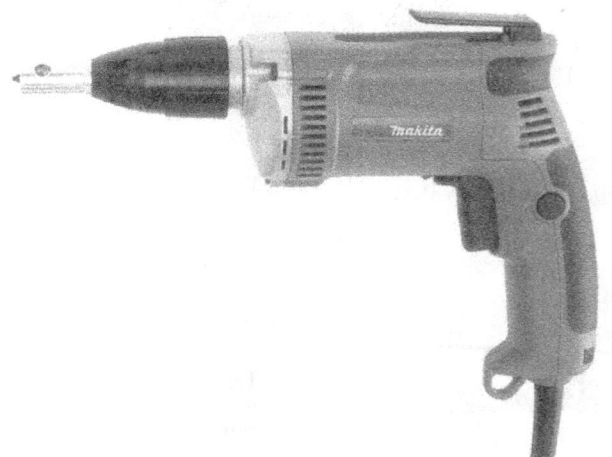

104F12.EPS

Figure 12 ◈ Gypsum board lifter.

104F15.EPS

Figure 15 ◈ High-speed drywall screw gun.

Figure 16 ◆ Drywall lift.

Figure 17 ◆ T-brace.

104F17.EPS

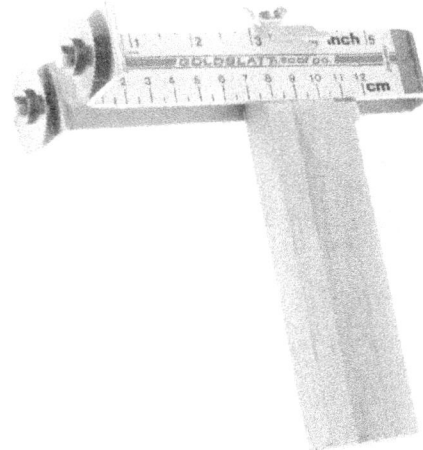

Cutting Small Strips

The drywall stripper shown here is designed to cut narrow strips of wallboard up to 4½". There is a sharp edge on each side of the tool, so it cuts both sides of the panel at once. The drywall stripper makes a cleaner cut than a utility knife when cutting long, narrow strips such as those that might be needed around a window or door.

104SA03.EPS

4.0.0 ◆ APPLICATION OF GYPSUM BOARD

Gypsum board panels can be applied over any firm, flat base such as wood or steel framing and furring. Gypsum can also be applied to masonry and concrete surfaces, either directly or to steel or wood furring strips. If the board is applied directly, any irregularities in the masonry or concrete surfaces must be smoothed or filled. Furring is a means to provide a flat surface for standard fastener application. It also provides a separation to overcome dampness in exterior walls.

The most common type of residential interior wall construction is the standard gypsum board system with joints between the panels and internal corners reinforced with tape and covered with joint treatment compound to prepare them for decoration.

The term *joint* is used to describe any point where two drywall panels meet. A butt joint is where two sheets of wallboard with untapered sides meet. A flat joint is the intersection of two bevel-edged wallboards. External corners are normally reinforced with corner bead, which, in turn, is covered with joint compound. Exposed edges

Locating Electrical Boxes

The Blind Mark™ system uses magnets to find electrical boxes under drywall. A target magnet is placed in the electrical box before the drywall is installed. After the drywall is up, a locator magnet is used to find the box containing the target.

(1) (2) (3) (4)

104SA04.EPS

are covered with metal or plastic trim. The result is a smooth, unbroken surface ready for final decoration. When architectural board is used, no further decoration is necessary but trim moldings or battens can be used to cover the joints. Special instructions for installing architectural panels will be given later in this module.

4.1.0 Single-Ply and Multi-Ply Construction

In light commercial and residential construction, single-ply gypsum board systems are commonly used (*Figure 18*). Generally, they are adequate to meet fire resistance and sound control requirements. Multi-ply systems, as shown in *Figure 19*, have two or more layers of gypsum board to increase sound isolation and fire-resistive performance. They also provide better surface quality because face layers are often laminated over base layers, thereby reducing the number of fasteners. As a result, the surface joints of the face layer are reinforced by the continuous base layers of gypsum board. Nail pop and ridging problems are less frequent, and imperfectly aligned supports have less effect on the finished surface.

Parallel vs. Perpendicular Installation

Perpendicular installation (at right angles to the framing members) is generally preferred when ceilings are normal height—8'-1" or less. A 54" board is available for use with 9' ceilings. Perpendicular installation has several advantages over parallel installation, including:

- There are fewer joints.
- There is less measuring and cutting required.
- Joints are at a convenient height for finishing.
- A single panel ties together more framing members to increase strength and hide framing irregularities.

For rooms with taller ceilings, parallel application is more practical. Parallel application may also be required for normal-height ceilings in order to meet fire ratings.

For ceiling application, select the method that results in the fewest joints.

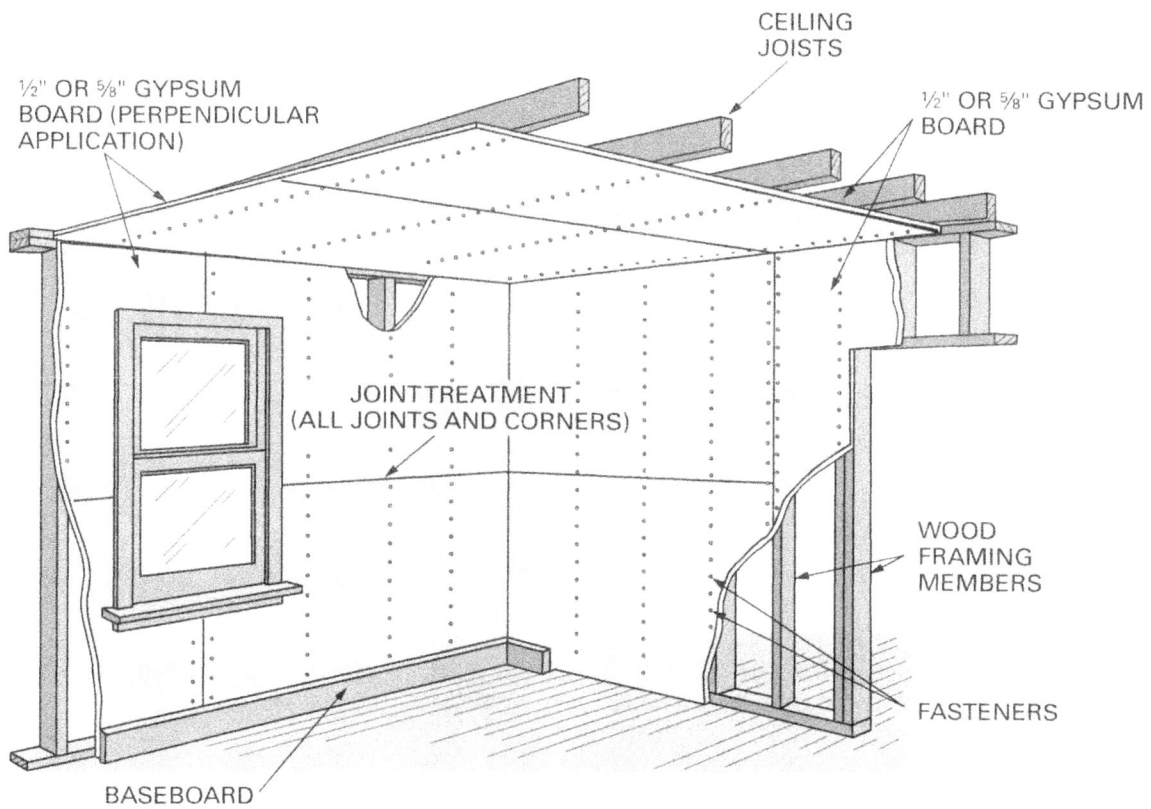

CEILING JOISTS

½" OR ⅝" GYPSUM BOARD (PERPENDICULAR APPLICATION)

½" OR ⅝" GYPSUM BOARD

JOINT TREATMENT (ALL JOINTS AND CORNERS)

WOOD FRAMING MEMBERS

FASTENERS

BASEBOARD

104F18.EPS

Figure 18   Single-ply construction.

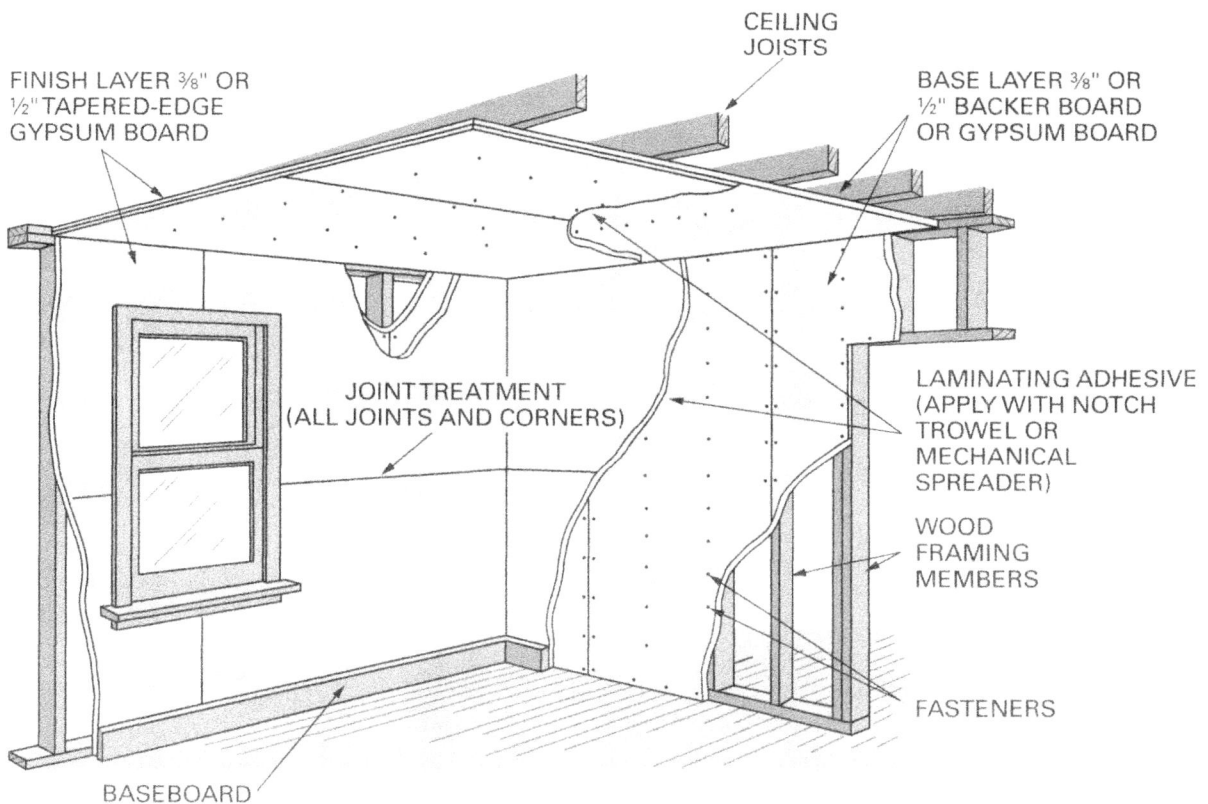

CEILING JOISTS

FINISH LAYER ⅜" OR ½" TAPERED-EDGE GYPSUM BOARD

BASE LAYER ⅜" OR ½" BACKER BOARD OR GYPSUM BOARD

JOINT TREATMENT (ALL JOINTS AND CORNERS)

LAMINATING ADHESIVE (APPLY WITH NOTCH TROWEL OR MECHANICAL SPREADER)

WOOD FRAMING MEMBERS

FASTENERS

BASEBOARD

104F19.EPS

Figure 19   Multi-ply construction.

Satisfactory results can be assured with either single-ply or multi-ply assemblies by requiring the following:

- Proper framing details, consisting of straight, correctly spaced, and properly cured lumber
- Proper job conditions, including controlled temperatures and adequate ventilation during application
- Proper measuring, cutting, aligning, and fastening of the board
- Proper joint and fastener treatment
- Special requirements for proper sound isolation, fire resistance, thermal properties, or moisture resistance

Single-ply and multi-ply installations will be discussed in more detail later in this module.

4.2.0 Job-Site Preparation

Job conditions such as temperature and humidity can affect the performance of joint treatment materials and the appearance of the joint. These conditions may also affect adhesive materials and their ability to develop an adequate bond. During the cold season, manufacturers usually specify that interior finishes should not be installed unless the building is maintained between 55°F and 70°F. These temperatures should also be maintained for at least 48 hours after the installation. All materials must be protected from the weather.

When ceilings are to receive water-based spray texture finishes, special attention must be given to the spacing of framing members, the thickness of the board used, ventilation, vapor barriers, insulation, and other factors, which can affect the performance of the system and cause problems, particularly sag of the gypsum board between framing members.

Lumber must be kept dry during storage and installation at the job site. Its moisture content should not exceed 15 percent at the time of gypsum board application. Green lumber should not be used for framing. Since lumber shrinks across the grain as it dries, it tends to expose the shanks of nails driven into the edges of the framing members. If shrinkage is substantial or the nails are too long, separation between the gypsum board and its framing member can result in nail pops.

The delivery of gypsum board should coincide with the installation schedule. Boards should be placed for convenience at the work location. Boards should be stored flat and under cover. The materials used as storage supports should be at least 4" wide. As the units are tiered, the supports should be carefully aligned from bottom to top so

that each tier rests on a solid bearing, as shown in *Figure 20*. Be careful to avoid excessive weight.

Stacking long lengths on short lengths should be avoided to prevent the longer boards from breaking. Leaning boards against the framing members for a prolonged period of time with the long edges horizontal is not recommended. Avoid leaning

104F20.EPS

Figure 20 ✦ Gypsum board storage.

INSIDE TRACK   **Moving Drywall**

A drywall dolly like the one shown here is specially designed to transport drywall panels at the job site.

104SA05.EPS

Storing and Handling Gypsum Drywall

Figure 20 shows gypsum drywall as it would be stored in a warehouse or building supply store. Just before installation, the drywall panels would be distributed along interior walls and stood on edge.

Drywall panels are sold in pairs. Two sheets of drywall are connected by a strip of paper tape, which can be stripped off to separate the panels. Drywall is heavy. Two ½" panels weigh about 110 pounds, while a pair of ⅝" panels weigh close to 150 pounds. The panels need to be handled carefully so they don't break under their own weight. They should be lifted and carried by the edges rather than the ends. Also, proper lifting procedures must be used by people handling drywall in order to prevent injury.

boards during periods of high humidity, as the boards could be subject to warping. All materials should remain stored in their original wrappers or containers until ready to use on the job site. When boards are moved, they should be carried, not dragged, so the edges are not damaged.

4.3.0 Cutting and Fitting Procedures

Any gypsum board installation should be carefully planned. Accurate measuring, cutting, and fitting are very important. In residential buildings with less than 8'-1" ceiling heights, it is preferred that the wallboard be installed at right angles to the supporting members because there are usually fewer joints to finish. On long walls, boards of maximum practical lengths should be used to minimize the number of end joints. Scored, scratched, broken, or otherwise damaged boards should not be used.

Measurements should be done accurately at the correct ceiling or wall location for each edge or end of the board. Accurate measuring will usually reveal any irregularities in framing or furring so corrective allowances can be made in cutting. Poorly aligned framing should be corrected before applying gypsum board.

Gypsum board should be cut by first scoring through the paper down to the core with a sharp utility knife, working from the face side. The board is then snapped back away from the cut face.

The back of the paper is cut with a utility knife. Gypsum board may also be cut by sawing. All cut edges and ends of the gypsum board should be smoothed to form neat, tight-fitting joints when installed. Ragged cut ends or broken edges can be smoothed with a rasp or sandpaper, or trimmed with a sharp knife. If burrs on the cut ends are not removed, they will form a visible ridge in the finished surface.

The practices listed below should be followed to ensure a sound application:

- Install the ceiling boards first, then the wall panels.
- The panels should fit easily into place without force.
- Always match edges and ends. For example, tapered end to tapered end and square-cut end to square-cut end.
- Plan to span the entire length of ceilings or walls with single boards, if possible, to reduce the number of end joints, which are more difficult to finish.
- Stagger end joints and locate them as far from the center of the ceiling or wall as possible so they will be inconspicuous.
- In a single-ply application, the board ends and edges parallel to the supporting members (framing) should fall on these members to reinforce the joint.
- Mechanical and electrical equipment, such as cover plates, registers, and grilles, should be installed to provide for the final wall thickness when applying the trim.
- Place a shim under the wallboard to keep it from absorbing moisture from the floor.

NOTE

The depth of electrical boxes should not exceed the framing depth, and boxes should not be placed back-to-back on the same stud. Electrical boxes and other devices should not be allowed to penetrate completely through the walls. This is detrimental to sound isolation and fire resistance. Make sure the wires are pushed back into the box.

4.4.0 Drywall Fasteners

Nails and screws are commonly used to attach gypsum board in both single-ply and multi-ply installations. Clips and staples are used only to attach the base layers in multi-ply construction. Special drywall adhesives can be used to secure single-ply gypsum board to framing, furring, masonry, and concrete, or to laminate a face ply to a base layer of gypsum board or other base material. Adhesives must be supplemented with mechanical fasteners.

Where fasteners are used at the board perimeter, they should be placed at least ⅜" from the board edges and ends. Fastening should start in the middle of the board and proceed outward toward the board perimeter. Fasteners must be driven as near to perpendicular as possible while the board is held firmly against the supporting construction.

Also, fasteners must be used with the correct shield, guard, or attachment recommended by the manufacturer. Nails should be driven with a crown-headed hammer, which forms a uniform depression or dimple that is not more than ¹⁄₃₂" deep around the nail head. See *Figure 21*.

4.4.1 Nails

Both annular and cupped-head nails are acceptable for gypsum board application (*Figure 22*). Preferably, the nails should have heads that are flat or concave and thin at the rim. The heads should be between ¼" and ⁵⁄₁₆" in diameter to provide adequate holding power without cutting the

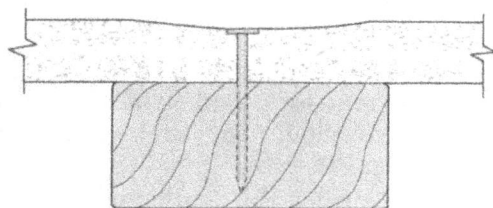

face paper when the nail is dimpled. Casing and common nails have heads that are too small in relation to the shank; they easily cut into the face paper and should not be used.

Nail heads that are too large are also likely to cut the paper surface if the nail is driven incorrectly at a slight angle. The nails should be long enough to go through the wallboard layers and far enough into the supporting construction to provide adequate holding power. The nail penetration into the framing member should be ⅞" for smooth shank nails and ¾" for annular ring nails, which provide more withdrawal resistance and require less penetration. For fire-rated assemblies, greater penetration is required (generally 1⅛" to 1¼" for one-hour assemblies).

Particular care should be taken not to break the face paper or crush the core by striking it too hard with the hammer.

Gypsum board can be attached by either a single nailing or a double nailing method. Double nailing produces a tighter board-to-stud contact. Whenever fire-resistive construction is required, the nail spacing specified in the fire test should be followed. Always check local codes for nailing requirements.

Figure 21 ◆ Uniform depression or dimple.

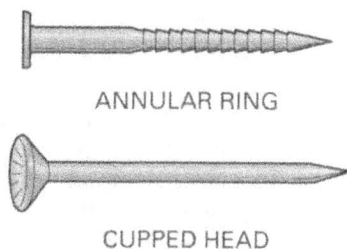

Figure 22 ◆ Nails.

INSIDE TRACK — Special Fasteners

Application of gypsum board requires special fasteners. Ordinary wood or sheet metal screws and common nails are not designed to penetrate the board without damage, to hold it tightly against the framing, or to permit correct countersinking for proper concealment. For high-end installations, sheets can be connected between studs using butt clips.

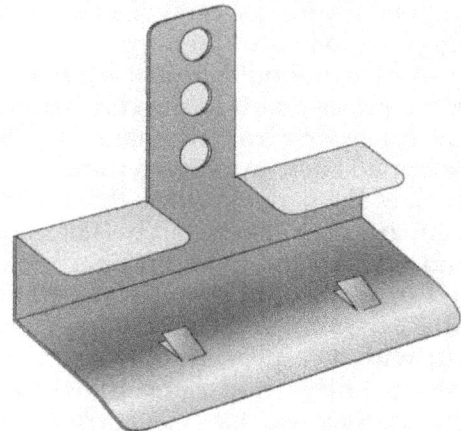

Single nails should be spaced at a maximum of 7" on center on ceilings and 8" on center on walls along framing members. See *Figure 23*.

Nails are first driven into the center or field of the board and then outward to the edges and ends. In single-ply installations, all ends and edges of gypsum board are placed over framing members or other solid backing, except where treated joints are at right angles to framing members.

In double nailing, the spacing of the first set of nails is 12" on center, with the second nailing 2" to 2½" from the first. See *Figure 24*.

The second set of nails is applied in the same sequence as the first set, but not on the perimeter of the board. The first nails driven should be reseated as necessary following the application of the second set. The general attachment procedure is as follows:

Step 1   Carefully measure and cut the board.

Step 2   Prior to nailing, mark the gypsum board to indicate the location of the framing.

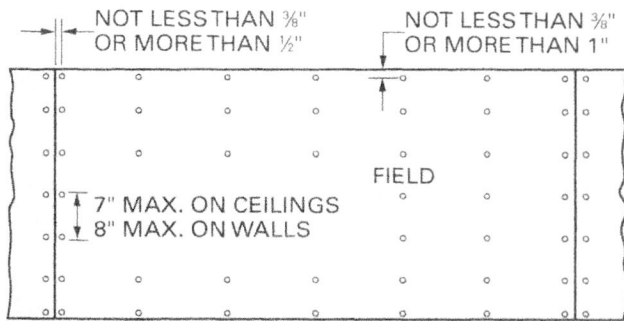

Step 3   To avoid nail pops or protrusions, hold the board firmly against the framing when nailing.

Step 4   Drive the nails straight into the framing member. Nails that miss the framing member should be removed, and the nail hole dimpled and covered with joint compound.

Step 5   Damage to the board caused by overdriving nails may be corrected by driving a new nail 2" away to provide firm attachment. Examples of correct and incorrect nailing are shown in *Figure 25*.

Other common causes of face paper fractures are misaligned or twisted supporting framing members and projections of improperly installed blocking or bracing, as shown in *Figure 26*.

Framing faults prevent solid contact between the gypsum board and framing members and hammer impact causes the board to rebound and rupture the paper. Defective supports should be corrected prior to the application of the gypsum board. Protruding framing members should be trimmed or reinstalled. Shims can be used, if

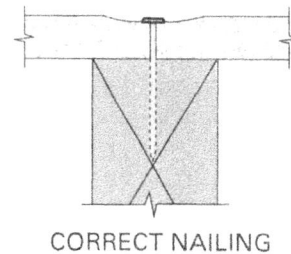

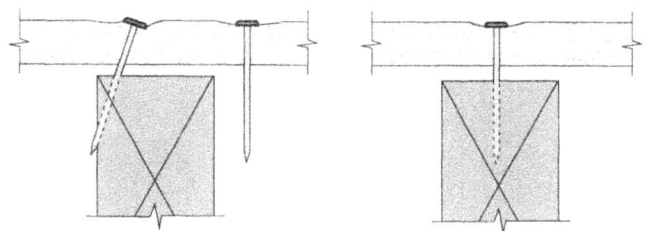

NOTE:
If screws are used in place of nails, spacing is as follows:

FRAMING SPACING	WALLS	CEILINGS
16" OC	16"	12"
24" OC	12"	12"

104F23.EPS

Figure 23   Single nail spacing.

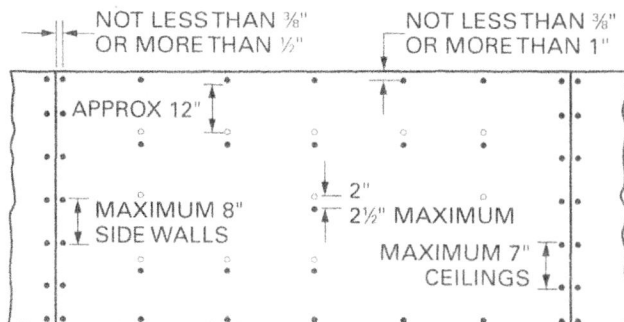

104F24.EPS

Figure 24   Double nail spacing.

CORRECT NAILING

INCORRECT NAILING

104F25.EPS

Figure 25   Correct and incorrect nailing.

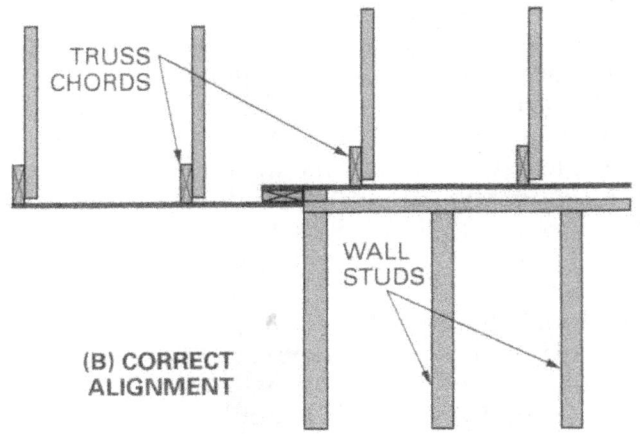

Figure 26 ❖ Incorrect and correct alignment.

necessary, for receiving framing members. The use of screws, adhesives, or two-ply construction will minimize problems resulting from these defects.

4.4.2 Screws

Drywall screws (*Figure 27*) are used to attach gypsum board to wood or steel framing or to other gypsum board. They have a Phillips head design that is intended to be used with a drywall power screwdriver. These screws pull the board tightly to the supports without damaging the board and minimize fastener and surface defects due to loose boards. The specially contoured head, when properly driven, makes a uniform depression that is free of ragged edges and fuzz.

Type W gypsum drywall screws are designed for fastening gypsum board to wood framing or furring. The Type W screw points are diamond-shaped to provide efficient drilling action through both gypsum and wood, and their specially designed threads provide both quick penetration and increased holding power.

The recommended minimum penetration into supporting construction is ⅝". However, in two-ply construction where the face layer is screw-attached, additional holding power is developed in the base ply, which permits reduced penetration into supports down to ½". Type S screws may be substituted for Type W screws in two-ply construction.

Figure 27 ❖ Drywall screws.

Selecting Drywall Screws

If you go to a building supply store to get drywall screws, you will find that there are many different types and sizes. It is important to determine exactly what you need before you make the trip.

Electric Screwdriver Attachment

This attachment holds a strip of 50 drywall screws. It has a depth control that allows the operator to set and lock in the correct depth for drywall screws.

104SA08.EPS

Type S gypsum drywall screws are designed for fastening gypsum boards to steel studs or furring. They are self-drilling with a self-tapping thread and generally a mill slot or hardened drill point that is designed to penetrate sheet metal with little pressure. Easy penetration is important because steel studs are often flexible. They tend to bend away from the screws, and the screws tend to strip easily.

Type G gypsum drywall screws are used for fastening gypsum board panels to gypsum backing boards. They are similar to Type W screws, but have a deeper, special-thread design. They are generally 1½" long, but other lengths are available. Gypsum drywall screws require a penetration of at least ½" of the threaded portion into the sup-

porting board. Allowing approximately ¼" for the point results in the minimum penetration of ¾".

Gypsum drywall screws should not be used to attach wallboard to ⅜" backer board because they do not provide sufficient holding strength. Nails or longer screws should be driven through both the surface layer and the ⅜" backer (base ply) to provide the proper penetration in the supporting wood or metal construction.

For best results, the screw gun should be kept perpendicular to the work surface. Adequate pressure must be exerted to engage the clutch and prevent the screw from slipping (also known as walking). The tool should be triggered continuously until each fastener is seated. A one-piece socket makes driving easier and more efficient than separate socket and extension pieces because it provides a more rigid base and firmer control. Depth gauges are useful to ensure proper penetration.

Because fewer fasteners are required when screws are used to attach gypsum board, the number of fasteners to be finished is reduced and possible application defects are minimized. Screws should be placed 12" on center on ceilings and 16" on center on walls where the framing members are 16" on center. Screws should be placed at a maximum of 12" on center on walls and ceilings where the framing members are 24" on center. Double screws are recommended in the latter case.

The required penetration for screws is as mentioned previously. Gypsum board should be attached to steel framing and furring with Type S screws spaced no more than 12" on center along supports for both walls and ceilings. Type S-12 screws are required for steel framing that is 20 gauge or heavier. A 12" screw spacing is also appropriate when gypsum board is mounted on resilient furring channels over wood framing.

4.5.0 Floating Interior Angle Construction

To minimize the possibility of fastener popping in areas adjacent to wall and ceiling intersections and to minimize cracking due to structural stresses, the floating angle method may be used for either the single-layer or double-layer application of gypsum board to wood framing.

Floating interior angle construction helps to eliminate nail popping and corner cracking by omitting fasteners at the intersections of walls and ceilings. This is applicable for single nailing, double nailing, and screw attachment. *Figure 28* shows a typical single-layer application. The same nail-free clearances at corners should be maintained in double nailing.

CEILING JOISTS

NAILS 7" OC

GYPSUM
WALLBOARD

FLOATING ANGLES
(OMIT NAILS)

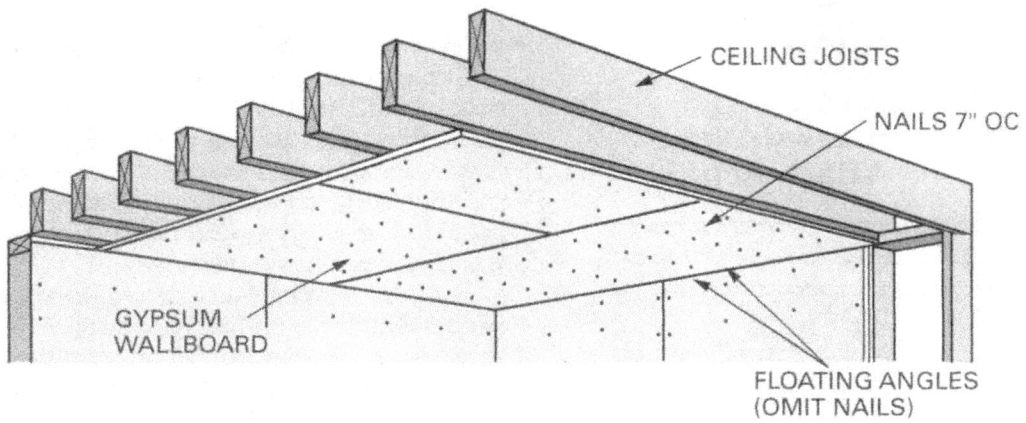

PERPENDICULAR CEILING APPLICATION
(SINGLE NAILING)

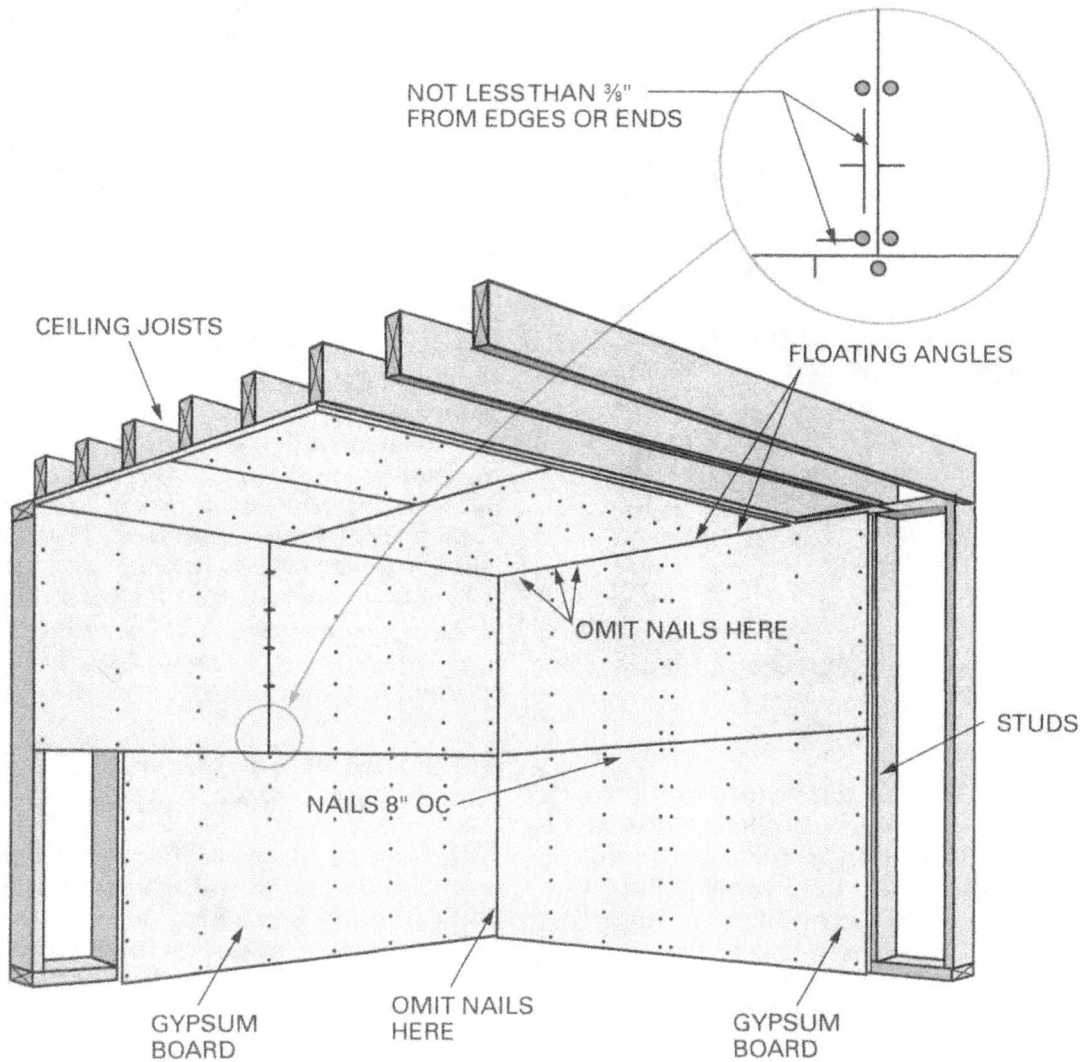

NOT LESS THAN ⅜"
FROM EDGES OR ENDS

CEILING JOISTS

FLOATING ANGLES

OMIT NAILS HERE

STUDS

NAILS 8" OC

GYPSUM
BOARD

OMIT NAILS
HERE

GYPSUM
BOARD

PARALLEL CEILING APPLICATION
(SINGLE NAILING)

104F28.EPS

Figure 28 ◆ Floating interior angle construction.

In floating interior angle construction where the ceiling framing members are perpendicular to the wall/ceiling intersection, the ceiling fasteners should be located 7" from the intersection for single nailing and 11" to 12" for double nailing or screw applications.

On ceilings where the joists are parallel to the wall intersection, nailing should start at the intersection. Gypsum board should be applied to the ceiling first and then to the walls. See *Figure 29*.

Gypsum board on side walls should be applied to provide a firm, level support for the floating edges of the ceiling board.

Apply the overlapping board firmly against the underlying board to bring the underlying board into firm contact with the face of the framing member behind it. The overlapping board should be nailed or screwed, and the fasteners should be omitted from the underlying board at the vertical intersection.

4.6.0 Adhesives

Adhesives are used to bond single layers of gypsum board directly to the framing, furring, masonry, or concrete. They can be used to laminate gypsum board to base layers of backer boards, sound deadening boards, rigid foam, and other rigid insulating boards. The adhesive must be used in combination with nails or screws, which provide supplemental support.

The adhesives used for applying wallboard finishes are classified as follows:

- Stud adhesives
- Laminating adhesives such as dry powder (including joint tape compound), special drywall laminating adhesives, and drywall contact and modified contact adhesives

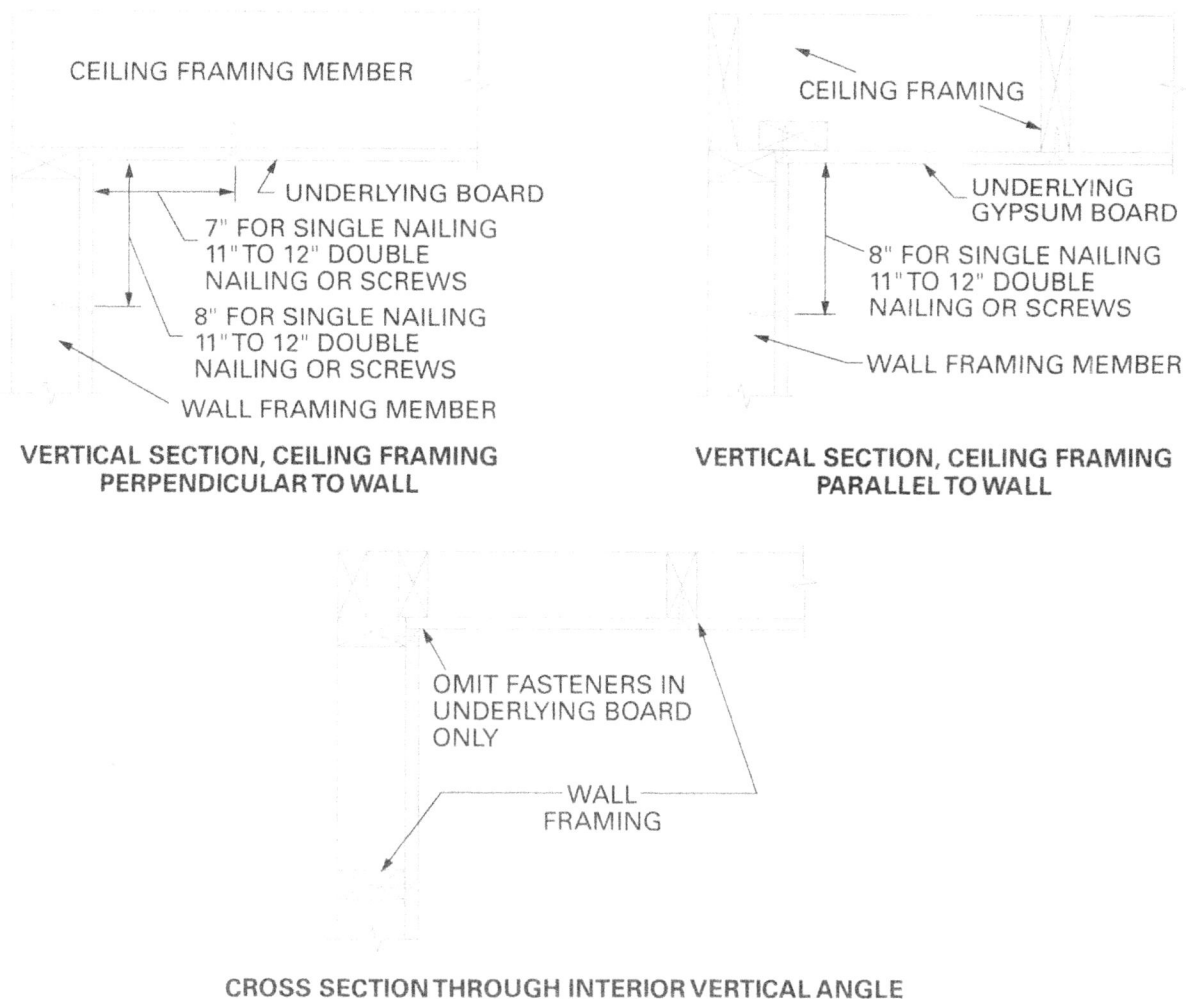

CEILING FRAMING MEMBER
UNDERLYING BOARD
7" FOR SINGLE NAILING 11" TO 12" DOUBLE NAILING OR SCREWS
8" FOR SINGLE NAILING 11" TO 12" DOUBLE NAILING OR SCREWS
WALL FRAMING MEMBER

VERTICAL SECTION, CEILING FRAMING PERPENDICULAR TO WALL

CEILING FRAMING
UNDERLYING GYPSUM BOARD
8" FOR SINGLE NAILING 11" TO 12" DOUBLE NAILING OR SCREWS
WALL FRAMING MEMBER

VERTICAL SECTION, CEILING FRAMING PARALLEL TO WALL

OMIT FASTENERS IN UNDERLYING BOARD ONLY
WALL FRAMING

CROSS SECTION THROUGH INTERIOR VERTICAL ANGLE

104F29.EPS

Figure 29   Fastener patterns for floating interior angle construction.

4.6.1 Stud Adhesives

Stud adhesives are specially prepared to attach single-ply wallboard to steel or wood studs and are generally used in conjunction with nails. Some permit a significant reduction in the use of mechanical fasteners, but they still require some fastening, at least at the board perimeters. These adhesives should be of caulking consistency so that they bridge framing irregularities. Stud adhesives should meet the requirements of the *Standard Specification for Adhesives for Fastening Gypsum Wallboard to Wood Framing (ASTM C557)*. This specification covers workability, consistency, open time, wetting characteristics, strength, bridging ability, aging, and freeze/thaw resistance. These adhesives are applied with an electric, pneumatic, or hand-operated gun (*Figure 30*) in a continuous or semi-continuous bead.

If the stud adhesive has a solvent base, it should not be used near an open flame, in poorly ventilated areas, or for lamination of architectural gypsum board. Special adhesives are available for architectural gypsum board.

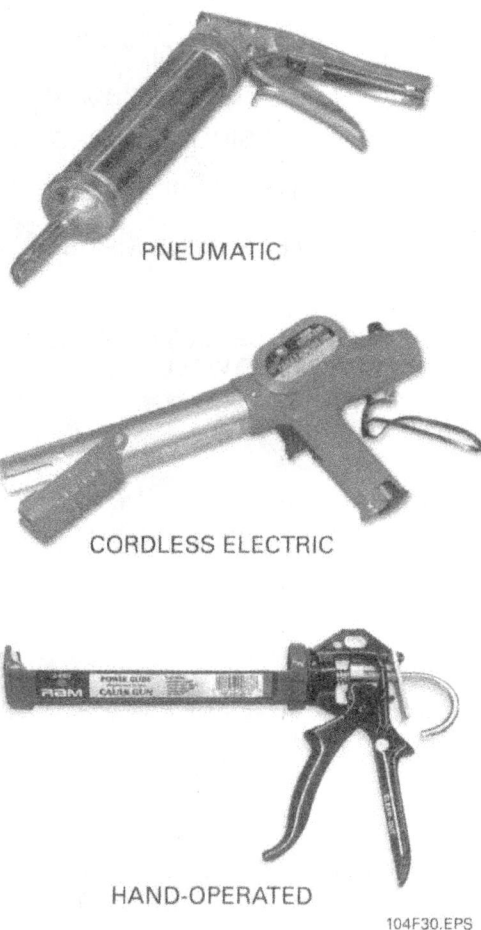

PNEUMATIC

CORDLESS ELECTRIC

HAND-OPERATED

104F30.EPS

Figure 30 ◆ Adhesive applicators.

NOTE

Check the temperature range for the adhesive you plan to use to make sure it is compatible with the expected operating temperatures of the building.

4.6.2 Dry Powder Adhesives

Dry powder laminating adhesives are generally gypsum drywall joint compounds used to embed joint-reinforcing tape. They are used to laminate gypsum boards to each other or to suitable masonry or concrete surfaces. Dry powder adhesives are not intended for use in bonding gypsum board to wood framing or furring, although special laminating adhesives, as well as some stud adhesives, can be used when recommended by the manufacturer.

Only as much laminating adhesive should be mixed as can be used within the working time specified by the manufacturer. The water used should be at room temperature and clean enough to drink. The adhesive may be applied over the entire board area with a suitable spreader, applied in spaced parallel ribbons, or applied in a pattern of spots, as recommended by the manufacturer. All dry powder laminating adhesives require permanent mechanical fasteners at the board perimeters.

If the boards are applied vertically on side walls, fasteners are placed at the top and bottom. Face boards may require temporary support or supplemental fasteners until the full bond strength is developed.

4.6.3 Drywall Contact Adhesives

Drywall contact adhesives require permanent mechanical fasteners at least at the perimeters of all boards applied to walls, ceilings, and soffits. If used to apply architectural gypsum board vertically on side walls, permanent fasteners are required only at the top and bottom of the boards, where they can be hidden by base and ceiling moldings or other decorative trim.

Contact adhesives may be used to laminate gypsum boards to each other or to steel studs. The adhesive is applied by roller, spray gun, or brush in a thin, uniform coating to both surfaces to be bonded. For most contact adhesives, some drying time is usually required before surfaces can be joined and the bond can be developed.

To ensure proper adhesion between surfaces, the face board should be impacted over its entire surface with a suitable tool, such as a rubber mallet. No temporary supports are needed while a contact adhesive sets and the bond forms.

One disadvantage of contact adhesives is their inability to fill irregularities between surfaces, which leaves some areas without an adhesive bond. Another disadvantage is that most of these adhesives do not permit moving of the board once contact has been made. A sheet of polyethylene film or tough building paper can be slipped between the surfaces so gradual bonding of surfaces will occur as the slip sheet is withdrawn. Extra care should be taken when contact adhesives are used. The manufacturer's recommendations should always be followed.

> **WARNING!**
> Observe all manufacturer's safety data sheet (MSDS) precautions for adhesives. Extreme caution must be taken when using contact cement as it is highly flammable. It should be used in a well-ventilated area as the fumes can quickly overcome a worker.

4.6.4 Modified Contact Adhesives

Modified contact adhesives provide a longer placement time. They have an open time (up to a half hour) during which the board can be repositioned, if necessary. They combine good long-term strength with a sufficient immediate bond to permit erection with a minimum of temporary fasteners.

In addition, these adhesives have enough bridging ability to cover up minor framing irregularities. Modified contact adhesives are intended for attaching wallboard to all types of supporting construction, such as solid walls, other gypsum boards, and various insulating boards, including rigid foam insulation.

Adhesives are also used for securing drywall materials and paneling to steel studs. The use of adhesives will eliminate some of the fasteners required. Use only the adhesives specified by the manufacturers or suppliers of the particular steel framing and sheathing material being used. Improper adhesives will not only fail to add to the structural integrity, but may be detrimental to system performance.

4.6.5 Application of Adhesives

Stud adhesives should be applied with a caulking gun in accordance with the manufacturer's recommendations. A straight bead, approximately ¼" in diameter, is applied to the face of the studs in the field (center) of the panel. See *Figure 31*. Where two gypsum panels join over a stud, two parallel beads of adhesive should be applied, one near each edge of the stud.

Single-ply gypsum board systems attached with stud adhesives require supplemental perimeter fasteners. The fasteners should be placed 16" on center along the edges or ends on boards that are perpendicular or parallel to the supports.

Ceiling installations require supplemental fasteners in the field as well as on the perimeter. They should be placed 24" on center. See *Figure 32*.

INSIDE TRACK   Using Adhesives on Drywall

When drywall is applied using adhesive, either drywall adhesive or construction adhesive may be used if it meets the requirements of *ASTM C557*. The adhesive should be allowed to dry for 48 hours before finishing the joints. Using adhesive does not eliminate the need for fasteners; it simply reduces the number of fasteners needed. Check the job specifications or local codes.

Adhesives cannot be used to attach drywall panels to studs if the building has an inside moisture barrier.

When laminating drywall panels for multi-layer installation, either lightweight or standard setting-type joint compound may be used in place of adhesive to laminate the panels.

104SA09.EPS

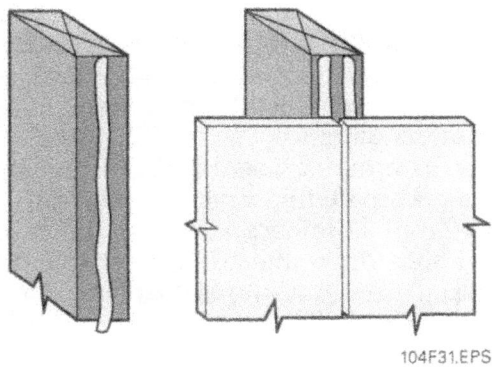

Figure 31 ❖ Adhesive applied to the edges of a stud.

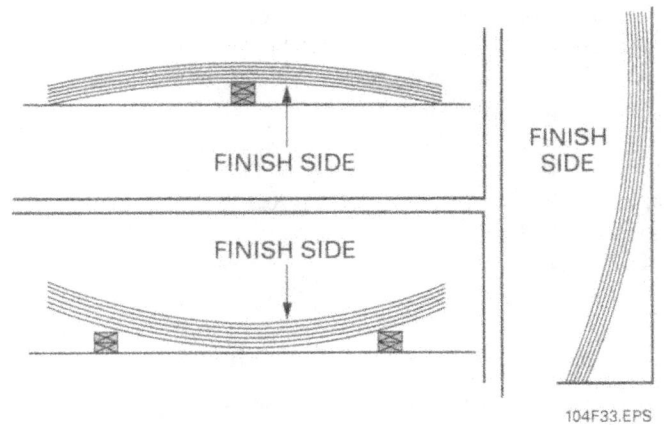

FIELD FASTENERS ON CEILINGS ONLY

PERIMETER SPACING SAME FOR WALLS AND CEILINGS

SUPPORT SPACING 16" OR 24" OC

HORIZONTAL APPLICATION

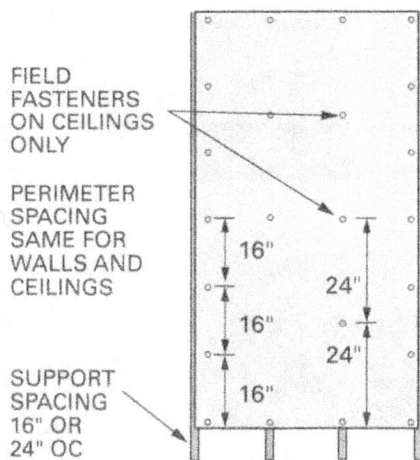

FIELD FASTENERS ON CEILINGS ONLY

PERIMETER SPACING SAME FOR WALLS AND CEILINGS

SUPPORT SPACING 16" OR 24" OC

VERTICAL APPLICATION

104F32.EPS

Figure 32 ❖ Supplemental wall and ceiling fasteners.

Adhesive is not required at top or bottom plates, bridging, bracing, or fire stops. Where fasteners at vertical joints are undesirable, gypsum panels may be prebowed, as shown in *Figure 33*.

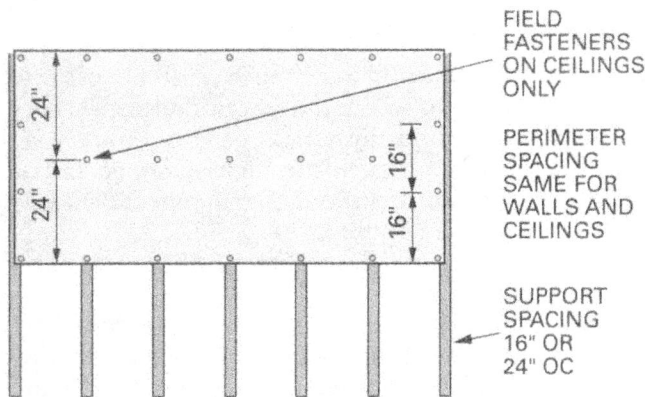

Figure 33 ❖ Prebowing of gypsum panels.

Prebowing puts an arc in the gypsum board, which keeps it in tight contact with the adhesive after the board is applied. Supplemental fasteners (placed 16" on center) are then used at the top and bottom plates.

Gypsum board may be prebowed by stacking it, face up, with the ends resting on 2 × 4 lumber or other blocks, and with the center of the boards resting on the floor. Allow it to remain overnight or until the boards have a permanent bow.

Architectural boards can also be installed using adhesive, but care should be taken to avoid adhesive contact with the decorated face. Position the boards within the open time specified for the adhesive and use a rubber mallet to tap the boards along the studs to ensure a continuous bond with the framing. Follow the manufacturer's specifications for architectural gypsum board.

4.6.6 Adhesive Application to Metal Framing

Some stud adhesives, such as those used with steel framing, require fasteners on intermediate supports as well as at the perimeters of gypsum panels. The framing spacing varies both according to the load and the type of board being used. See *Table 1*.

4.6.7 Adhesive Application to Concrete and Masonry

Gypsum board panels can be laminated directly to above-grade interior masonry and concrete wall surfaces if the surface is dry, smooth, clean, and flat. Gypsum board can be laminated directly to exterior cavity walls if the cavities are properly insulated to prevent condensation and the inside face of the cavity is properly waterproofed.

Table 1 Maximum Spacing of Ceiling Framing

Gypsum Board (Thickness)		Application to Framing		Maximum OC Spacing of Framing
Base	Face	Base	Face	
3/8"*	3/8"	Perpendicular	Perpendicular or Parallel	16"
1/2"*	3/8" or 1/2"	Perpendicular or Parallel	Perpendicular or Parallel	16"
5/8"*	1/2"	Perpendicular or Parallel	Perpendicular or Parallel	16"
5/8"*	5/8"	Perpendicular or Parallel	Perpendicular or Parallel	24"

*Adhesive between plies should be dried or cured prior to any decorative treatment. This is especially important when a spray-applied, water-based texture finish is to be used.

Sidewalls — For two-layer application with adhesive between the plies, 3/8", 1/2", 5/8" gypsum board may be applied perpendicularly (horizontally) or parallel (vertically) on framing spaced a maximum of 24" OC.

104T01.EPS

Prefinished gypsum board with a surface that is highly resistant to water vapor should not be laminated to concrete or masonry, because moisture may become trapped within the gypsum core of the board. The base surface must be made as level as possible. Rough or protruding edges and excess joint mortar should be removed and any depressions filled with mortar to make the wall surface level.

Base surfaces should be cleaned of all form oil, curing compound, loose particles, dust, and grease in order to ensure an adequate bond. Concrete should be allowed to cure for at least 28 days before gypsum board is laminated directly to it.

Exterior below-grade walls or surfaces should be furred and protected with the installation of a vapor barrier and insulation in order to provide a suitable base for attaching the gypsum board. This is also true for any surface that cannot be prepared readily for direct lamination.

Supplemental mechanical fasteners spaced 16" on center may be used to hold the gypsum board in place while the adhesive is developing a bond.

> **NOTE**
> An alternative to anchoring furring to a masonry wall is to build a 1⅝" metal stud wall.

5.0.0 ◆ RESURFACING EXISTING CONSTRUCTION

Gypsum board may be used to provide a new finish on existing walls and ceilings of wood, plaster, masonry, or wallboard. If the existing surface is structurally sound and provides a sufficiently smooth and solid backing without shimming, ¼"

gypsum board can be applied with adhesives, nails, or screws. Drywall nails should penetrate the framing by ⅞". When power-driven screws are to be used, the threaded portion of the screw must penetrate the framing by at least ⅝".

Existing surfaces that are too irregular to receive gypsum board directly should be furred and shimmed to provide a suitable fastening surface. The minimum gypsum board thickness for various support spacing and installation methods should be as previously recommended for new construction over furring. Any surface trim for mechanical and electrical equipment, such as switch plates, outlet covers, and ventilating grilles, should be removed and saved for reinstallation. Electrical boxes should be reset prior to the installation of new gypsum board.

6.0.0 ◆ DRYWALL TRIMS

Trim comes in a variety of shapes and sizes, each one having a particular function. It can be made of metal or vinyl (see *Figure 34*). Corner trim, or corner bead as it is usually called, is used to protect and reinforce exposed outside corners of wallboard, as well as provide a straight guide for finishing.

One type of corner bead is nailed or clinched onto the framing members through the drywall panels. Other types might have a bullnose (a metal corner bead with rounded edges) and/or metal mesh flanges, either in regular or expanded widths. Still another type has paper flanges attached to the corner bead, which is applied with joint compound. It usually receives a three-coat finishing process to obtain a smooth surface and to conceal any fasteners. The exposed nose of outside corners or the edge of inside corner bead provides a guide for making the finish a flush surface.

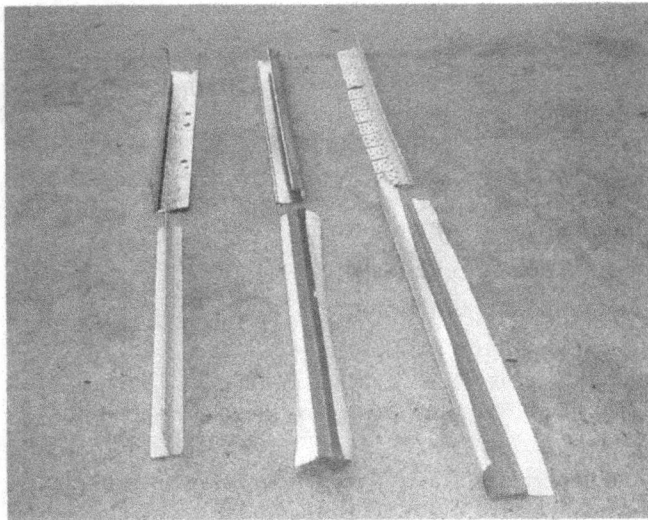

Figure 34 ◆ Drywall trims.

104F34.EPS

L-bead and J-bead (sometimes called casings) are metal or plastic pieces shaped like Ls or Js. They provide maximum protection and neat finished edges to wallboard at window and door jambs and other abutments. They are available in sizes to accommodate all thicknesses of gypsum board. They are especially useful and attractive when gypsum board abuts dissimilar objects.

Some types of trim are designed to be finished with compound in the same manner as other drywall joints. The USG 200 series trims are an example. These are known as wet trim. Other types, called dry trims, do not require finishing. The USG 400 series trims are an example of this type.

Flex tape is another type of drywall trim designed to reinforce edges and corners that serve as decoration. This material might be used as border or edging on arches, splayed angles, door and window frames, or to reinforce in a decorative way various odd intersections of ceiling or wall panels.

Reveal trim is another type of finishing or decorating material often classified as decorative trim or architectural molding. In general, reveal trim works like decorative spacing between gypsum board panels. It may also function as decorative batten strips or borders or moldings around room perimeters, door frames, windows, archways, and other architectural features.

There are many different types of decorative trims, but all of them generally attempt to eliminate the need for the usual joint finishing procedures. In place of taping and topping, a decorative piece can be installed by drywall finishers. However, some reveal trims need to be installed at the time the wallboard is hung. Such trims usually have a design feature that allows them to stay covered up until the rest of the room or project is finished. Then the protective covering is removed to show off the reveal trim between the wallboard panels.

Achieving a Rounded Appearance

A smooth, rounded finish appearance can be obtained by using a bullnose corner molding and cap such as the ones shown here.

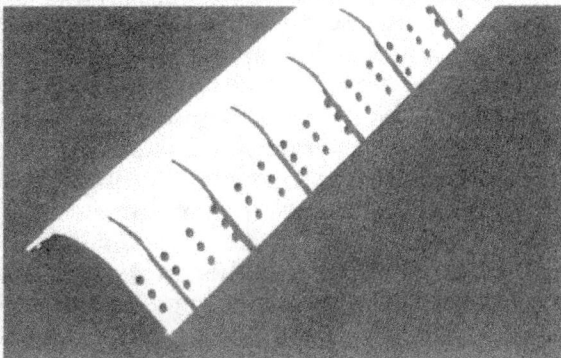

ARCH BEAD

CORNER BEAD

CORNER CAP

104SA10.EPS

7.0.0 ◆ FIRE-RATED AND SOUND-RATED WALLS

The construction of walls and partitions is driven by the fire and soundproofing requirements specified in local building codes. In some cases, a frame wall with ½" gypsum drywall on either side is satisfactory. In extreme cases, such as the separation between the offices and the manufacturing spaces in a factory, it may be necessary to have a concrete block (CMU) wall combined with fire-resistant gypsum board, along with rigid and/or fiberglass insulation, as shown in *Figure 35*. This is especially true if there is any explosion or fire hazard.

While they are only occasionally used in residential construction, steel studs are the standard for framing walls and partitions in commercial construction.

Once the studs are installed, one or more layers of gypsum board and insulation are applied. The type and thickness of the wallboard and insulation depend on the fire rating and soundproofing requirements. Soundproofing needs vary from one use to another and are often based on the amount of privacy required for the intended use. For example, executive offices, medical examination rooms, high-rise condos, hospitals, and homes for the elderly may require more privacy than general offices.

The requirements for sound reduction and fire resistance can significantly affect the thickness of a wall. For example, a steel stud wall with a high sound transmission class (STC) and fire resistance might have a total thickness of nearly 6½", while a low-rated wall might have a thickness of only 3½" using steel studs.

7.1.0 Fire-Rated Construction

Every wall in a building is rated for its fire resistance, as established by building codes. The rating is measured in terms of hours, such as one-hour wall or two-hour wall. It indicates the length of time an assembly can withstand fire and provide for evacuation of occupants, as determined under laboratory conditions (*Figure 36*). The greater the fire rating, the thicker the wall is likely to be.

In multi-family residential construction, the walls and ceilings dividing the occupancies must meet special fire and soundproofing requirements. The code requirements will vary from one location to another and may even vary within areas of a jurisdiction.

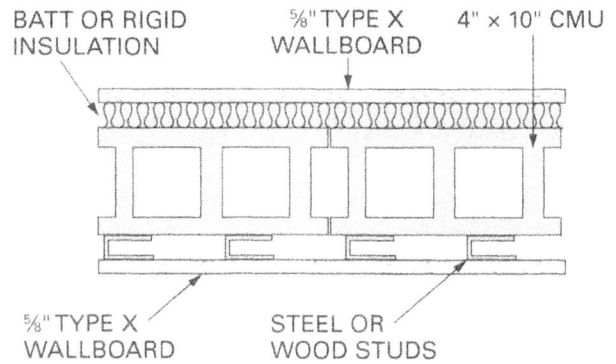

Figure 35 ◆ High fire/noise resistance partition.

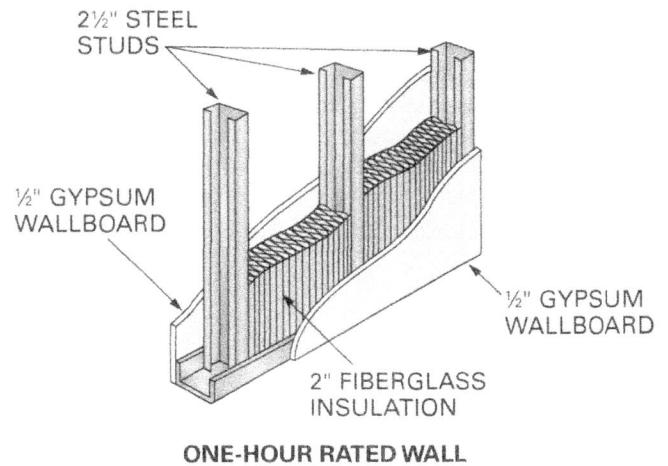

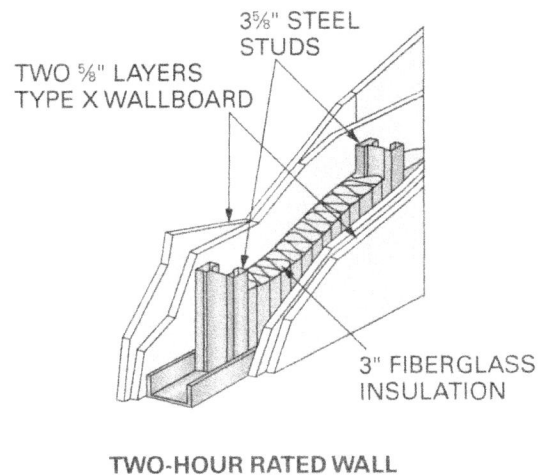

ONE-HOUR RATED WALL

TWO-HOUR RATED WALL

104F36.EPS

Figure 36 ◆ Partition wall examples.

Local codes specify the fire ratings that must be achieved in different occupancies and uses. They may also specify the nailing pattern to be used on drywall panels. Check the local codes before proceeding. Also, keep in mind that electrical and plumbing installations must be inspected in order for the building to receive a certificate of occupancy. Before covering these installations with drywall, make sure the inspection has been performed. An inspection sheet should be at the site.

There are many different construction methods for so-called party walls. Each is designed to meet different fire and soundproofing standards. The wall is generally at least 3″ thick and contain several layers of gypsum board and insulation. A fire-rated wall may abut a non-rated partition or wall. Ensure that the rated wall is carried through to maintain the fire rating. *Figure 37* shows an example of this.

When gypsum drywall is used in a party wall, the architect's plans must be followed precisely. If the inspector finds flaws in the construction, a certificate of occupancy will not be issued. This photo illustrates why building codes place so much emphasis on properly constructed party walls.

104SA11.EPS

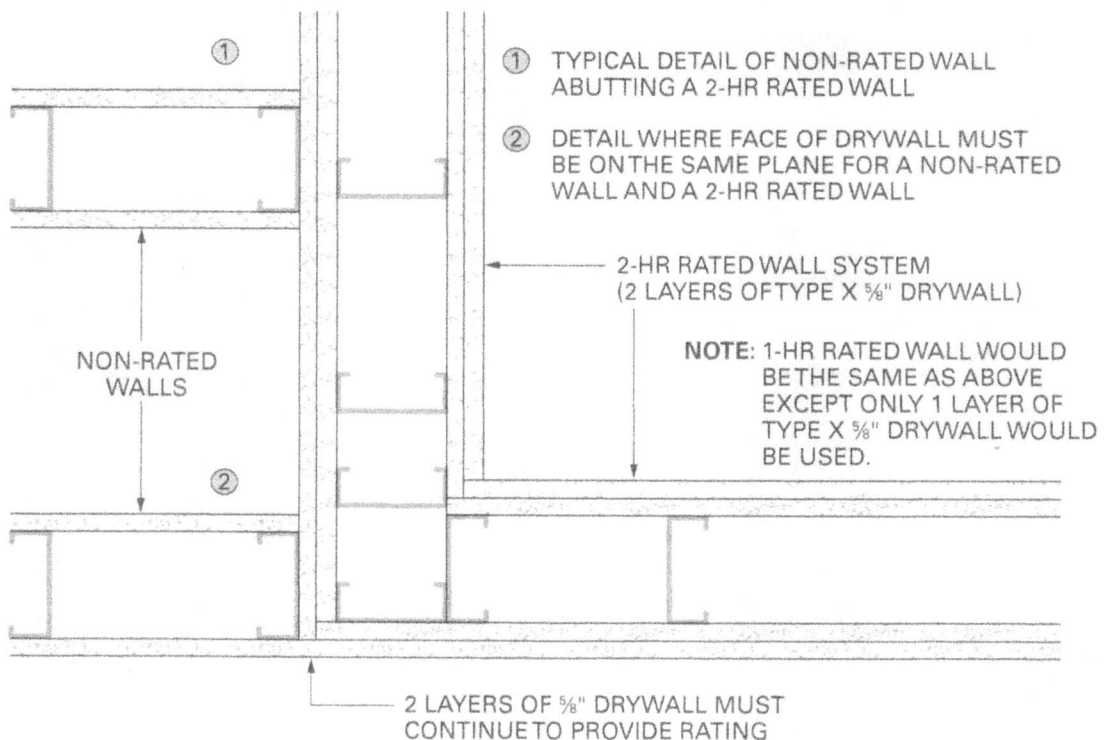

① TYPICAL DETAIL OF NON-RATED WALL ABUTTING A 2-HR RATED WALL

② DETAIL WHERE FACE OF DRYWALL MUST BE ON THE SAME PLANE FOR A NON-RATED WALL AND A 2-HR RATED WALL

2-HR RATED WALL SYSTEM (2 LAYERS OF TYPE X ⅝″ DRYWALL)

NOTE: 1-HR RATED WALL WOULD BE THE SAME AS ABOVE EXCEPT ONLY 1 LAYER OF TYPE X ⅝″ DRYWALL WOULD BE USED.

NON-RATED WALLS

2 LAYERS OF ⅝″ DRYWALL MUST CONTINUE TO PROVIDE RATING

104F37.EPS

Figure 37 ❖ An example of a fire-rated wall abutting a non-rated wall.

Drilling Fire-Rated Walls

There are many wall variations, so you must first establish the type of construction before undertaking any invasive action, such as drilling. If drilling is permitted, you may have to use firestopping materials to seal off the opening.

FIRESTOPPING
COMPOUND

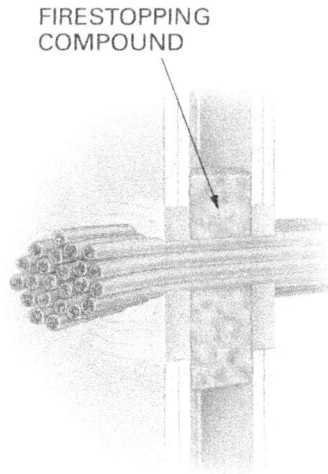

104SA12.EPS

7.1.1 Firestopping

Firestopping means cutting off the air supply so that fire and smoke cannot readily move from one location to another.

In frame construction, a firestop is a piece of wood or fire-resistant material inserted into an opening such as the space between studs. This firestop acts as a barrier to block airflow that would feed and carry a fire to the upper floors.

In commercial construction, firestopping material is used to close wall penetrations such as those created to run conduit, piping, and air conditioning ducts. If such openings are not sealed, fire will travel through the openings.

In order to meet the fire rating standards established by the building and fire codes, the openings must be sealed. The firestopping methods used for this purpose are classified as mechanical and nonmechanical.

Mechanical firestops are devices such as the one shown in *Figure 38* that mechanically seal the opening.

Nonmechanical firestops are fire-resistant materials, such as caulks and putties, that are used to fill the space around the conduit or pip-

Avoid Back-to-Back Fixtures

Medicine cabinets; electrical, telephone, television, and intercom outlets; and plumbing, heating, and air conditioning ducts should not be installed back-to-back. Any opening for such fixtures, piping, and electrical outlets should be carefully cut to the proper size and caulked.

104F38.EPS

Figure 38 ◆ Mechanical firestop device.

ing. You may be required to install various non-mechanical firestopping materials when working with fire-rated walls and floors. Holes or gaps affect the fire rating of a floor or wall. Properly filling these penetrations with firestopping materials maintains the rating. Firestopping materials are typically applied around all types of piping, electrical conduit, ductwork, electrical and communication cables, and similar devices that run through openings in floor slabs, walls, and other fire-rated building partitions and assemblies.

Nonmechanical firestopping materials are classified as intumescent or endothermic. Both are formulated to help control the spread of fire before, during, and after exposure to open flames. When subjected to the extreme heat of a fire, intumescent materials expand (typically up to three times their original size) to form a strong insulating material that seals the opening for three to four hours. Should the insulation on the cables, pipes, etc., passing through the penetration become consumed by the fire, the expansion of the firestopping material also acts to fill the void in the floor or wall in order to help stop the spread of smoke and other toxic products of combustion.

Firestopping Materials

There are a wide variety of firestopping materials on the market. Shown here are just a few examples. Firestopping and fireproofing are not the same thing. Firestopping is intended to prevent the spread of fire and smoke from room to room through openings in walls and floors. Fireproofing is a thermal barrier that causes a fire to burn more slowly and retards the spread of fire.

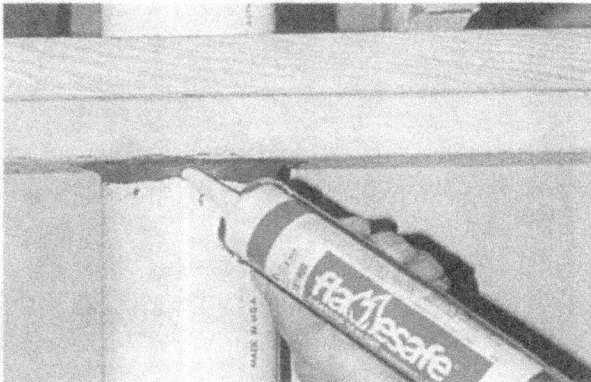
FIRESTOP SEALANT

CABLE PROTECTION SPRAY

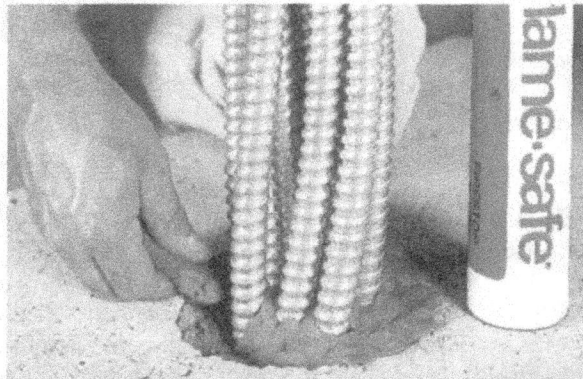
INTUMESCENT PUTTY

104SA13.EPS

Endothermic materials block heat by releasing chemically bound water, which causes them to absorb heat.

Firestopping materials are formulated in such a way that when activated, they are free of corrosive gases, reducing the risks to building occupants and sensitive equipment.

Firestopping materials are made in a variety of forms including composite sheeting, caulks, silicone sealants, foams, moldable putty, wrap strips, and spray coatings. They come in both one-part and two-part formulations. The installation of these materials must always be done in accordance with the applicable building codes and the manufacturer's instructions for the product being used. Depending on the product, firestopping materials can be applied via spray equipment, conventional caulking guns, pneumatic pumping equipment, or a putty knife.

Any firestopping materials used must meet the criteria of standard *ASTM E814, Fire Test*, as tested under positive pressure. They must also have an hourly rating that is equal to or greater than the hourly rating of the floor or wall being penetrated. Based on *ASTM E814/UL 1479* tests, one of two ratings, measured in time, may be applied to firestopping materials and systems. These ratings are as follows:

- *F rating* – A firestopping system meets the requirements of an F rating if it remains in the opening during the fire test for the rating period without permitting the passage of flames through the opening or the occurrence of flaming on any element of the unexposed side of the assembly.
- *T rating* – This rating measures thermal conductivity. It is the amount of time that it takes a thermocouple on the unexposed side of an assembly (the side away from the fire) to increase in temperature by 325°F. It does not measure passage of flames so much as dangerous levels of heat that could help spread a fire should one start. A low T-rating indicates that hot surfaces could readily conduct a fire, or spread a fire by coming into contact with combustible materials.

Firestopping is required at the head of wall joints, which are the joints that occur where a wall assembly meets a floor or roof assembly. The firestopping must be done in accordance with an approved design. Many different designs are needed because of the variety of materials that may be used in the abutting structures.

7.2.0 Sound-Isolation Construction

The first step for airborne sound isolation of any assembly is to close off air leaks and flanking paths. Since noise can travel over, under, or around walls, through the windows and doors adjacent to them, through air ducts, and through floors and crawl spaces, these paths must be correctly treated.

Buildings are generally required to meet a sound transmission class (STC) rating. The STC is a numeric rating representing the effectiveness of the construction in isolating airborne sound

transmission. The higher the STC rating, the better the sound absorption. Hairline cracks and other openings can have an adverse effect on the ability of a building to achieve its STC rating, particularly in higher-rated construction. Where a very high STC performance is needed, air conditioning, heating, and ventilating ducts should not be included in the assembly. Failure to observe special construction and design details can destroy the effectiveness of the best assembly. Improved sound isolation is obtained by the following:

- Separate framing for the two sides of a wall
- Resilient channel mounting for the gypsum board
- Using sound-absorbing materials in wall cavities
- Using multi-layer gypsum board of varying thicknesses in multi-layer construction
- Caulking the perimeter of gypsum board partitions, openings in walls and ceilings, partition/mullion intersections, and outlet box openings
- Locating recessed wall fixtures in different stud cavities

The entire perimeter of sound-isolating partitions should be caulked around the gypsum board edges to make it airtight, as detailed in *Figure 39*. The caulking should be a non-hardening, non-shrinking, non-bleeding, non-staining, resilient sealant.

Sound-control sealing must be covered in the specifications, understood by all related tradespeople, supervised by the appropriate party, and inspected carefully as the construction progresses.

7.2.1 Separated Partitions

A staggered wood stud gypsum partition placed on separate plates will provide an STC between 40 and 42. The addition of a sound-absorbing material between the studs of one partition side can increase the STC by as much as 8 points. With ⅝" Type X gypsum board on each side, an assembly has a fire resistance classification of one hour.

THINK ABOUT IT

STC-Rated Construction

What types of buildings require the highest STC-rated construction?

DOUBLE-SOLID SYSTEM   METAL STUD SYSTEM

UNDER AND OVER PARTITIONS

CAULK METAL STUD   CAULK DOUBLE STUD   CAULK WOOD STUD

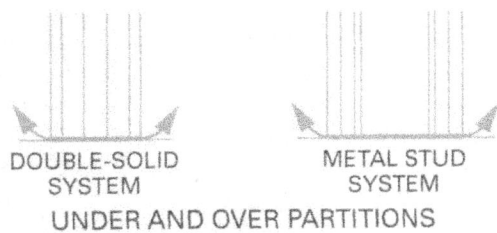

NORMAL CONSTRUCTION

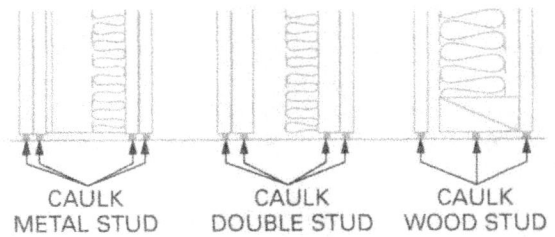

SELECT CONSTRUCTION

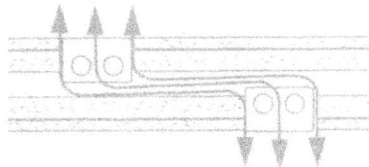

THROUGH PARTITIONS
• OPENINGS
• OUTLET BOXES

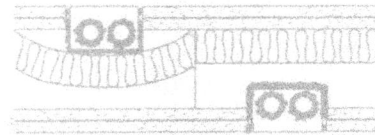

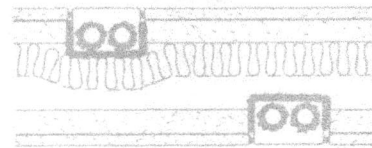

CAULKING OF OPENINGS THROUGH PARTITIONS

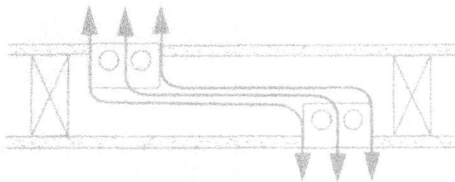

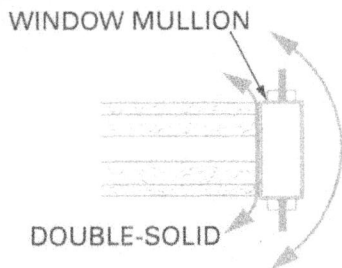

THROUGH PARTITIONS
• OPENINGS
• OUTLET BOXES

CAULKING OF OPENINGS THROUGH PARTITIONS

WINDOW MULLION

DOUBLE-SOLID

AROUND – FLANKING PARTITION ENDS

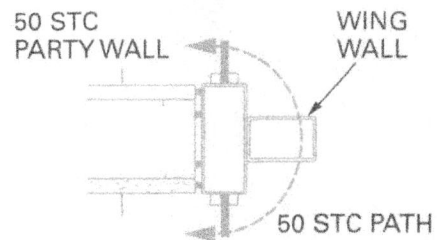

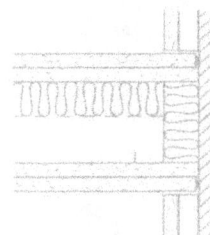

50 STC PARTY WALL   WING WALL

50 STC PATH

TYPICAL PARTITION-MULLION INTERSECTION

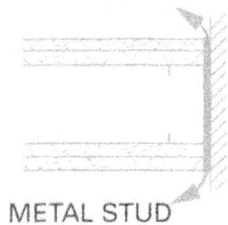

METAL STUD

AROUND – FLANKING PARTITION ENDS

INTERSECTION WITH INTERIOR WALL

104F39A.EPS

Figure 39 ◆ Caulking of sound-isolation construction (1 of 2).

PRE-DESIGN CONSTRUCTION
SIMULATING LABORATORY CONDITIONS

¼" PERIMETER RELIEF AND CAULKING
TO SEAL AGAINST LEAKS

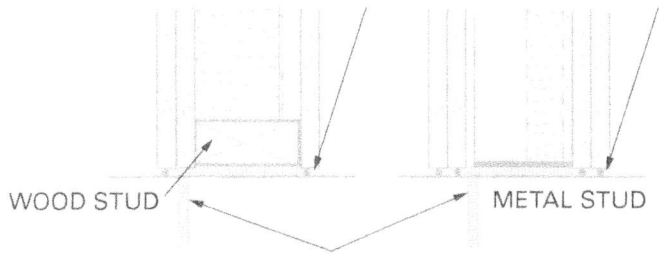

WOOD STUD

METAL STUD

GASKET IMPEDES STRUCTURAL
FLANKING THROUGH FLOOR

TYPICAL FLOOR-CEILING OR ROOF DETAIL

VOID BETWEEN BOX AND
WALLBOARD CAULKED

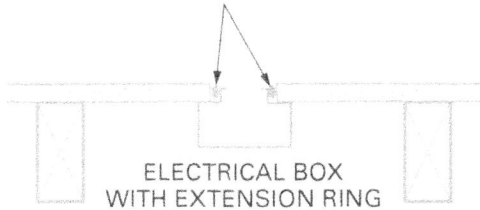

OFFSET BOXES MINIMUM OF ONE
STUD SPACE AND CAULK OPENING

ELECTRICAL BOX
WITH EXTENSION RING

OUTLET BOX DETAIL

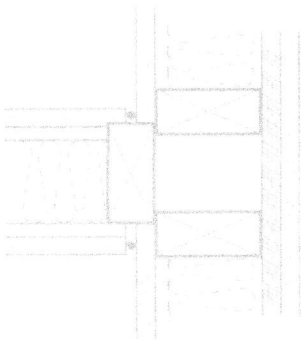

OUTLET BOX DETAIL

WOOD OR METAL STUD

INTERSECTION WITH EXTERIOR WALL

TYPICAL PARTITION INTERSECTIONS

104F39B.EPS

Figure 39   Caulking of sound-isolation construction (2 of 2).

Separated walls without framing can also be constructed by using an all-gypsum, double-solid, or semi-solid partition.

Steel or wood tracks fastened to the floor and ceiling hold the partitions in place. For the attachment of kitchen cabinets, lavatories, ceramic tile, medicine cabinets, and other fixtures, a staggered stud wall rather than a resilient wall is recommended. The added weight and fastenings may short circuit the construction acoustically.

7.2.2 Resilient Mountings

Resilient attachments acting as shock absorbers reduce the passage of sound through the wall or ceiling and increase the STC rating. Further STC increases can result from more complex construction methods incorporating multiple layers of gypsum board and building insulation in the wall cavities.

Resilient furring channels are attached with the nailing flange down and at right angles to the wood stud, as shown in *Figure 40*.

To install furring channels, drive 1¼" Type W screws or 6d coated nails through the prepunched holes in the channel flange. With extremely hard lumber, ⅞" or 1" Type S screws may be used. Locate the channels 24" from the floor, within 6" of the ceiling line, and no more than 24" on center. Extend the channels into all corners and fasten them to the corner framing. Attach ½" × 3" gypsum board filler strips to the bottom plate directly over the studs by overlapping the ends and fastening both flanges to the stud. Apply the gypsum board horizontally with the long dimension parallel to the resilient channels using 1" Type S screws spaced 12" on center along the channels. The abutting edges of boards should be centered over the channel flange and securely fastened.

7.2.3 Sound-Isolating Materials

Sound-isolating materials include:

- Mineral fiber (including glass) blankets and batts used in wood stud assemblies
- Semi-rigid mineral or glass fiber blankets for use with steel studs and laminated gypsum partitions
- Mineral (including glass) fiberboard
- Rigid plastic foam furring systems
- Lead or other special shielding materials

Mineral wool or glass fiber insulating batts and blankets may be used in assembly cavities to absorb airborne sound within the cavity. They should be placed in the cavity and carefully fitted behind electrical outlets and around any cutouts necessary for plumbing lines. Insulating batts and blankets may be faced with paper or another vapor barrier and may have flanges or be of the unfaced friction-fitted type.

Gypsum board may be applied over rigid plastic foam insulation. It is applied on the interior side of exterior masonry and concrete walls to provide a finished wall and to protect the insulation from early exposure to fire originating within the building. Additionally, these systems provide the high insulating values needed for energy conservation.

In new construction or for remodeling, these systems can be installed with as little as 1" dimension from the inside face of the framing or masonry (½" insulation and ½" Type X gypsum board).

When applying gypsum board over rigid foam insulation, the entire insulated wall surface should be protected with the gypsum board, including the surface above ceilings and in closed, unoccupied spaces.

Single-ply or double-ply, ½" or ⅝" gypsum board should either be screw-attached to steel wall furring members attached to the masonry or nailed directly into wood framing, as shown in *Figure 41*. Follow the insulation manufacturer's instructions.

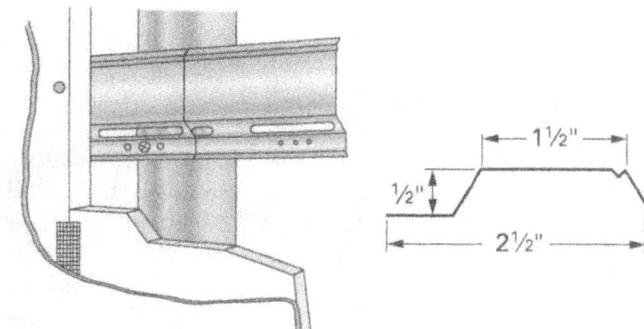

SINGLE NAILING

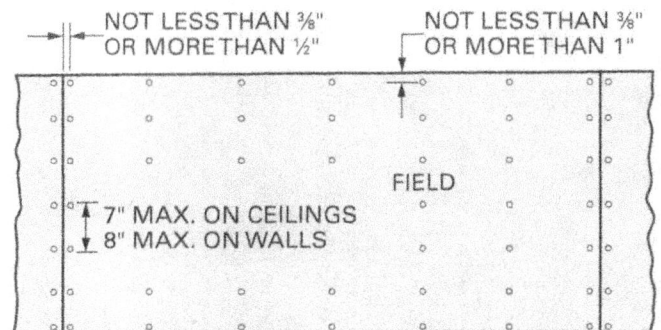

DOUBLE NAILING

206F51.EPS

Figure 41 ◆ Nailing patterns for installation over rigid foam insulation.

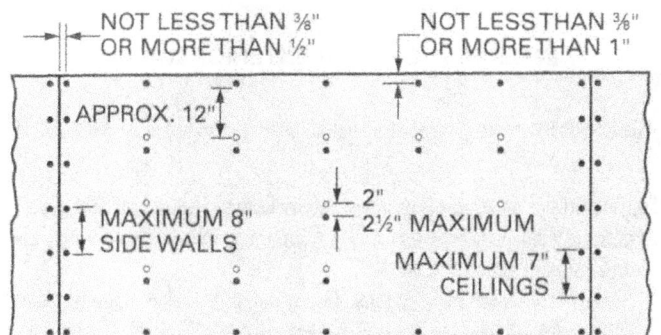

104F40.EPS

Figure 40 ◆ Attached resilient furring channel.

Furring members should be designed to minimize thermal transfer through the member and to provide a 1¼" minimum width face or flange for screw application of the gypsum board.

Furring members should be installed vertically and spaced 24" on center. Blocking or other backing as required for attachment and support of fixtures and furnishings should be provided. Furring members should also be attached at floor/wall and wall/ceiling angles (or at the termination of the gypsum board above suspended ceilings) and around door, window, and other openings. Single-ply gypsum board should be applied vertically, with the long edges of the board located over furring members. The installation should be planned carefully to avoid end joints. The fastener spacing should be as required for single-ply application over framing or furring.

In double-ply applications, the base ply should be applied vertically. In horizontal face ply applications, the face ply and end joints should be offset by at least one framing or furring member space from the base ply edge joints.

The fastener spacing should be as required for two-ply application over framing or furring, as discussed previously.

In wallboard applications, mechanical fasteners should be of such a length that they do not penetrate completely to the masonry or concrete. In single-layer applications, all joints between gypsum boards should be reinforced with tape. In addition, gypsum board joints should be finished with joint compound. In two-ply applications, the base layer joints may be concealed or left exposed.

Adhesive should not be used to apply vinyl-faced gypsum board face layers over a wall insulated with rigid foam.

7.3.0 Control Joints

Control joints (*Figure 42*) should be installed in gypsum board systems wherever expansion or control joints occur in the base exterior wall and not more than 30' on center in long wall furring runs. However, this increases to 50' if perimeter relief is provided. Wall or partition height door or window frames may be considered control joints.

Control joints should also be used over window and door openings. The gypsum board should be isolated from structural elements such as columns, beams, and loadbearing interior walls, and from dissimilar wall or ceiling finishes by control joints, metal trim, or other means. Refer to the specifications and drawings for the locations of control joints.

Movement of the structure can impose severe stresses and cause cracks, either at the joint or in the field of the board. Cracks are more prevalent at an archway or over a door, because this is usually the weakest point in the construction. In new construction, it is wise to wait until at least one heating season has passed before repairing or refinishing.

A source of cracking in nonbearing walls of high-rise or commercial buildings is the modern trend toward less rigid structures. Larger deflections in structural members and greater expansion and contraction of exterior columns can impose unexpected loads on nonbearing walls and lead to cracking. Detail designs for perimeter relief of nonbearing partitions are available to improve this condition. One solution is to use relief runners to attach nonbearing walls to ceiling and column members (*Figure 43*).

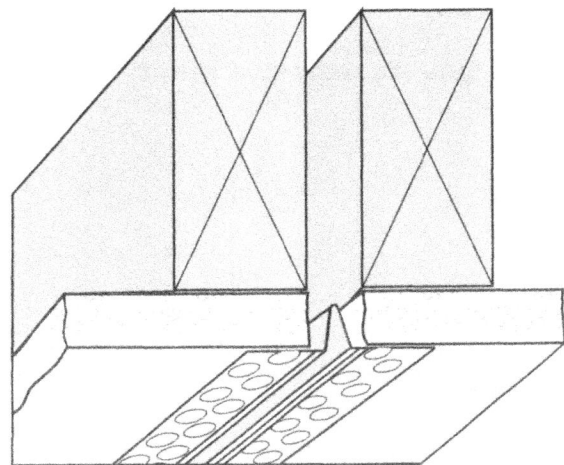

206F52.EPS

Figure 42 Typical control joint.

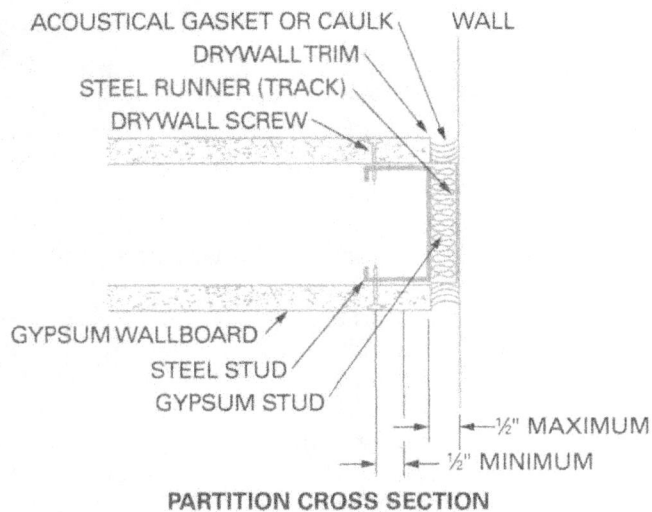

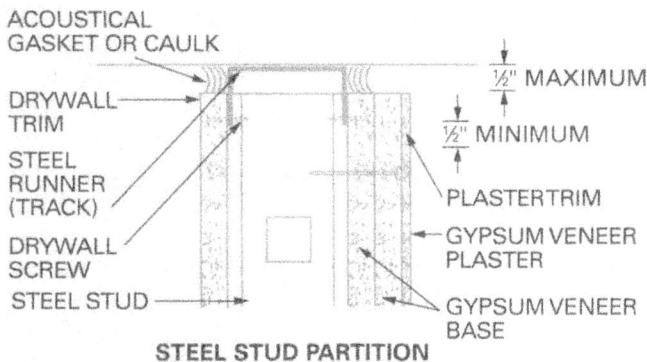

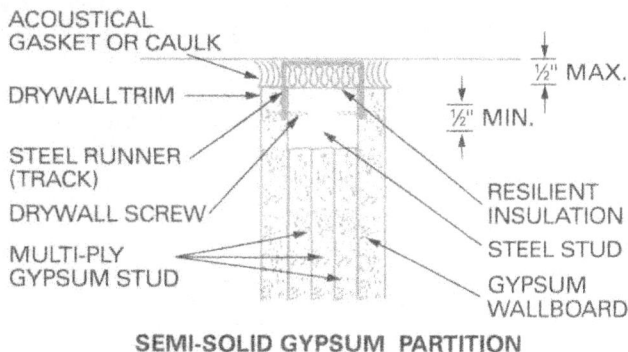

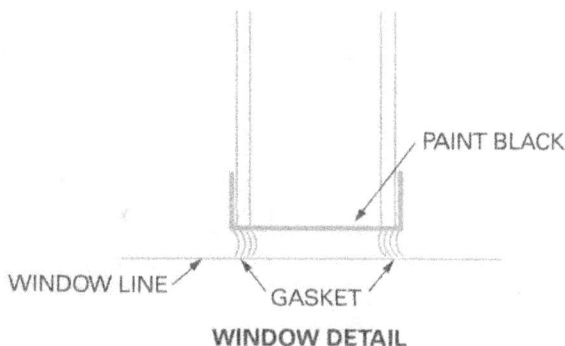

PARTITION CROSS SECTION

ACOUSTICAL GASKET OR CAULK — WALL
DRYWALL TRIM
STEEL RUNNER (TRACK)
DRYWALL SCREW
GYPSUM WALLBOARD
STEEL STUD
GYPSUM STUD
½" MAXIMUM
½" MINIMUM

STEEL STUD PARTITION

ACOUSTICAL GASKET OR CAULK
DRYWALL TRIM
STEEL RUNNER (TRACK)
DRYWALL SCREW
STEEL STUD
½" MAXIMUM
½" MINIMUM
PLASTER TRIM
GYPSUM VENEER PLASTER
GYPSUM VENEER BASE

SEMI-SOLID GYPSUM PARTITION

ACOUSTICAL GASKET OR CAULK
DRYWALL TRIM
STEEL RUNNER (TRACK)
DRYWALL SCREW
MULTI-PLY GYPSUM STUD
½" MAX.
½" MIN.
RESILIENT INSULATION
STEEL STUD
GYPSUM WALLBOARD

WINDOW DETAIL

PAINT BLACK
WINDOW LINE
GASKET

104F43.EPS

Figure 43 ● Designs for perimeter relief.

INSIDE TRACK   **Control Joints**

Control (expansion) joints are used in large expanses of wall or ceiling drywall to compensate for the natural expansion and contraction of a building. Control joints help prevent cracking and joint separation. They are common in commercial construction, especially where exterior concrete walls contain expansion joints.

If the control joint is installed in a space where fire rating and/or sound control are important, a seal must be used behind the control joint. The control joint has a ¼" slot that is covered by plastic tape. The tape is removed after the joint is finished, leaving a small recess.

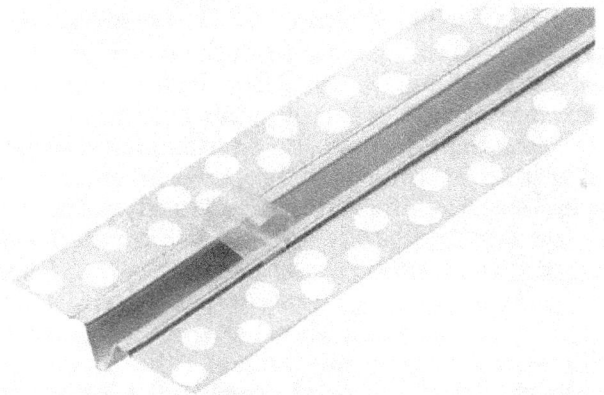

104SA14.EPS

8.0.0 ● MOISTURE-RESISTANT CONSTRUCTION

Special consideration must be given when finishing bathrooms, laundries, kitchens, and other areas subject to moisture. Although water-resistant gypsum board can be used in these applications, a waterproof tile backer such as cement board should be used. Unlike water-resistant gypsum board, cement board can also be used in areas of high moisture and humidity such as saunas and gang showers.

As mentioned earlier, gypsum board that will be subjected to moisture should not be foil backed or applied directly over a vapor barrier, as the vapor barrier will trap moisture within the board.

In moisture-resistant construction, the tile backer or gypsum board should be applied horizontally, with the factory-bound edge spaced a minimum of ¼" above the lip of the shower pan or tub. Shower pans or tubs should be installed prior to the installation of the board. Shower pans should have an upstanding lip or flange located at a minimum of 1" higher than the entry wall to the shower. It is recommended that the tub be supported. If necessary, the board should be furred away from the framing members so the upstanding leg of the pan (*Figure 44*) will be on the same plane as the face of the board.

> **NOTE**
>
> Different types of waterproof boards have different applications and limitations, so it is always necessary to check the manufacturer's product data sheets and installation instructions for the type of board being used.

An additional board extending the full height from floor to ceiling is required for a fire-rated or sound-rated construction (*Figure 45*).

Suitable blocking should be provided approximately 1" above the top of the tub or pan. Between-stud blocking should be placed behind the horizontal joint of the board above the tub or shower pan. For ceramic tile applications, use studs that are at least 3½" deep and placed 16" on center. Appropriate blocking, headers, or supports should be provided for tub plumbing fixtures and to receive soap dishes, grab bars, towel racks, and similar items.

Tile backer boards should be applied with nails or screws spaced not more than 8" on center. When ceramic tile more than ⅜" thick is to be applied, the nail or screw spacing should not exceed 4" on center. When it is necessary for joints and nail heads to be treated with joint compound and tape, either use waterproof, non-hardening caulking compound or seal-treat joints and nail heads with a compatible sealer prior to the installation.

> **NOTE**
>
> The caulking compound or sealer must be compatible with the adhesive to be used for the application of the tile. Follow the adhesive manufacturer's instructions.

Interior angles should be reinforced with supports to provide rigid corners. The cut edges and openings around pipes and fixtures should be caulked flush with a waterproof, non-hardening, silicone caulking compound or adhesive complying with the *American National Standard for Organic Adhesives for Installation of Ceramic Tile*. The directions of the manufacturer of the tile, wall panel, or other surfacing material should also be followed.

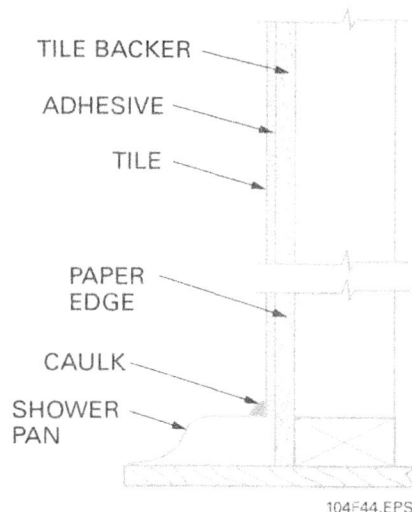

Figure 44   Pan is on the same plane as the face of the board.

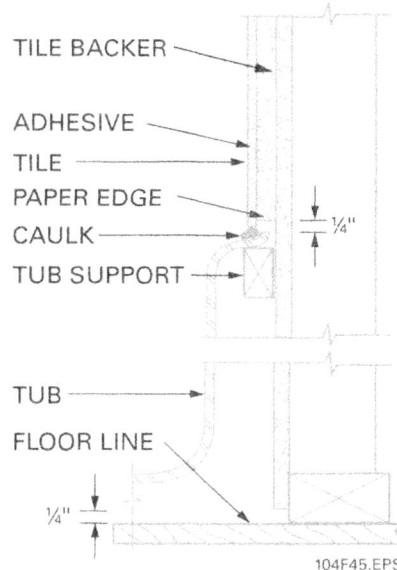

Figure 45   Sound-rated construction.

INSIDE TRACK — Tile Application

Ceramic wall tile application to gypsum board should meet the *American National Standard Specifications for Installation of Ceramic Tile with Water-Resistant Organic Adhesive*. The adhesives used should meet the *American National Standard for Organic Adhesives for Installation of Ceramic Tile*.

The surfacing material should be applied down to the top surface or edge of the finished shower floor, return, or tub, and installed to overlap the top lip of the receptor, subpan, or tub.

9.0.0 ◆ ESTIMATING DRYWALL

This section covers the guidelines for estimating wallboard and fasteners. *Table 2* shows rules of thumb for ordering different types of drywall nails and screws.

9.1.0 Gypsum Board

To estimate how many sheets of wallboard you will need for a job, first determine how many square feet of space the room contains. Multiply the length of each wall by two, add the results, then multiply the total wall space by the wall height to get the total square footage of wall space. If the ceiling is to be covered with wallboard, its square footage would be included as well. Then divide the total square footage by 32, which is the square footage of a 4' × 8' sheet, or 36, which is the square footage of a 4'-6" × 8' panel. For example, a 10' × 12' room with 8' high walls would contain 352 square feet of wall space. Converting this number into sheets (352 ÷ 32) yields 11. Door and window openings are usually figured solid unless there is a large picture window or door. This creates a built-in allowance for waste.

9.2.0 Fasteners

The types and amounts of fasteners will vary from job to job. It is important to consult the job specifications as well as the manufacturer's guidelines for this information. The following will serve as a rule-of-thumb guide to estimating fasteners.

For single-layer application 16" OC, approximately 1,000 screws are needed per 1,000 square

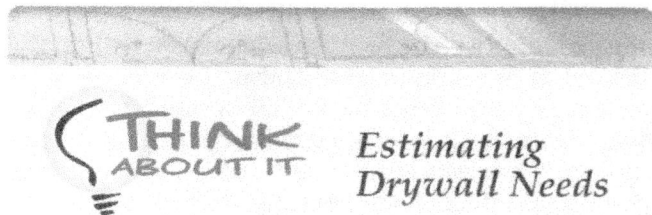

Estimating Drywall Needs

How many square feet of drywall are required for the walls and ceiling for a room that is 12' × 16' with 8' ceilings? How many pounds of 1¼" annular ring nails would be required for a single-layer installation?

Table 2 Fastening Materials Required for 1,000 Square Feet of Drywall

Material	Amount Required for 1,000 Square Feet of Drywall
1¼" annular-ring nails	6¼ lb
1⅜" annular-ring nails	6¾ lb
1" drywall screws	3 lb
1¼" drywall screws	4¼ lb
1⅜" drywall screws	5½ lb

104T02.EPS

feet of wallboard. If the framing is 24" OC, about 850 screws per 1,000 square feet will be needed.

If nails are used, you need to calculate the amount required in pounds. To install a single layer of ⅝" wallboard, you will need about 5 pounds of nails for every 1,000 square feet of wallboard. The size of the nail, and therefore the total weight, depends on the thickness of the wallboard. It can range from 4½ pounds of nails for ¼" wallboard to 7 pounds of nails for 1¼" (two ⅝" sheets laminated) wallboard.

Selecting the Right Size Panel

Drywall panels come in several common sizes, including 8', 9', 10', and 12'. The 12' size is likely to be more economical for some spaces because 12' panels result in fewer joints. Therefore, less finishing labor and finishing materials are required. For example, if you have an 11' wide, 8' high wall, two 12' panels would span the entire wall, leaving only one wall joint plus the corner joints to finish. With 8' panels, the same wall would have three wall joints to finish.

1. Gypsum board retards the spread of fire because _____.
   a. it is coated on both sides with a fireproofing substance
   b. its core contains asbestos
   c. the paper covering is fireproof
   d. the core releases water as steam when exposed to high heat

2. The tool commonly used to cut gypsum board panels to size is a _____.
   a. circular saw
   b. cross-cut saw
   c. utility knife
   d. hook-bill knife

3. The point at which the edges of two panels of drywall meet is known as a _____.
   a. bedding seam
   b. joint
   c. gypsum lath
   d. dimple

4. As a general rule, gypsum drywall should only be installed when the building temperature is greater than _____.
   a. 55°F
   b. 60°F
   c. 70°F
   d. 80°F

5. Fasteners should be applied to wallboard working from _____.
   a. top to bottom
   b. edge to edge
   c. center to edge
   d. corner to corner

6. When you are installing fire-rated wallboard, the nails should penetrate the studs by at least _____.
   a. ½"
   b. ¾"
   c. ⅞"
   d. 1⅛"

7. In a double-nailing system, the second set of nails is driven about _____ from the first.
   a. 2"
   b. 4"
   c. 8"
   d. 12"

8. In single-nailing floating interior angle construction where the framing is perpendicular to the wall/ceiling intersection, the ceiling fasteners should be located _____ from the intersection.
   a. 1"
   b. 7"
   c. 10"
   d. 12"

9. When covering existing walls with structurally sound surfaces, _____ gypsum board should be used.
   a. ⅛"
   b. ¼"
   c. ⅜"
   d. ⅝"

10. What is measured by a T-rating?
    a. Speed of flames spreading
    b. Thermal conductivity
    c. Thickness of gypsum board
    d. Sound insulation

11. Each of the following is a construction method used to control noise *except* _____.
    a. caulking around outlet box openings
    b. placing air conditioning ducts back-to-back
    c. using separate framing for the two sides of a wall
    d. mounting gypsum board in resilient channels

12. A low STC rating indicates _____.
    a. excellent fire resistance
    b. excellent sound isolation
    c. poor fire resistance
    d. poor sound isolation

13. Which of the following is true regarding the use of water-resistant gypsum board?

    a. It is recommended for use in saunas and steam rooms.
    b. It is not recommended for showers and tubs.
    c. It should have a foil backing.
    d. It should be applied directly over a vapor barrier.

14. When you are installing cement board, the factory-bound edge should be placed at least _____ above the lip of the shower pan or tub.

    a. ¼"
    b. ½"
    c. ¾"
    d. 1"

15. In moisture-resistant construction, the tile backer or gypsum board should be applied _____.

    a. vertically
    b. horizontally
    c. diagonally
    d. as a multi-ply system

Summary

This module covered gypsum drywall and its installation. Once the drywall is properly installed, the joints must be finished to create a smooth surface for application of the final decorative finish, such as paint or wallcovering.

Gypsum drywall panels installed over wood or steel framing are the most common method of finishing walls. In residential applications, drywall panels are commonly used to finish ceilings as well.

There are different types and sizes of drywall designed for different applications. It is therefore important that you know the different types, their applications, and their methods of installation bits, convey a positive, cooperative attitude to those around you, and practice good safety habits every day. During drywall installation, you must also be aware of how to comply with specific fire and sound ratings.

Notes

Robert J. Pelletier

Owner
Pelletier Construction Inc.
Albuquerque, NM

Bob Pelletier worked his way up the ladder from drywall installer/finisher to foreman, and eventually to project manager. In 1993, he decided to start his own contracting business. That business has grown from a one-man operation to a successful enterprise that employs 85 people.

How did you get started in the construction industry?
I came by it naturally. My grandfather was a home builder in Central Massachusetts, where I grew up. Three of his sons—my uncles—followed him into the industry as builders and carpenters, as did other family members. One of my uncles became a building inspector. Most of my training came from on-the-job experience working with my family. I learned early in life on that construction work paid well for anyone who was willing to work hard and maintain a high quality standard.

What kinds of work have you done in your career?
I started out doing drywall installation and finishing and eventually became foreman of a drywall crew. From there, I advanced to working as project manager and project superintendent for general contractors. In 1993, I decided to start my own drywall contracting business. My wife Kathy and I started out using our savings to get the business rolling. I did the site work and she ran the office. We started out working from home and using storage units for materials and equipment in order to keep our costs down. It wasn't easy. We always made sure our employees were paid first, so in the early days there were weeks when they got a bigger paycheck than we did.

What factors have contributed most to your success?
I have worked hard to establish and maintain a good reputation for honesty and quality work. When I make a commitment, I do whatever is necessary to keep it, even if that means working around the clock. Ask anyone who employs contractors what they value most and they'll tell you they just want someone who will do what they say they are going to do, when they say they're going to do it—in short, someone who keeps their word.

What advice would you give to someone just entering the field?
A good work ethic is critical to success. You have to show up for work and be on time. You need to accept responsibility for your actions and, above all, honor your commitments. One thing that everyone in the construction industry needs to realize is that project schedules depend on everyone doing their part of the job correctly and on time. If they fail, then everyone downstream from them is affected. For example, if a crew is scheduled to do the interior painting on a certain day, and the drywall work isn't finished on time, that day is wasted for the painting crew. They may have other jobs scheduled for the rest of the week, so then the carpets can't be installed on schedule. This could means the buyers' moving van might show up and not be able to unload.

Tell us some interesting career-related facts or accomplishments.
I have been president of the American Subcontractors Association of New Mexico, and I presently serve on the board of the Associated Builders and Contractors (ABC), a national organization. I am proud of the fact that I started my own business and grew it into a company that has about 85 employees. I believe that one of my most important accomplishments as a business owner is that I never borrowed money to run the company or buy equipment.

Corner bead: A metal or plastic angle used to protect outside corners where drywall panels meet.

Floating interior angle construction: A drywall installation technique in which no fasteners are used at the edge of the panel in order to allow for structural stresses.

Gypsum board: A generic term for paper-covered gypsum core panels; also know as gypsum drywall.

Joint: A place where two pieces of material meet.

Nail pop: The protrusion of a nail above the wallboard surface that is usually caused by shrinkage of the framing or by incorrect installation. Also applies to screws.

Slurry: A thin mixture of water or other liquid with any of several substances such as cement, plaster, or clay.

Substrate: The underlying material to which a finish is applied.

Additional Resources and References

Additional Resources

This module is intended to be a thorough resource for task training. The following reference work is suggested for further study. This is optional material for continued education rather than for task training.

Gypsum Construction Handbook. Chicago, IL: United States Gypsum Company, 2000.

Figure Credits

NCCER CURRICULA — USER UPDATE

NCCER makes every effort to keep its textbooks up-to-date and free of technical errors. We appreciate your help in this process. If you find an error, a typographical mistake, or an inaccuracy in NCCER's curricula, please fill out this form (or a photocopy), or complete the online form at **www.nccer.org/olf**. Be sure to include the exact module ID number, page number, a detailed description, and your recommended correction. Your input will be brought to the attention of the Authoring Team. Thank you for your assistance.

Instructors – If you have an idea for improving this textbook, or have found that additional materials were necessary to teach this module effectively, please let us know so that we may present your suggestions to the Authoring Team.

NCCER Product Development and Revision
13614 Progress Blvd., Alachua, FL 32615

Email: curriculum@nccer.org
Online: www.nccer.org/olf

❏ Trainee Guide   ❏ Lesson Plans   ❏ Exam   ❏ PowerPoints   Other _____

Craft / Level: Copyright Date:

Module ID Number / Title:

Section Number(s):

Description:

_____

_____

_____

_____

Recommended Correction:

_____

_____

_____

_____

Your Name:

Address:

_____

Email: Phone:

45105-07

Drywall Finishing

45105-07
Drywall Finishing

Topics to be presented in this module include:

Overview

When gypsum drywall is first installed, there are visible seams between the drywall sheets and at corners. These seams must be closed in a way that makes them invisible to anyone looking at the finished wall. Look around your home. Although the walls were most likely finished with 4 × 8 sheets of drywall, it should look as if it was done with one solid sheet. The seams are covered with paper or fiberglass tape, which is embedded with a compound commonly called mud. When the mud has dried, the joint is sanded flat. This process is usually done three times for each seam. It sounds simple, but like other areas of construction, it is not. There are many types of compound, a variety of finishing tools, and some specialized skills that can only be learned with practice.

Objectives

When you have completed this module, you will be able to do the following:

1. State the differences between the six levels of finish established by industry standards and distinguish a finish level by observation.
2. Identify the hand tools used in drywall finishing and demonstrate the ability to use these tools.
3. Identify the automatic tools used in drywall finishing.
4. Identify the materials used in drywall finishing and state the purpose and use of each type of material, including:
   - Compounds
   - Joint reinforcing tapes
   - Trim materials
   - Textures and coatings
5. Properly finish drywall using hand tools.
6. Recognize various types of problems that occur in drywall finishes; identify the causes and correct methods for solving each type of problem.
7. Patch damaged drywall.

Trade Terms

All-purpose compound	Ridges
Bullnose	Skim coat
Feathering	Tape
Joint compound	Tapered joint
Lightweight compound	Taping compound
Mud	Topping compound

Required Trainee Materials

1. Pencil and paper
2. Appropriate personal protective equipment

Prerequisites

Before you begin this module, it is recommended that you successfully complete *Core Curriculum*; and *Drywall Level One*, Modules 45101-07 through 45104-07.

This course map shows all of the modules in the first level of the *Drywall* curriculum. The suggested training order begins at the bottom and proceeds up. Skill levels increase as you advance on the course map. The local Training Program Sponsor may adjust the training order.

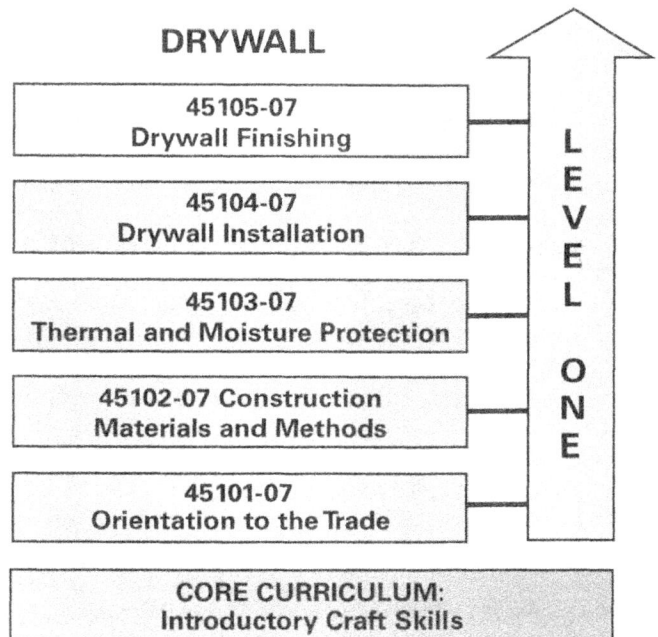

DRYWALL

LEVEL ONE
45105-07 Drywall Finishing
45104-07 Drywall Installation
45103-07 Thermal and Moisture Protection
45102-07 Construction Materials and Methods
45101-07 Orientation to the Trade

CORE CURRICULUM: Introductory Craft Skills

105CMAP.EPS

1.0.0 ◆ INTRODUCTION

In some situations, you may have to install and finish gypsum drywall. A remodeling project is one example. In other cases, it may be up to you to repair damaged drywall or drywall that was improperly installed by someone else. Therefore, it is essential to be thoroughly familiar with the tools, materials, and procedures used in drywall finishing and repair.

2.0.0 ◆ FINISHING STANDARDS

Drywall requires different levels of finish depending on its location, lighting conditions, and decorative treatment. A hidden surface, such as an attic area, requires far less finishing than one in full view, such as a living room wall. In addition, even minor flaws are apt to be more evident when the surface is exposed to strong lighting conditions or decorated with certain finishes, such as gloss or semi-gloss paints or thin wallcoverings. Generally, the more visible a surface, the more likely its lighting or decoration are to show surface defects, and the more finishing work it requires.

These factors are addressed in *A Recommended Specification for Levels of Gypsum Board Finish*, which was jointly developed by the Painting and Decorating Contractors of America, the Association of the Wall & Ceiling Industries–International, the Gypsum Association, and the Ceilings & Interior Systems Construction Association.

This specification is designed to serve as a standard reference for architects, specification writers, contractors, building owners, and others. It provides them with a specific description of the final appearance of gypsum walls and ceilings finished to different levels before the application of a decorative coating of paint, texture material, or wallcovering.

The specification describes the following six levels of finish and typical applications for each of them:

- *Level 0* – No taping, finishing, or accessories required. This level might be used for temporary construction or where final decoration is undetermined.
- *Level 1* – All joints and interior angles shall have tape embedded in joint compound (also referred to as mud or taping compound). Surface shall be free of excess joint compound. Tool marks and ridges are acceptable. A Level 1 finish might be specified for attics, areas above ceilings, service corridors, and other areas not generally seen by the public. It provides some degree of smoke and sound control. In some areas, it is called fire-taping.

- *Level 2* – One separate coat of joint compound shall be applied over all joints, angles, fastener heads, and accessories. The surface shall be free of excess joint compound. Tool marks and ridges are acceptable. All joints and interior angles shall have tape embedded in joint compound. Joint compound applied over the body of the tape at the time of tape embedment shall be considered a separate coat of joint compound and shall satisfy the conditions of this level. A Level 2 finish might be recommended for garages, warehouses, and other areas where surface appearance is not critical. It is specified where water-resistant gypsum backing board (*ASTM C 630*) is used as a substrate for tile.
- *Level 3* – All joints and interior angles shall have tape embedded in joint compound and one separate coat of joint compound applied over all joints and interior angles. Fastener heads and accessories shall be covered with two separate coats of joint compound. All joint compound surfaces shall be smooth and free of tool marks and ridges. A Level 3 finish might be specified for surfaces to be finished with a medium or heavy texture before painting or with heavy-grade wallcovering. It is not recommended for light or medium weight wallcoverings or for smooth painted surfaces.
- *Level 4* – All joints and interior angles shall have tape embedded in joint compound and two separate coats of joint compound applied over all flat joints and one separate coat of joint compound applied over interior angles. Three separate coats of joint compound shall be applied over all fastener heads and accessories. All joint compound shall be smooth and free of tool marks and ridges. Light textures or wallcoverings require this level of finish. The specification notes that in critical lighting areas, flat paint over light textures reduces shadowing of finished joints through the surface decoration, but that gloss, semi-gloss, and enamel paints are not recommended for this level of finish. It also notes that the type of wallcovering applied over this level should be chosen carefully to properly conceal joints and fasteners.
- *Level 5* – All joints and interior angles shall have tape embedded in joint compound, two separate coats of joint compound applied over all flat joints, and one separate coat of joint compound applied over interior angles. Three separate coats of joint compound shall be applied over all fastener heads and accessories. A thin skim coat of joint compound, or a material manufactured especially for this purpose, shall be applied to the entire surface. The surface shall be smooth and free of tool marks and ridges.

Level 5 is recommended for gloss, semi-gloss, enamel, or non-textured flat paints or severe lighting conditions. This is the highest level of finish and provides the best protection against joints or fasteners being visible through the decorative coating.

The specification, which is available from any of the associations that developed it, notes that for Levels 3, 4, and 5, it is recommended that the prepared surface be coated with a drywall primer prior to the application of finish paint. It also notes that the effects of severe lighting on a surface can be minimized by skim coating the drywall, by decorating it with medium to heavy textures, or by using window coverings to soften shadows.

3.0.0 ◆ DRYWALL FINISHING TOOLS

Proper finishing of walls and ceilings would be nearly impossible without a variety of finishing tools. These hand-operated and mechanical tools not only speed up the finishing process, but also help create walls and ceilings that are smooth and flat. This module will introduce you to many tools that are used specifically in finishing procedures.

3.1.0 Tool Safety

Although they appear simple and harmless, finishing tools can be as dangerous as any other kind of tool when handled improperly. Finishing knives have very sharp edges and corners. Automatic tapers have gears and chains that can catch fingers or hair. Automatic finishers have sharp blades, and some have spring-loaded hatches that can catch fingers, hair, and clothing.

The best way to avoid accidents and tool-related injuries is to treat your tools with respect. Never horse around when tools are being used by you or anyone else. Do not handle any tool unless you have been thoroughly trained in its use. If you do not know how to use a tool, ask your instructor, job supervisor, or another drywall mechanic to teach you.

Always keep your finishing tools clean and free of rust or excess compound, both of which can be poisonous if they get into a cut. Do not use knives with chipped blades or broken handles. Not only can they hurt you, but they can also ruin your finishing work.

Treat automatic tools with special care. They must be cleaned and maintained according to the manufacturer's directions. You can hurt yourself trying to force a mechanical tool to work when the tool is jammed with dried joint compound. Inspect automatic tools before every use; if a tool looks damaged or in bad repair, report the problem to your supervisor.

3.2.0 Hand Tools

The following hand tools are used to cut, hang, and finish drywall:

- 4' straightedge or T-square
- Utility knife with plenty of blades
- Drywall saw
- Circle cutter
- Drywall hammer
- Caulking gun
- Screwdriver
- Broad knife
- Joint trowel
- Corner tool
- Mud pan or hawk
- Sandpaper/drywall screen
- Sanding block, pole sander, or electric sander
- Sponge sander

An easy way to cut drywall is to place the straightedge on the finished side of the panel and score the drywall with a utility knife, cutting through the paper facing and into the gypsum. Then snap the panel apart along the score line by applying pressure from the back of the board. If the backing paper is still intact, use the knife to finish cutting through it.

An alternative method of cutting through drywall is to use a drywall saw. It is good for making straight as well as curved cuts. A saw with a long, thin, pointed blade (*Figure 1*) works well for limited or detail cuts such as when cutting out a hole for an electrical box. The saw can cut in a straight line or in an arc. Use a drill to make a starter hole for the blade.

Finishing knives range in type and size from the 1¼" putty knife (*Figure 2*) to the extremely wide 24" taping knife (*Figure 3*). They are all similar in function, however, because they bed and feather the taping or topping compound. Each knife is designed for use in a different situation. For example, the smallest putty knife is used in hard-to-reach spaces, for patching, and for working the compound around windows, cabinets, and doors. The 4", 6", 8", 10", and 12" widths are the most common.

Taping knives generally have blades made of blue steel, which flexes under the pressure of bedding and feathering taped joints. They range in size according to blade width.

A long-handled broad knife is used to wipe excess compound from freshly taped joints. The broad knife's blade is made of stainless steel or blue steel and may range in width from 7" to 9". Typically, a broad knife's handle is about 10" to 12" long, but handles up to 28" long are available. These long knives are handy tools for cleaning up messy joints and spatters on high walls and ceilings.

Finishing trowels (*Figure 4*) are similar to cement and plaster trowels, but they are available with either a flat or concave finishing surface (also called a blade). The concave blade is very useful for shearing the joint taping compound. These types of trowels are often preferred over taping knives for finishing wallboard joints. Blades range from 10" to 18" in length and are usually 4½" wide. Blades maintain their bow shape due to the spring steel or flexible stainless steel from which they are made. Flat-blade trowels may be used for applying textured finishes.

105F02.EPS

Figure 2 ◈ Putty knife.

105F03.EPS

Figure 3 ◈ Finishing knives.

105F01.EPS

Figure 1 ◈ Drywall saw with thin, pointed blade for detail cuts.

INSIDE TRACK

Dress Properly

Always wear the proper clothing and safety gear when finishing drywall. Wear goggles and a dust mask to keep dripping compound and dust out of your eyes, nose, and mouth. Adequate clothing can help protect you from cuts and falling objects.

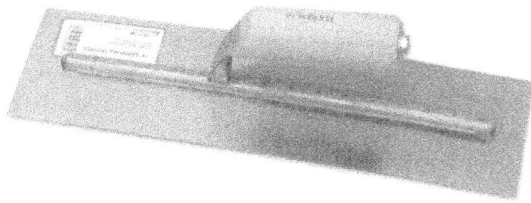

CURVED BLADE TROWEL

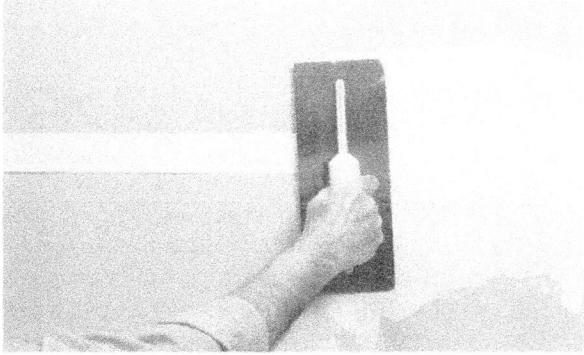

TROWEL IN USE

105F04.EPS

Figure 4 Finishing trowel.

INSIDE
CORNER
TROWEL

OUTSIDE
CORNER
TROWEL

105F05.EPS

Figure 5 Corner trowels.

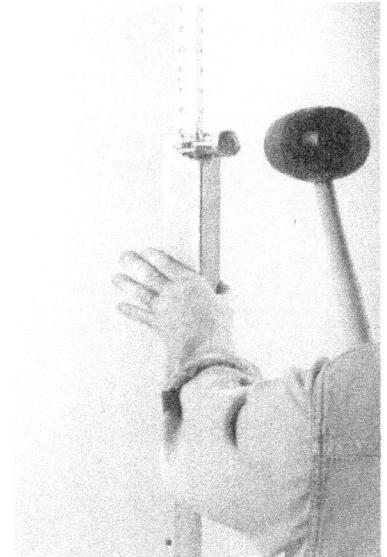

Cornering tools enable you to finish both inside and outside corners. They include not only troweling tools, but sanding and bead-attaching devices as well. Corner trowels are made for finishing inside and outside corners. While inside and outside corner trowels are shaped alike, their handles are on opposite sides (*Figure 5*). These tools are generally not used by professional drywall finishers.

Some corner systems have a paper-faced corner bead that can be coated with joint compound for direct application to a corner. Corner beads can be screwed on, clinched, or applied with staples.

For attaching corner bead to outside corners, the corner clinching tool may also be used. The clincher is struck with a rubber mallet once the corner bead is positioned and the tool is set over it. When struck, the corner clinching tool centers the bead and clinches it to the corner by crimping each side into the wallboard at four points. The entire length of corner bead can be attached by moving the clincher up or down the corner and striking it several times (*Figure 6*).

A mud pan (*Figure 7*) is simply a long container that holds joint compound (mud) to be knifed or troweled over the wallboard. The pan is commonly made of steel, aluminum, or plastic. It also may have replaceable steel edges mounted on the long sides for scraping excess compound off the knife or trowel. A typical mud pan looks just like a baker's bread pan.

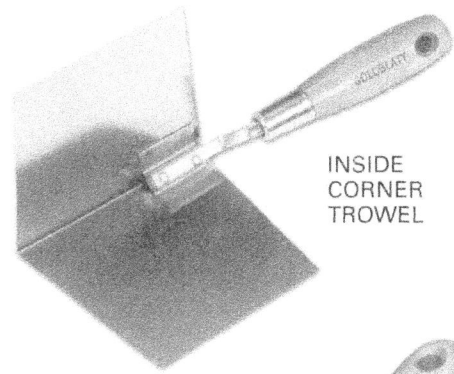

105F06.EPS

Figure 6 Corner clinching tool with a rubber mallet.

A hawk (*Figure 8*) is a handheld metal sheet, similar to a mortar board, on which a supply of plaster or compound is placed until used. This tool was originally used by plasterers, and it is still known by the name given to it by workers in that trade.

Figure 7 ◆ Mud pan.

Figure 8 ◆ Hawk.

Figure 9 ◆ Pole sander.

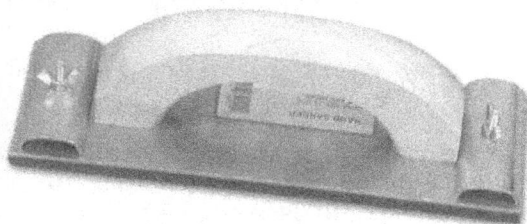

Figure 10 ◆ Commercial hand sander.

Sanding is an important part of the finishing process, and tools such as sanders are available to help you with this task. There are two basic types of sanders: the pole sander (*Figure 9*) and the hand sander (*Figure 10*). Both use the same principle of fastening down a strip of sandpaper by means of clamps and wing nuts.

With the hand sander, you are better able to sand dried finished joints and fastener heads that are within normal reach. With a pole sander, you can do the same job over your head without relying on scaffolds or stilts.

Hand sanders, also known as sanding blocks, may be purchased or made on the job site. A commercial sanding block consists of a wooden or metal base, around which a sheet of sandpaper is wrapped. A second block is pressed against one side of the base and tightened down with a wing nut to hold the sandpaper in place around the base. The sander allows you to apply even pressure over the sandpaper's entire face while saving your fingers from abrasions and other injuries.

Electric sanders, which will be discussed later, are also available.

3.3.0 Mixing Tools

Both hand tools and attachments to power drills can be used to mix joint compound and other liquified materials at the job site. For hand mixing, you can use a mud masher, which is a lot like a potato masher (*Figure 11*). The mud is mashed in a pail until it is mixed to a smooth consistency.

If you need to mix a full bucket of compound, it is faster and easier to use a power drill equipped with a long-stemmed mud mixer (*Figure 12*). There are several types of spinning mud mixers.

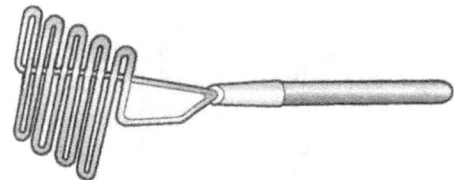

Figure 11 ◆ Hand-operated mud mashers.

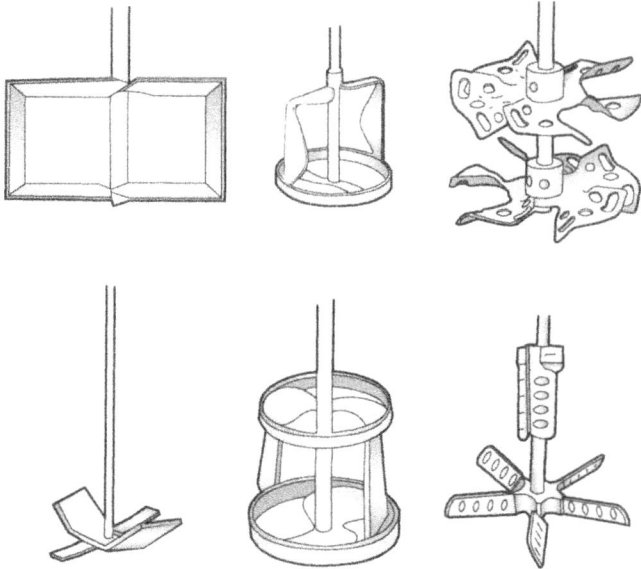

105F12.EPS

Figure 12   Mud mixers used with a power drill.

The mixer end of the device is placed in the mud once its shaft has been secured in the power drill chuck. The drill is used like a kitchen hand mixer to stir the mud.

3.4.0 Tape Dispensers

A 500' roll of joint tape can be awkward to carry around while trying to apply compound or tape a long joint. Simple tape dispensers and tape holders solve the problem. Many of these lightweight tape-holding devices are designed to hang from belts or shoulder slings, leaving your hands free (*Figure 13*). Some dispensers even crease the tape for application in corners.

A banjo (*Figure 14*) is a large tape dispenser. It is loaded with a full 500' roll of joint tape and a supply of mud. The banjo applies the mud to the tape as the tape is pulled out. The banjo can be adjusted to change the amount of mud applied to the tape. The banjo may get its name from its similarity in shape to the musical instrument of the same name.

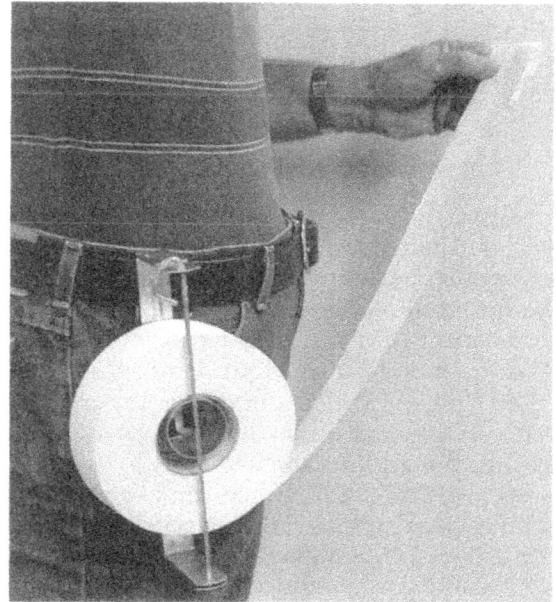

105F13.EPS

Figure 13   Tape dispenser used with paper or fiberglass mesh tape.

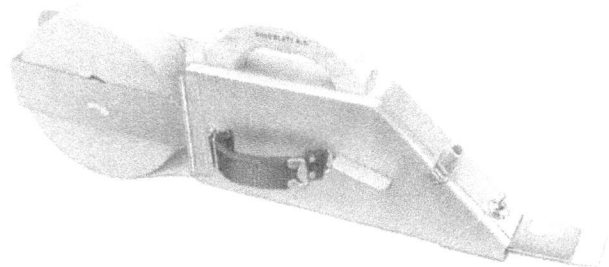

105F14.EPS

Figure 14   Banjo.

3.5.0 Automatic Finishing Tools

Not only are tools available to help you apply compound and tape and finish the joints, but there are also tools that do many of these operations at the same time—automatically. These tools are operated by hand, but they are referred to as automatic finishing tools or mechanical finishing tools. This is because they use intricate mechanisms to do work that otherwise would have to be done with hand tools and fingers. Although they are operated by hand, the hand operation usually involves simply pushing the tool along a path.

3.5.1 Automatic Taping Tools

One of the most popular mechanical finishing tools is the automatic taping tool (BAZOOKA®), which applies mud and tape to joints (*Figure 15*).

The automatic taper uses gears, rollers, pulleys, a piston, and levers to quickly guide the tape, coat it with a measured layer of mud, and dispense the tape along the joint. The automatic taper even cuts the tape at the end of a pass and can crease the tape for application in corners.

Using the BAZOOKA®, ceilings up to 10' high can be finished without using a ladder, scaffolding, or stilts. An extension can be mounted to the taper's base to reach even higher ceilings. For closets and other tight spaces, miniature tapers are also available.

You may need a lot of practice to master the use of an automatic taper, but you will probably find it time well spent. Using automatic tools, an experienced mechanic can tape an entire room faster than using only hand tools.

Automatic tapers are made primarily of aluminum, plastic, and other rust-resistant materials. This makes it easy to keep the unit clean, which is necessary to guarantee proper operation.

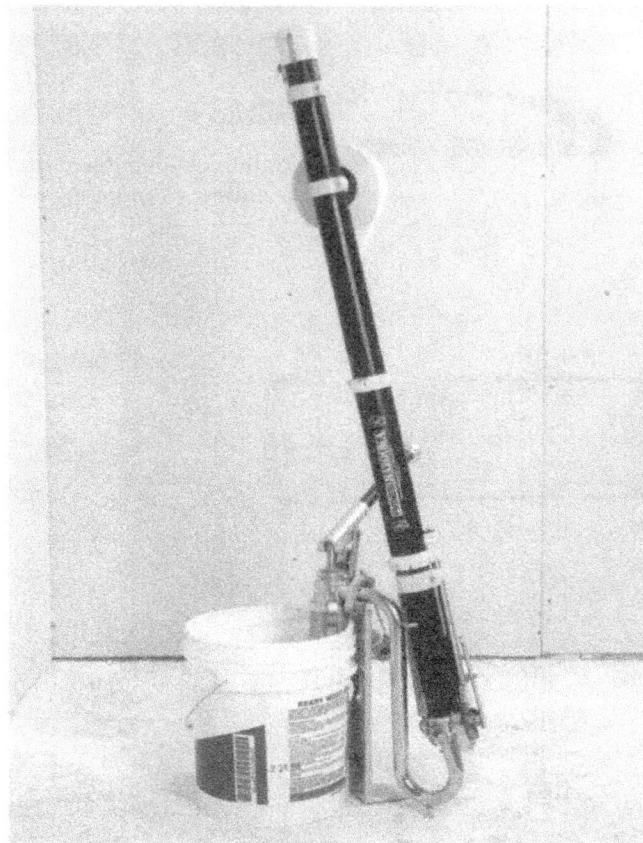

105F15.EPS

Figure 15 ✦ BAZOOKA® automatic taping tool.

The taper is cleaned by filling the empty joint compound chamber with water. Water is then forced out of the unit through the valve that distributes the compound by moving the floating piston up and down inside the compound chamber. The gears and valve openings at the head of the taper should be sprayed clean with a high-pressure water hose.

Automatic taping tools use a small razor blade to cut the joint tape at the end of a pass. This blade must be changed occasionally, as must the cable that operates the unit's piston. The piston moves up and down inside the taper's tube, forcing out compound. You can change the blade or cable in just a few minutes, following the instructions that come with the taping unit. Always follow the manufacturer's instructions for maintaining the taper to avoid breakdowns and lost time on the job.

3.5.2 Nail Spotters

A much simpler automatic finishing tool is the nail spotter (*Figure 16*). The nail spotter quickly applies compound over nail and screw dimples in fastened wallboard.

INSIDE TRACK

Keeping Your Compound Fresh

To prevent the compound from drying out in the taper's head during use, always stand the unit headfirst in a pail of clean water whenever you must stop working for a short time. At the end of the work day, pump all the compound out of the taper before cleaning it out with water.

The spotter is simply a metal box on a swiveling pole. The pole's swivel allows you to use the spotter at any angle and to push the device along a wall or ceiling while standing still.

The pole is attached to a hinged plate on top of the spotter. As you push the spotter along, the pressure you exert on the pole is transferred to the plate. As the plate sinks into the metal box, it forces compound out through a small opening in the bottom of the box. The compound fills the dimples and excess mud is automatically scraped off by a blade mounted in the trailing edge of the box.

Nail spotters are commonly made in 2" and 3" widths. The pole lets you reach high ceilings and walls without a ladder or stilts. The mechanism is small and light enough to be used in closets and other cramped spaces.

The nail spotter is much easier to use than the automatic taper. Once it is mastered, an entire room can be spotted in just a few minutes. Since the tool automatically scrapes off excess mud and feathers the spot, each dimple must be gone over only once.

The spotter's blade is bowed slightly to leave a small crown of compound over each dimple. The blade is very hard, but can be damaged or broken by an exposed nail or screw head. Broken or worn blades must be replaced. The blade is usually mounted to the unit with a clamp and one or two wing nuts. By removing the nuts, you can back off the clamp and slide out the blade. The

new blade is simply slid into place and held down by the wing nuts.

The nail spotter should be cleaned thoroughly after every use by flushing with a high-pressure water hose. The units are usually made of rust-resistant aluminum or stainless steel.

3.5.3 Flat Finishers

The flat finisher is also known as a box. You will generally use more than one box because they come in different sizes (*Figure 17*). The idea of a flat finisher is to apply topping and finishing coats to taped drywall joints. As each successive layer is applied, you use a wider box. The flat finisher works on the same principle as the nail spotter and dispenses an even, measured strip of compound along any flat joint. Boxes come in 7", 10", and 12" widths. Each applies topping coats of those dimensions. Use the smallest box for the first topping coat after taping; use the 10" or 12" box for the second topping coat if two topping coats are all that are required; or use the 10" for the second coat and the 12" for a third coat.

Like the nail spotter, the flat finisher is a metal box with a hinged lid. When you push the box along the joint, pressure on the handle forces the hinged lid down, squeezing mud out of the box through a small opening at the bottom. The opening can be adjusted to change the amount of mud released. Automatic-feed versions supply compound to the finisher, reducing the amount of effort required.

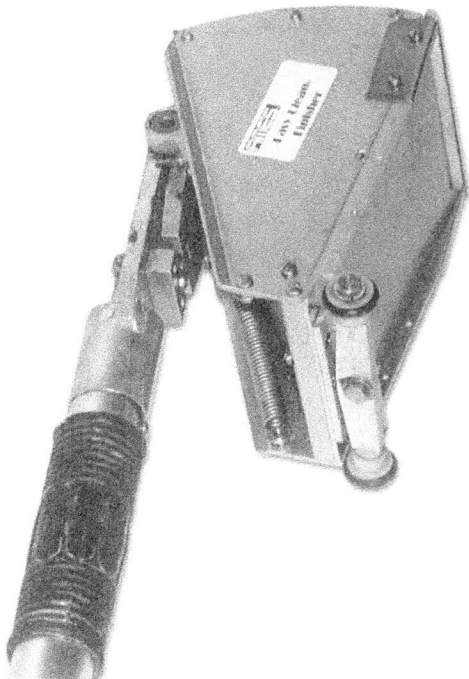

105F16.EPS

Figure 16   Nail spotter.

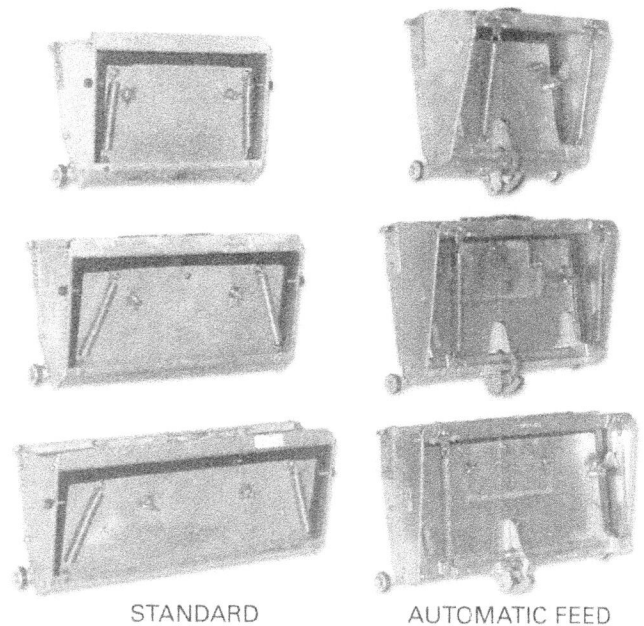

STANDARD   AUTOMATIC FEED

105F17.EPS

Figure 17   Flat finishers.

Once flat joints have been taped either by hand or with an automatic taper, they can be finished using an automatic finisher. This tool applies a smooth, even coat of mud over the taped joint, automatically feathering the edges and crowning the center.

The flat finisher operates on the same principle as the nail spotter. The device is pushed along, and pressure on the handle forces compound out of a metal box. As you guide the unit along the joint, a metal blade scrapes off excess compound and leaves the desired crown height. The finisher can be adjusted for different crown heights.

Flat finishing tools are available in three widths: 7" for applying a topping coat to a taped joint; 10" for applying a third or finish coat; and 12" for applying a fourth or skim coat.

Flat finishers can be used on wall and ceiling joints running in any direction. The tool is not only easy to operate, but allows you to get the job done in a small fraction of the time it would take to finish joints by hand.

Adjustable handles enable the finisher to reach ceilings up to 12' high. The device is compact enough to use in closets and other cramped spaces. Like most other automatic finishing tools, the flat finishing tool must be kept clean in order to work properly. These applicators must be cleaned thoroughly after each use by squeezing the remaining compound out of the reservoir and flushing the unit with water from a high-pressure hose.

3.5.4 Corner Applicators and Finishers

The corner applicator (*Figure 18*) works in conjunction with a corner finisher, or plow (*Figure 19*). By attaching the plow to the ball/cone end of the corner applicator with a locking retainer clip, the operator can put a finish coat on both sides of an angle at the same time. The plow operates on the same principle as the flat finisher or nail spotter, but its dispensing surface is shaped in a 90-degree angle to fit into interior corners. The applicator is available in 2" or 3" widths and can be used to apply bedding coats of compound prior to taping or to apply finish coats that are feathered and smoothed with other tools.

Once a bed of compound and tape has been applied in interior corners, the tape is smoothed and embedded into the mud using a corner roller (*Figure 20*). This device consists of four stainless steel rollers mounted on a swiveling pole.

For best results start from the middle of the angle joint, and use light pressure to roll toward both ends. Make a second pass, again from the middle, working toward both ends with firm

Figure 18 ❖ MudRunner™ corner applicator.

Figure 19 ❖ Corner finisher (plow).

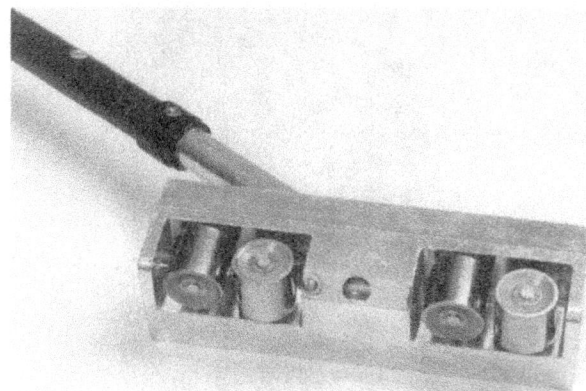

Figure 20 ❖ Corner roller.

pressure. This will force excess compound from under the tape. To maintain the corner roller, spray it with a high-pressure hose after use.

3.5.5 Automatic Loading Pumps

Automatic taping and finishing tools are filled with compound through automatic loading pumps (*Figure 21*). These pumps look a lot like the old-fashioned hand-cranked water pumps used before the invention of the faucet, and they work on the same principle.

The pump's intake nozzle is placed in a five-gallon pail of compound. To stabilize the device, place your foot on the foot plate on the pump's main body, which sticks out of the bucket and sits on the floor.

As you operate the pump's handle, mud is forced out through a J-shaped outlet nozzle. Several attachments can be used to adapt the nozzle to feed various automatic tools. The attachments let you fill the tools quickly and without any overspill or mess.

Like all automatic tools, the pump must be maintained properly. At the end of the day, remove the device from the pail and force out any compound remaining inside the pump. Flush the pump with water from a high-pressure hose or by placing the intake nozzle in a bucket of fresh water and pumping the water through the compound.

To prevent clogging the pump, make sure the joint compound is properly mixed and free of lumps. Remix the mud periodically if it must stand unused for a while.

The pump's intake nozzle features a small screen that stops lumps or debris from being drawn into the pump. Following the manufacturer's directions, remove and clean the screen after each use. You will probably need to replace the screen occasionally.

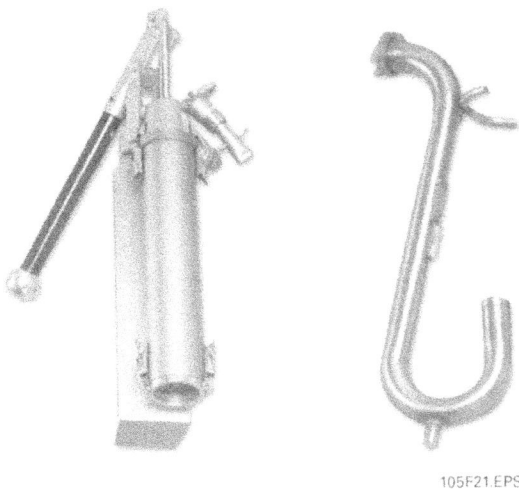

3.5.6 Vacuum Sanders

The hand-operated vacuum sander works like a hand-operated pole sander. The vacuum sander's pole, however, is actually a rigid hose connected to a powerful vacuum cleaner. A piece of screen-backed sandpaper or a tough sanding mesh is stretched across the hose's flat, hollow head. As the sanding head is pushed back and forth across the dried joint compound, the vacuum sander pulls dust and chunks through the sanding mesh and into the hose. The dust is collected in a large filter bag, as in an ordinary household vacuum cleaner.

There are also power-driven versions of these sanders (*Figure 22*). They provide a fast, clean way of finishing drywall jobs. The vacuum hose is connected to a shop vacuum. The sander uses foam-backed sanding pads that are specially

105F21.EPS

Figure 21   Automatic joint compound loading pump and gooseneck.

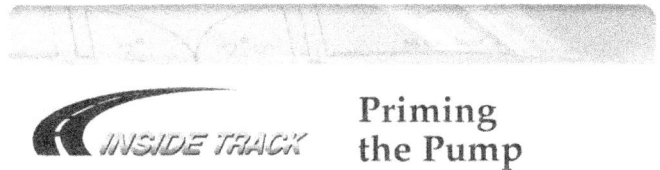

105F22.EPS

Figure 22   Power-driven vacuum sander.

designed for drywall work. The machines are safe, effective, and easy to use. Hoses and attachments enable the finisher to sand normal-height ceilings and walls without stilts or ladders.

4.0.0 ◆ DRYWALL FINISHING MATERIALS

This section covers various types of drywall finishing materials.

4.1.0 Joint Reinforcing Tapes

Three kinds of tape may be used in drywall finishing. They are paper tape, fiberglass mesh tape, and metal edge tape. Each kind may further be divided into those pre-coated with adhesive and those without. The tape most frequently used is paper tape without adhesive.

Standard paper tape is used to cover and reinforce seams, joints, and patchwork. It is a strong paper with feathered edges. There are two types of paper tape available: plain paper and perforated paper.

Paper tape (*Figure 23, left*) can vary in width, but it is generally about 2" wide. The specific width required by most automatic taping tools is 2$\frac{1}{16}$". The tape usually comes in rolls that range in length from 60' to 500'.

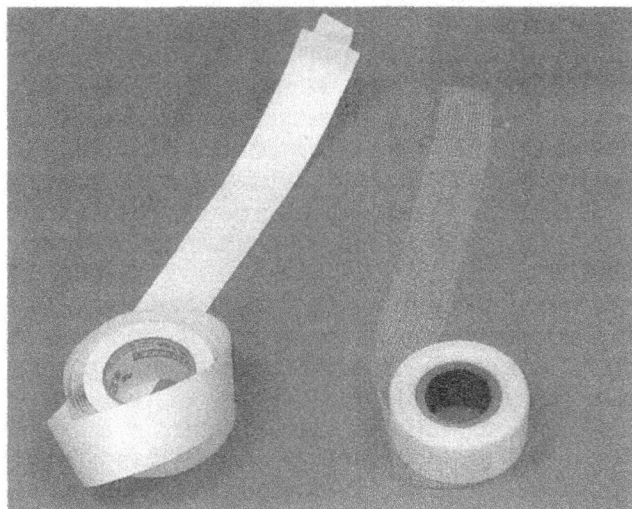

105F23.EPS

Figure 23 ◆ Joint reinforcing tape.

The paper tape may also contain fibers, crisscrossed or woven into the material, and it often comes tapered or feathered at the edges. The overall surface may also be scuffed or roughened in order to allow better bonding with the taping compound. Paper tape is considered to be superior to fiberglass tape for many applications because paper tape resists stretching and distortion. It also

provides a more consistent bond between the face papers of the gypsum boards on each side of most joints.

Perforated paper tape has larger, more visible holes which are designed to allow more compound to ooze through them, producing a better bond once dry. These holes may vary in size, depending upon the manufacturer, but they are usually about $\frac{1}{16}$" in diameter.

Fiberglass mesh tape (*Figure 23, right*) is generally self-sticking fiberglass joint tape made of fabric-woven filaments that do not decay. This type of tape is also available without adhesive. Fiberglass mesh makes a strong tape and may be more durable than paper tape under certain conditions, such as in high-moisture areas and when used with moisture-resistant drywall. Fiberglass mesh tape is much more costly than paper tape, but is good for repair work, veneer taping, and other specialized applications. Its use can increase productivity enough to offset the extra cost of the tape.

Metal edge tape (*Figure 24*) is a type of paper tape with the added feature of two galvanized strips of steel down the center, with a small gap between them to allow for the crease. The metal strips are typically $\frac{1}{2}$" wide. This tape is sometimes referred to as flexible metal corner reinforcing tape. How-

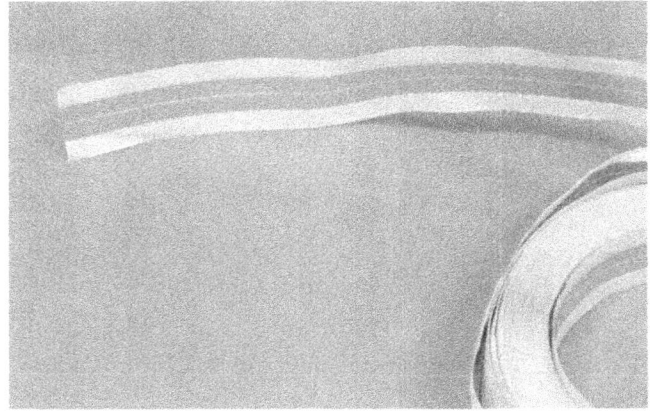

105F24.EPS

Figure 24 Metal edge tape.

ever, the tape is still applied and finished just like regular paper tape.

Metal edge tape is best used for corners with other than 90-degree angles. It is also used for corners formed by radius wall and ceiling intersections, arches, drops, splays, and wherever wallboards need to join in unusual configurations. Metal edge tape makes any outside angled corner straight and sharp, with some reinforcing qualities. The edges beyond the metal strips are feathered like those of standard paper tape.

Using Fiberglass Tape

INSIDE TRACK

Fiberglass mesh tape pre-coated with adhesive is designed to be installed without first applying a bedding coat of taping compound. In other words, adhesive-backed tape is applied directly to the joint and pressed into place with a taping knife or trowel. After the tape is stuck into place, it is covered with layers of compound. Fiberglass mesh tape without adhesive backing is sometimes installed using staples.

Some manufacturers recommend that you do not use fiberglass mesh tape with the usual ready-mixed or powder joint compounds. Instead, special powder compounds, such as quickset compounds, have been developed that work much better with fiberglass tape. Quickset compounds are discussed in more detail later in this module.

An automatic taping tool specifically designed for placing fiberglass tape is shown here.

105SA01.EPS

Corner bead is not very flexible, so something else is needed for odd corners made by curved or angled wallboard intersections. The principal idea of metal edge tape is to provide, as much as possible, the same strength and finishing quality that corner bead provides.

Metal edge tape is generally 2⅛" wide and comes in 100' rolls. The gap between the metal strips, allowing for the paper crease, is usually only ¹⁄₁₆". The tape is designed so that the metal side is applied to the wall.

4.2.0 Finishing Compounds

Joint compound is more commonly known as mud. It comes in both wet and powder forms. Wet compound has been premixed by the manufacturer, while powder compound is mixed on the job site by adding the proper amount of cold, clean water. Each of these products is designed to accomplish a specific result, depending on the job. Their ingredients are often different, so you should never mix wet compound with dry compound.

There are two basic types of joint compound: setting-type and drying-type. Drying-type compounds (*Figure 25*) have a consistency that makes them easy to sand. However, they typically take 24 hours or more to fully dry so they can be sanded and finished. Drying-type compound is commonly used in high-end work. Drying-type compounds are available in powder and pre-mix forms.

Setting-type compounds (*Figure 26*), also known as quickset compounds, contain chemicals that cause them to harden quickly. Although they are convenient because of their quick-dry qualities, these compounds do not sand easily. They must be smoothed before they completely harden. Quick-set compounds are often used for embedding tape, followed by finish coats of drying-type

compound. Because these compounds begin to harden as soon as they are exposed to air, they are only available in powder form.

Several types of joint compound (mud) are used in finishing drywall. One popular approach is the two-step system in which the tape is embedded in the joint using a taping compound to obtain a strong bond. The joint is then finished with a topping compound, which is much easier to sand. All-purpose compounds combine the characteristics of the two-step system into a single compound.

Taping compound is designed for its bonding qualities and strength in bedding and reinforcing taped joints. It is also used as a first coat on metal corner bead or trim, nail or screw heads, and other fasteners. Taping compound is also used as prefill and fill coats, and for repairing surface drywall cracks and cracks in plaster. Taping compound is generally the most likely to shrink. It is also the strongest bonding and most difficult to sand.

Topping compound is used for the second and third coats of the finishing process. This type of compound is softer and easier to sand. It also produces less shrinkage. You must never use topping compound for taping because topping compound is not designed to embed and bond a taped joint. A joint bedded with topping compound will crack with the first slight movement or settling of the wall. Topping compound is designed to be molded and sanded flat on top of a joint that has already been fastened together.

All-purpose compound combines the features of both taping and topping compounds, but in so doing, it gives up some of the qualities of each. For example, it loses some of the bonding qualities of taping compound and some of the soft and smooth drying capabilities of topping compound.

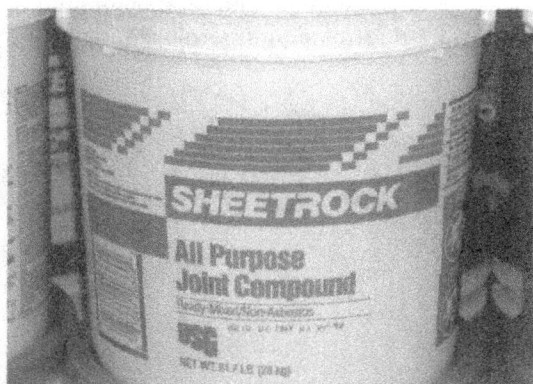

105F25.EPS

Figure 25 ❖ All-purpose drying-type compound.

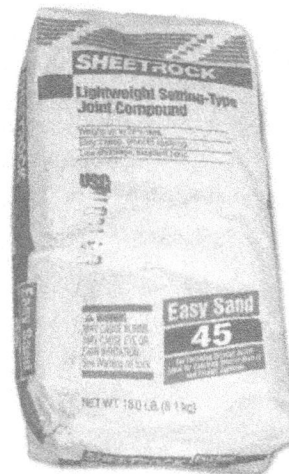

105F26.EPS

Figure 26 ❖ Setting-type compound.

However, all-purpose compound is excellent for use in textured finish applications. This type of compound is often used to finish walls with various interesting effects. In that respect, you are almost practicing the art of plastering.

Lightweight compound (*Figure 27*) has the advantages of an all-purpose compound and is 25 to 35 percent lighter. It also has less shrinkage and sands as easily as topping compound. Lightweight compound can be used to laminate gypsum panels, coat interior concrete ceilings and above-grade columns, and patch cracks in plaster. It can also be used for texturing.

105F27.EPS

Figure 27   Lightweight compound.

4.2.1 Powder Compounds

Compounds packaged in dry powder form store better than other forms of compound. Dry powder can be stored at any temperature and in any storage area or warehouse that is kept dry and free from moisture. Since warehouses are rarely heated, powder compounds are best for winter storage. However, it is recommended that powder compounds be moved to a warm mixing room a full day before they are mixed.

All powder compounds must be mixed with clean water in exactly the proportions specified on packaging instructions. Generally, these proportions depend upon the amounts of compound you need for the job. The amounts needed are also determined by the square footage of the joint areas you intend to cover. Once mixed, the compound may be kept in tightly covered containers in storage for many days, as long as the storage area is kept at room temperature. If the powder has been properly mixed to start with, it will not clump up or harden in stored closed containers, but you should always stir or mash it again before use.

Both powder and premixed forms of compound shrink by drying out. As water evaporates from the compound, the compound shrinks to fill in space vacated by the water. This process is different from the chemical hardening process used by special quickset compounds (discussed later), which is more like glue drying.

When Is Five Gallons Not Five Gallons?

The traditional five-gallon pail is no longer five gallons. Currently, these full plastic pails contain only 4½ gallons of premixed compound.

Taping, topping, and all-purpose compounds are generally available in dry powder form. They may be packaged in bags or cartons. The bags generally contain 25 pounds of powder. Cartons may be measured in gallons or pounds and usually contain more powder than bags.

Current building codes and standards generally forbid any use of asbestos in construction materials. Always make sure that the powder you are going to use is right for the job and complies with local regulations.

4.2.2 Premix Compounds

In general, premix (ready-mix) compounds are formulated, mixed, and packaged by their manufacturers. They are vinyl-based and require little or no mixing. This feature reduces the need for readily available water on the job site. Premix compounds also reduce the waiting or soak times after application, and they offer good crack resistance after drying.

These products will freeze, however, so you need to take precautions in cold weather. If vinyl premix does freeze, thaw it only at room temperature, and do not apply any additional heat. Always use premix products in their packaged state of consistency as much as possible to minimize shrinkage. Follow the specific instructions on the label.

Premix compounds are available in both plastic pails and sealed cartons. They are also quite heavy because, as the name implies, the water is already mixed in. The full pails weigh over 60 pounds; full cartons might be 50 or 60 pounds.

The advantage of premix compounds is that they can be used at job sites where there is no supply of fresh water, which must be available in order to use powders. Of course, powders can be mixed where there is fresh water and then transported in closed pails to sites without water. Generally, however, most contractors simply use the premix in these situations.

The disadvantage of premix is that it must be stored where it will not freeze during the winter. Finishing compounds must be used only in room-temperature environments. In colder climates in winter, you cannot do drywall finishing until the job site interiors are closed up and heat is installed and working.

Even though the premix is already mixed at the factory and ready to use on the job, you will want to mix or mash it again before you use it. Hand mashing is done by forcing the masher down repeatedly in the center and around the sides of the pail. Use smooth, complete downward strokes. On the upward stroke, scrape the masher along the sides of the pail to loosen any compound sticking to the edge.

Only two or three minutes of mixing is usually enough to ensure a smooth, even consistency all through the compound.

You will need to dilute the premix compound slightly for use with automatic finishing tools. Generally, add ½ cup of clean, cold water to 4½ gallons of premix compound. For the automatic taper, mix in two cups of clean, cold water per 4½ gallons of compound. Note that some manufacturers have compounds specifically designed for use with their line of taping and finishing tools.

The full range of compounds are available in premix form: taping, topping, all-purpose, and specialty compounds. Specialty compounds are all-purpose compounds that offer enhanced bonding, shrinking, and sanding capabilities. They often eliminate the need for a third topping coat, and they are good for laminated applications. Specialty compounds are also useful for texturing applications.

4.2.3 Setting-Type Compounds

Quickset compound, commonly called 20-minute mud, 30-minute mud, or hot mud, is a compound that sets up very quickly in comparison to other compounds because it hardens chemically instead of by water evaporation. Shrinkage is reduced considerably, so quickset makes an excellent filler for metal trim, repairs, and around pipes.

Quickset compound is available only in powder form; it needs to be mixed with water on the job. It is absolutely essential that the water be clean and cold. It is also very important to mix only as much as you can apply in the allotted set-up time.

Quickset compound is good for small jobs, prefill, corner beads, and finishing bathrooms and other high-moisture areas. It is especially good in humid weather. Because it sets so quickly, you need to wipe off excess compound immediately. Sanding the dried compound is also difficult, so take care to apply and wipe it as smooth as possible before it dries. Accelerators are available to make quickset compounds set even faster for special needs.

Quickset compound is packaged and sold by its setup time, which generally ranges from 20 minutes to 360 minutes (6 hours). (Note that the compound shown in *Figure 28* sets up in 90 minutes.) Because it works chemically, once it is hardened it will not shrink, even though it is not dry. This allows a strong bond to form and remain in high-humidity environments.

Quickset compounds are also very good for laminating applications, especially for laminating wallboard layers together. Quickset compounds can be used for coating concrete walls and ceilings (above ground), for filling in cracks and holes, for skim coating, and even for surface texturing. They

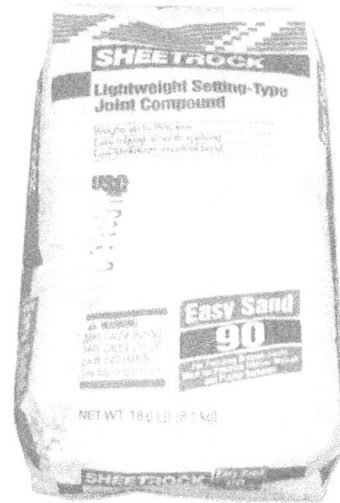

105F28.EPS

Figure 28 Quickset compound.

are also preferred for finishing exterior ceiling boards and for presetting joints of veneer finish systems.

> **NOTE**
> Don't overmix setting compounds. Overmixing will cause air bubbles in the finish.

4.2.4 Dry-Mix Safety

Whenever you mix dry compounds, be aware of the dust level and try to keep it to a minimum. You must also use the proper respirator when mixing dry powder compounds. In addition:

- Make sure all mixing containers are clean and free of residue.
- Use only clean, drinkable water for mixing.

- Make sure all tools and mixing blades are clean.
- Mix these compounds only according to the directions on their labels.
- If you use a power mixer or mixers operated by a power drill, use a slow speed such as 300 rpm to 500 rpm.
- Do not try to mix different types of compounds together. Their chemical makeup often differs from manufacturer to manufacturer. Even with the same manufacturer, the different types of compound are not compatible with each other.

4.2.5 Estimating Joint Treatments

The approximate quantities of materials needed for 1,000 square feet of wallboard are as follows:

- *Joint tape* – 370'
- *Joint compound* – 83 pounds of conventional powder or 138 pounds of conventional ready-mix

These figures can be used to calculate the requirements for other square footages by reducing the square footage to a decimal percentage of 1,000 and multiplying the area required by the above amounts. For example, 2,000 square feet would require twice the amount listed above and 1,400 square feet would require 1.4 times the amount listed above.

4.3.0 Sanding Materials

Sanding operations may be done with sponge sanders, hand sanders, pole sanders, or power sanders, which were described earlier. The sanders require sheets or strips of sandpaper to be fitted to them. Sanding is an important part of the finishing process, and tools are available to help you with this task. With the hand sander, you are

INSIDE TRACK

Wet Sanding

If only minimal sanding is required, a wet sponge can be used. This method produces no dust and will not scuff the paper. Use a polyethylene sponge, which looks something like carpet padding. Wet the sponge with clean water that is cool to lukewarm. Wring out the sponge enough to prevent dripping, then rub the joints to remove high spots using as few strokes as possible. Clean the sponge frequently during use.

better able to sand dried finished joints and nail spots that are within normal reach. With the pole or power sander, you can do the same job over your head without having to bother with scaffolds or stilts.

Sandpaper is rated by coarseness, called grit, which varies by degree of fineness. The lower the grit number, the coarser the paper. Coarse paper is used for first sanding jobs where the surface is very rough and you are trying to smooth it out fairly quickly. Fine paper is used for finish sanding where you are making the surface as smooth as possible, usually in preparation for painting.

Sanding is typically done with 180-grit sandpaper. Sandpaper is also available in grit numbers from 80 to 150 or more. It is packaged in sheets or rolls for hand sanders and in discs for power sanders. The precut sheets are designed to fit specialized tools such as angle sanders and wall sanders (sanding poles). Precut sheets are usually sold in packages of 100.

Sandpaper rolls are designed for you to cut off pieces of whatever length you need to fit onto a hand sander or pole sander. These rolls may vary in total length from 30' to 150' and in width from 3⅓" to 12".

Other sanding materials include sanding cloth, abrasive mesh cloth, and open mesh cloth, which are used chiefly for cutting joint mud. These materials may also be packaged in rolls or as individual sheets. Sanding sponges have become popular for sanding work. Sanding sponges have abrasive material bonded to one or both sides of a sponge. Some of them are beveled to make it easier to sand inside corners.

Mesh cloth, commonly called sanding screen, is preferred by many finishers for use on pole sanders. For first sanding of a joint or seam, this cloth prevents raising the nap (tiny hair-like fibers) on the face paper of the drywall. Do not raise this nap at all, if possible, because it makes for better painting if the nap is smooth. An advantage of mesh cloth is that it does not load.

Film-backed drywall abrasives are also effective. A major advantage is that they do not raise the face of recycled paper.

4.4.0 Textures

A texture is any wall or ceiling coating that serves as its own finish. It may also serve to hide taping and other finishing so that the surface need not be sanded smooth or otherwise prepared for painting, wallpapering, or other treatments. *Figure 29* shows examples of texture finishes.

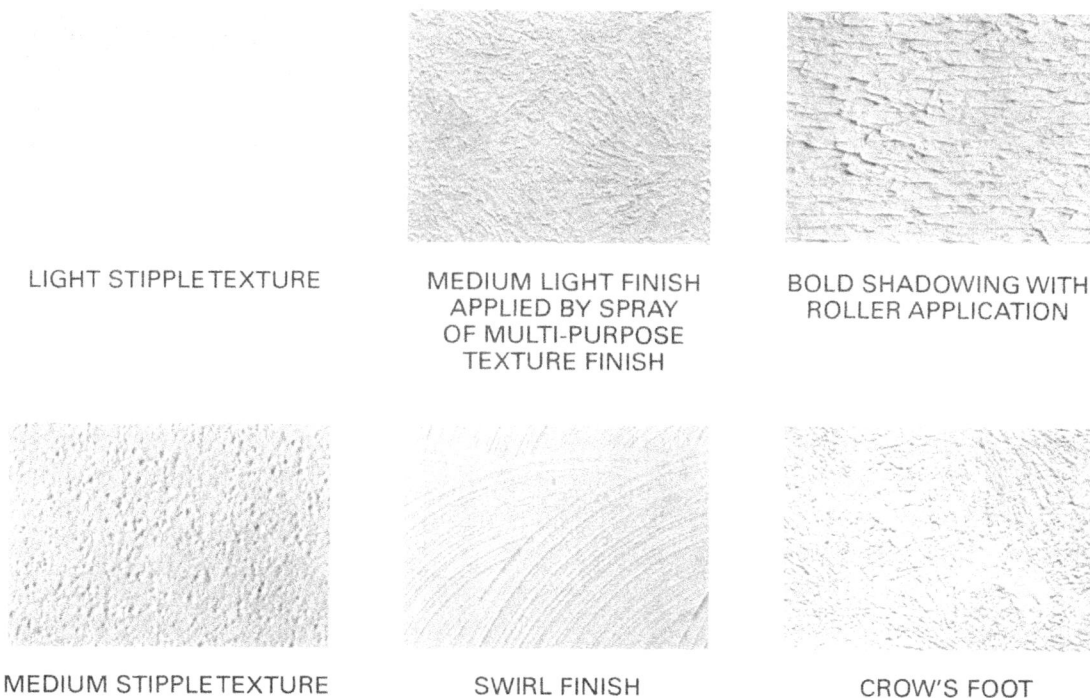

LIGHT STIPPLE TEXTURE

MEDIUM LIGHT FINISH
APPLIED BY SPRAY
OF MULTI-PURPOSE
TEXTURE FINISH

BOLD SHADOWING WITH
ROLLER APPLICATION

MEDIUM STIPPLE TEXTURE

SWIRL FINISH

CROW'S FOOT

105F29.EPS

Figure 29   Examples of texture patterns.

Decorative textures are very popular in both residential and commercial construction. Interesting patterns, simulated acoustical effects, and light or heavy finishes may be applied with rollers, brushes, stencils, sprayers, and other tools and equipment. *Figure 30* shows a roller designed to apply a particular texture finish pattern, along with some of the many roller patterns available. Some of the more common equipment used in texture finishing is shown in *Figure 31*.

1. *Glitter gun* – Used to embed glitter in wet texture ceilings. The hand-crank model shown is most economical, but not as efficient as an air-powered type.

2. *Drywall mud paddle* – Used with an electric drill at less than 400 rpm to mix drywall mud. It is designed to reduce the entrapment of air bubbles in the mixture.

3. *Stucco brush* – Used to create a variety of textures from stipple to swirl. Other variations can be achieved with thicker application and deeper texturing.

4. *Texture brush* – Available in many sizes and styles; tandem-mounted brushes cover a large area to speed a texturing job.

5. *Wipe-down blade* – Has a hardened steel blade and long handle to speed cleaning of walls and floors after application of texture materials. The blade has rounded corners to avoid gouging.

6. *Long-handled roller* – A standard paint roller adapted to the particular type of finish required. Available roller sleeves include short nap, long nap, looped, foam stipple, and carpet types in professional widths.

7. *Texture roller pan* – Used with rollers. Some models can hold up to 25 pounds of mixed texture.

8. *Flat blade knife* – Used to apply texture material and for troweled finishes.

9. *Circular patterned sponge* – Used to achieve patterned swirl finishes.

10. *Sea sponge* – Used to achieve free-form texture finishes.

11. *Whisk broom* – Similar to a stucco brush but stiffer. It can be used to produce a bolder brush pattern.

12. *Short-handle roller* – Same as the long-handled roller with the same sleeves. A looped texture roller sleeve is shown.

13. *Trowel* – Used to apply texture materials and for achieving troweled or knockdown finishes.

14. *Texture paddle* – Similar to the drywall mud paddle, but designed for mixing texture materials at 300 rpm to 600 rpm.

POINSETTIA

CROW'S FOOT

VINE

BASKET WEAVE

MONTERREY

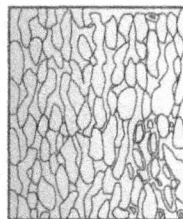
TREE BARK

PALM LEAF

105F30.EPS

Figure 30 ⬦ Texture rollers.

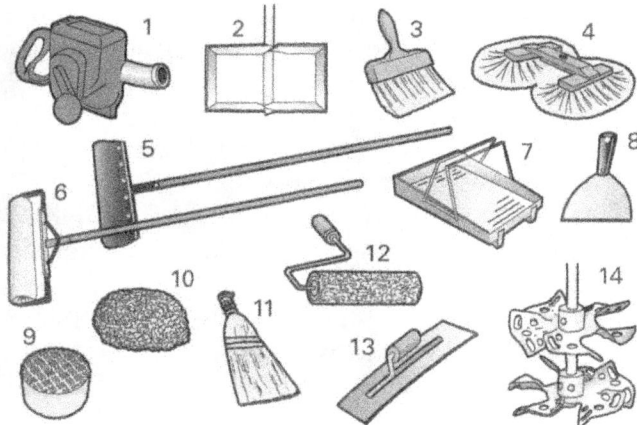

105F31.EPS

Figure 31 ⬦ Typical texture finishing equipment.

Texture Materials

Texture materials are similar to joint compounds in that they are made in both powder and premixed forms. In fact, some joint compounds can be easily used for texturing.

Texture materials may be manufactured in both powder and ready-mixed forms. The four types are described as follows:

• *Powder textures* – These textures may be either aggregated or unaggregated, which means there may or may not be other particles mixed into the powder. In aggregated products, particles of such substances as perlite, vermiculite, and polystyrene are used to make textured effects on primed surfaces, particularly ceilings. Aggregated powder products are designed for spray application. They also have a good solution time, only minimum-to-moderate fallout, and good bonding power and crack resistance. When properly sprayed on, these textures hide substrate imperfections very well. Unaggregated powder products may either be sprayed or hand applied to primed walls or ceilings. Several of these products are limited to hand applications only, so that you can produce crow's foot, stipple, or other pattern texture effects. Crow's foot is a design produced by a roller, which makes a kind of random bird-track pattern across the textured surface.

• *Powder joint compound textures* – Generally, powder joint compound textures are the same as the topping or all-purpose powder compounds used for normal joint finishing. For texturing with these products, you use a brush, roller, or trowel to produce light and medium hand-formed textures on walls or ceilings. Typically, you use swirling motions to make random patterns. Powder joint compounds are easily mixed in the usual way. They are smooth working and easy to texture by hand. They also hide surface imperfections very well, produce good bonding, and resist cracking. The color is white after drying, which can be left as is or painted another color, depending upon the finish specifications.

Spray Texturing Machines

Spray texturing machines are available for large jobs. They range from a hopper with a pneumatic spray nozzle to a self-contained machine like this one with its own built-in compressor.

105SA02.EPS

- *Premixed textures* – These include thick, heavy-bodied, vinyl-based materials, which are able to produce smooth, very deep textures. They generally dry to hard white finishes that are often left unpainted, especially on ceilings. These textures may be sprayed, troweled, or applied with a roller or brush. They go well over concrete and are able to fill voids and cracks and cover surface blemishes. Premixed texture offers good resistance against cracking on walls and ceilings. These textures are factory-mixed to a smooth consistency. They are easy and fast to apply and generally produce favorable results.
- *Premixed joint compound textures* – These textures consist of topping or all-purpose compounds, which are able to produce light to medium textures on ceilings and walls. These compounds are well suited for small jobs that need only

brushes, rollers, or trowels. However, they can also be spray applied. You can use these materials to produce a great variety of patterns and designs. These textures dry white, offer good crack resistance, and can be painted when dry. They are factory mixed to be smooth and free from lumps. These compounds go on quickly and easily for low-cost, yet good quality results.

5.0.0 DRYWALL FINISHING PROCEDURES

Drywall finishing involves taping, topping (also known as buttering), and sanding the wallboard seams and joints, whether on walls or ceilings, so these surfaces can be made ready for final decorating. Some jobs use texturing and have a reduced need for taping and sanding. Other jobs require a large amount of butt and seam taping, three or more topping or skim coats, and a lot of sanding to produce the expected results.

5.1.0 Site Conditions

Job site conditions such as temperature and humidity affect the performance of most finishing materials. During winter conditions, drywall finishing should not be attempted unless the building has heat in a somewhat controllable range between 50°F and 80°F. Furthermore, all materials must be protected from the weather at all times. If the humidity is excessive, ventilation must be provided. Windows should be kept open to provide air circulation. In enclosed areas without natural ventilation, fans should be used to move the air. When drying is slow, additional drying time should be allowed between applications of joint compound. During hot, dry weather, drafts should be avoided so the joint compound will not dry too rapidly.

5.1.1 Drywall Inspection

The professional always inspects installed drywall before finishing it. This determines if the drywall is ready for joint treatment. Improperly installed drywall is difficult to finish.

Examine the hung drywall. Nail and screw heads should all be dimpled. This means that each fastener head should have been driven into the wallboard so that a slight depression is made (*Figure 32*). No part of the fastener head should be above the rest of the board's surface. You can check this fairly easily by running your bare hand over the rows of fasteners.

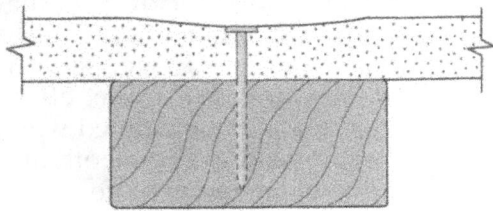

105F32.EPS

Figure 32 ◆ Dimple or uniform depression.

105F33.EPS

Figure 33 ◆ Butt joint misalignment (low side on right).

Determine if the fasteners are holding the wallboard panels tightly against the framing members. You can detect loose fasteners by placing a finger over the fastener head. Press the adjoining drywall area in toward the framing with your other hand. When loose, movement will be felt through the fastener head. If you discover this, install an extra fastener above and below the original fastener to make the drywall more secure.

Check the butted joints and outside corners. Remove loose paper and broken board from the drywall edges. Cut the paper back to where it still adheres; do not pull it.

Examine the board fields and the butted joints. Check for torn areas and large gaps. Mark all areas with a lead pencil only; ink or crayon will bleed through. These damaged areas must be repaired.

Examine the inside and outside corners. How well do the wallboard panels butt together? Make sure the panels are properly aligned. When one board sticks out farther than the other, the joint is difficult to tape and finish smooth. Another way of saying this, especially for butt joints, is that there might be a high side and a low side. A high side is produced by a butted panel that sticks out too far from the framing. A low side is then produced in the other abutting panel, which does not stick out as far. This is shown in *Figure 33*. This condition might be caused by a twisted stud. Part of the panel rests against a part of the stud that is not even with the part of the stud to which the other board is attached.

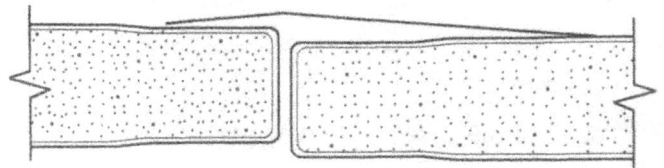

Since at this stage it is too late to correct the framing, the best you will be able to do is hide the offset. You can hide it by applying a little more compound and finishing the joint a little wider than usual. In fact, any butt joint at all is finished wider than a tapered joint because there is no tapered depression to hold the compound. The best you can do with any bad butt joint is to camouflage it.

All these inspections help you determine which procedures to use first in any given situation. If a major repair is required, inform your supervisor. If drywall needs serious correction, such as reframing or rehanging, it is much better to get it done before you begin taping and trying to hide everything with compound. However, if the repair is not that serious, you should be able to fix it.

5.2.0 Overview of the Finishing Process

As a general statement, the ideal drywall finishing process requires five steps. Depending on the drying time, it may take five different trips to the job site:

Step 1 Complete repairs, cut-outs, and prefill. Apply bedding tape at all joints and seams. Complete corner bead and trim installations. Top outside corners. Use an 8" knife on headers and spot fastener heads.

INSIDE TRACK

Maintaining the Proper Temperature

If drywall finishing is done during cold weather, the building must be heated to 55°F minimum, and the heat must be maintained during the entire finishing process and until the finish is dry. Avoid sudden changes in temperature, which can cause cracking due to thermal expansion.

Finishing compounds lose strength if they are subjected to freeze-thaw cycles. If a compound has been frozen, it should be discarded.

Drying Time

Atmospheric conditions and other factors always play a part in how fast the taping and topping coats dry. Another factor might even be the wallboard face paper itself. New and recycled paper might well have different drying rates, and different compound materials will vary as to how fast they dry. Therefore, drying times might be longer than just overnight.

Step 2 Apply first topping coat over taped joints and seams. Use a 10" box. Make double-wide topping coats at all butt joints. Top all corners and angles.

Step 3 Sand lightly, and scuff angles and flats. Apply a second (flash) topping coat to fill in all pits, gaps, depressions, or shrinkage. Apply a straddle coat on the butt joints. Apply another coat on the fastener heads.

Step 4 Perform light sanding and scuffing. Apply a third (skim) topping coat. Use an 18" knife on flat seams and butts. Use a 3" plow for angle topping. Apply a flash coat on headers, seams, and outside corners.

Step 5 Complete pole and hand sanding.

5.3.0 Automatic Taping and Finishing Procedures

Taping and finishing is a multi-step process that varies between three and five different stages of finishing of each board joint or seam. Joints are generally considered to be any place where two edges of wallboard come together. Seams are places where tapered board edges meet each other. Butt joints are places where square (nontapered) board edges meet each other. Flat joints are places where two beveled edges meet.

Generally speaking, joint taping includes pre-filling the joint, taping and bedding, topping and skim coats, and sanding. After each step, the compound is allowed to dry, usually overnight.

Joint compound and tape shrink as they dry. This shrinkage results in slight depressions that need to be filled out again by applying topping, skim, or finishing coats of compound. Using an automatic taping tool system, the actual processes at each joint or seam are greatly speeded up. The whole idea is to make every joint or seam as flat as the rest of the wall or ceiling. The idea is also to make these joints and seams undetectable once the decorating is complete.

Large jobs may require drywall finishing equipment. These specialized tools, which were discussed earlier in this module, enable drywall finishers to tape and finish drywall uniformly and efficiently. Basic tool components of such a system are the loading pump, nail spotter, automatic taper, corner roller, flat finisher, corner finisher, and corner applicator. The finishing process using automatic tools is outlined as follows:

Step 1 Apply the tape using the automatic taper.

Step 2 Press the tape into corners using the corner roller.

Step 3 Wipe down excess compound and embed the tape using a broad knife or taping knife.

Step 4 After the bedding coat has dried, apply a topping coat using a flat finisher for seams and butts. Use a corner applicator/finisher for angles.

Thinning the Compound

Experience has shown that best results come from adding ½ cup of clean, cold water to 4½ gallons of compound for all automatic tools except the automatic taper. For the automatic taper, mix in two full cups of clean, cold water per 4½ gallons of compound. Use the compound full strength for all hand tool applications, nail spotting, prefill, and skim coating.

Step 5 After the first topping coat has dried, apply a second topping coat using a wider flat finisher for seams and butts, and a corner applicator and finisher for angles.

Step 6 After this topping coat has dried, another skim coat may be applied using taping knives.

Step 7 After all coats have dried, sand the areas treated with compound. Wipe them down with a damp sponge after sanding. This helps the paper fibers to lie down.

5.3.1 Pump Loading Procedures

The loading pump (*Figure 34*) has nozzles of different sizes. Nozzle selection depends upon the equipment and material being used. The pump has a replaceable screen at the loading pump intake. This screen prevents large particles from passing through the pump. The pail holder is designed for a standard five-gallon pail. A gooseneck attachment mounts on the pump to fill the automatic taper. The loading pump without the gooseneck attachment fills the nail spotter, flat finisher, and corner applicator.

The pump is simple and rugged. It is designed to fill automatic taping and finishing tools with compound. It requires very little training for use and fills the application tools quickly and surely without pumping air.

Be sure to mix all compound thoroughly before using and especially before pumping into any automatic tool. Be sure to mix out any lumps, especially when using powder compounds. Use a power drill with mixer attachment, if available, or a mud masher. With a mud masher, you should plunge it down into the center of the pail and bring it up, scraping against the sides. Rotate around the pail as you mix. This keeps lumpy residues from forming on the sides of the pail. Also, be sure to remix any compound that has been left standing for a period of time.

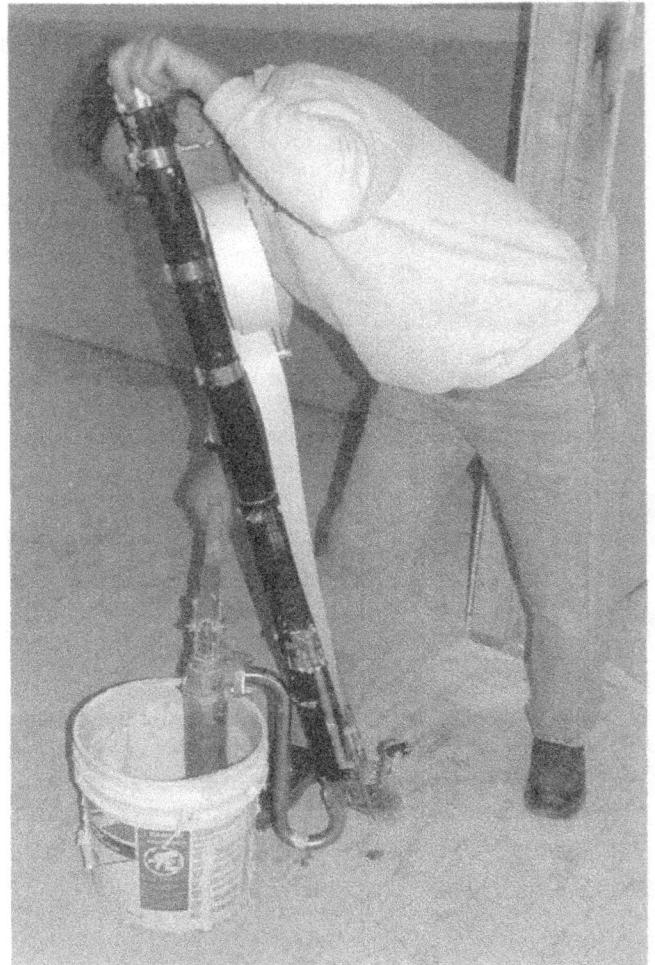

105F34.EPS

Figure 34 ❖ A BAZOOKA® being loaded with an automatic loading pump.

Keep the pump screen clean and free of lumps or dried compound. It may need to be replaced quite often, along with the O-ring on the gooseneck.

A final recommendation is to have two loading pumps at each job site, one with the gooseneck attachment, and the other with a fill adapter. This will speed up the operations of filling the different types of automatic tools.

INSIDE TRACK

Neatness Counts

Always set up your mixing pails and other equipment on a large scrap sheet of wallboard. This helps keep splashes off the floor and will greatly simplify your cleanup efforts. You should also keep at least one full bucket of water handy in this area to soak automatic tool heads when not in use.

5.3.2 Pumping

Set the pump into a full standard-sized premix compound pail, step on the pump's footplate outside the pail, and pump the compound using the pump handle. Before you pump any compound into an automatic tool, however, you may need to add water to improve the consistency of the pre-mixed compound. Taping compound should have a thinner consistency for use with the automatic taper, and a thicker consistency for topping applications. Always follow the recommended mixing instructions on the labels of the products you are using.

Also, before pumping any compound into an automatic tool, be sure to pump the handle a few times to clear out any air. Without the gooseneck attachment, you will simply be pumping compound back into its own pail. With the gooseneck attached, you should use another container to catch these first few pumps. Never attach a taper to the gooseneck until you are sure all the air is out. Then attach the taper upside down to the fitting and pump it full.

5.3.3 Automatic Nail Spotter Procedures

You may use a nail spotter (*Figure 35*) to fill countersunk nail and screw heads with joint compound. A complete row of fastener heads can be filled in one pass with this tool, which is normally available in 2" and 3" widths. The tool allows you to fill rows of fastener head dimples with compound, on both walls and ceilings, while working from the floor.

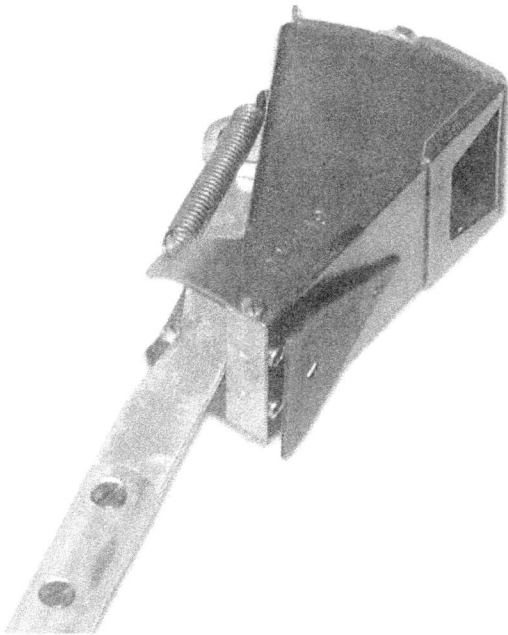

Use the loading pump to fill the nail spotter with compound. Set the pump into the pail, making sure the compound is well mixed and free of lumps. The pump needs only to be pumped full of compound; no priming is needed. Use the adapter spout (not the gooseneck) to fill the nail spotter at its opening.

Each row of dimples can be filled in one pass. Be sure to make positive contact with the wallboard surface at the beginning of each row. Draw the tool smoothly along the entire row, applying some pressure to force the compound out into the surface. The blade skims off excess compound and leaves a slight crown over each dimple as the tool floats along on the rocking skid.

After you pass over the last dimple, gradually break contact with the surface by using a sweeping motion. This procedure will fill the dimples without leaving excess compound that needs to be removed by hand.

The nail spotter is a simple tool to learn to use. Generally, you will have the procedure down by the end of your first day of using it.

5.3.4 Using the BAZOOKA® Automatic Taper on Ceiling Joints

Fill the automatic taper with joint compound using the gooseneck adapter with the loading pump. Close the gate valve and turn the tool upside down to fit its opening over the gooseneck opening. Put one to two fingers of your free hand into the end of the taper. Pump in the compound and stop pumping when your fingers feel the piston. This is to avoid overfilling. If you do happen to overfill, relieve pressure by depressing the filler valve stem with a nail.

Figure 35 Nail spotter.

105F35.EPS

WARNING!

Start slowing down at about six or seven pumps so you will not injure your fingers against the piston if it rises too quickly.

You will likely find that it takes about 9½ pumps to fill an empty taper.

Step 1 After loading, open the gate valve and turn the key counterclockwise until joint compound covers the leading edge of the tape. This is only necessary the first time you tape after each filling.

Step 2 Hold the taper with one hand on the control tube, called the slide, which is similar in operation to a shotgun pump (see *Figure 36*). Put your other hand on the bottom of the mud tube so you can operate the creaser control lever. You might even put several fingers into the end of the mud tube if you find this gives you greater control.

Step 3 Start by taping ceiling butt joints, and then tape the ceiling flat joints. Use both drive wheels for the first 4" to 6" in order to secure the tape to the ceiling. Start at one end of the seam and work to the other end. After taping the first 6", tilt the taper at about a 20- or 30-degree angle away from the plane you are working in so that only one drive wheel is pressed to the board surface. This helps limit the amount of compound and prevents air bubbles. Walk backwards as rapidly as possible, leading with the head of the tool.

Step 4 As you approach the end of the joint, gradually bring the tool back to alignment with your vertical working plane. At about 3" to 4" from the end, stop and pull down sharply on the control tube, which will cut the tape. You will need to slow down and stop completely in order to do this because the tape or blade will jam if you do not stop to make this cut.

Step 5 Return the slide to its neutral position and bring back the other drive wheel so both wheels press against the surface once again. Keep both wheels rolling to maintain the continuous buttering of compound needed to press the last bit of tape onto the end of the joint. At this ending sequence (when both wheels are rolling), push the slide forward to eject the end of a new tape. It will be buttered with the correct amount of compound that allows you to start directly into the next run.

Step 6 To finish the ceiling taping procedure, close the gate valve on the automatic taper and put it head first into a pail of clean water. Take a long-handled broad knife and wipe down all the ceiling butt and flat joints. The bedding joint compound should be wiped down while it is still wet, so that it is easily workable and excess moisture does not soak into the wallboard.

INSIDE TRACK

Using the Automatic Taper

Except for beginnings and ends of tape runs, always hold the automatic taper at about a 20- or 30-degree angle to the wall or ceiling and operate on one wheel only. It is important to have at least one wheel pressing against the board surface at all times while you are taping. These wheels control compound flow onto the tape. If you simply push the tape along without engaging a wheel, you will produce air pockets or bubbles, which are gaps in the tape where compound is missing. You should correct this immediately by going back, tearing out the unbuttered tape, and re-taping. If you do not, the topping will dry over the air pocket and will eventually crack apart and crumble or fall off the surface completely.

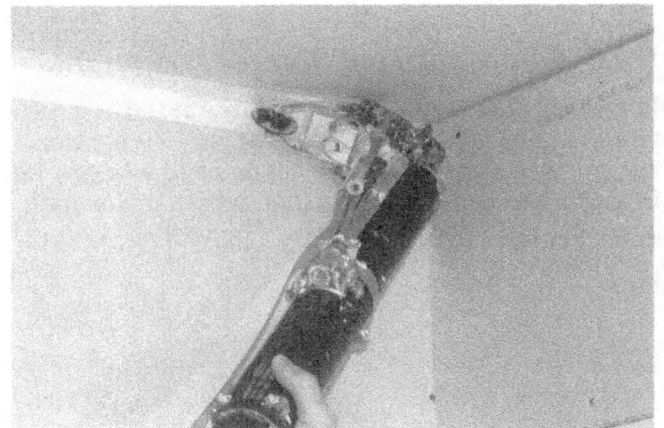

105F36.EPS

Figure 36 ◆ Using an automatic taping tool (BAZOOKA®).

Step 7 Use firm pressure and hold your knife blade at about a 45-degree angle to the surface. Wipe along each taped joint, laying the tape flat and forcing out excess compound from underneath. Start in the middle of each seam and work first toward one end, then the other. This helps to avoid wrinkles and bunching up of the tape. Be sure to catch all the excess compound you squeeze out on your knife and scrape it off on your hawk or into a mud pan. The whole process is meant to embed the tape, fill the joint, and leave a generally flat surface.

5.3.5 Using the Automatic Taper on Wall Joints

After the ceiling is wiped down, the next process is to tape the horizontal and vertical wall joints. Again, start with the butt joints. Remember to open the gate valve on the taper.

Step 1   For vertical wall joints, place the automatic taper at the bottom of the joint, about parallel to the floor. Push the control tube forward one or two inches to make a tape leader. Start the vertical taping with the leader overlapping the floor a little. As you proceed upwards, the tape will be drawn up as you go. With a little practice, you will know exactly how much of a leader to use in order to get the tape to end up exactly at the floor line after you have taped that joint. As soon as you can, when moving upwards off the floor, maneuver the taper so you are leading with the head. Also, shift so that you are tracking with just one wheel in contact with the wall.

Step 2   At 3" or 4" from the top of the wall, pull back the control tube to cut the tape and continue rolling to the ceiling intersection on both wheels. To start the next joint, roll the wheels slightly against the surface, starting the flow of compound while ejecting a new leader with the control tube.

Step 3   For taping horizontal joints, push forward on the control tube to produce a 2" or 3" tape leader. Place the leader, again with a slight overlap, at the beginning of the horizontal seam. Except for the start and end of a joint, always hold the taper at an angle to the wall (base of the tool angled downward) so only the bottom drive wheel is rolling against the surface. At the beginning of each seam, however, you need to push both wheels against the wall for about 6".

Step 4   As you come to within 3" or 4" of the end, pull back on the slide tube to cut the tape and continue on both drive wheels to the end of the joint while pushing forward on the slide. This applies the last several inches of tape while feeding out another leader for the next joint.

Step 5   For outside corners (where you are not using corner bead), simply follow the same procedure as explained above for vertical wall joints, but this time only apply tape to one side of the corner. Let the other side remain exposed to the air. When you have completed the vertical run, close the gate valve and fold the tape over the corner using your broad knife.

Step 6   At ceiling angle or wall intersection joints, you need to use the creaser wheel, which is extended by pulling on the trigger near the end of the automatic taper. You can also use the creaser wheel to help roll the tape against a flat seam, which is critically important when taping inside corners. Bisect the angle with your tape and make sure both wheels press equally on each side of the angle as you roll. Be sure to track in a straight line. Avoid twisting the automatic taper as you move. Again, start with a 1" or 2" leader to allow for the tape to be pulled toward the joint end (sometimes called "creeping"). You may have to push the leader into position in the angle using your fingers before you are able to proceed. Otherwise, taping inside corners and ceiling intersections is the same process that is used for vertical wall joints.

NOTE

Be sure you do not continue using the automatic taper for more than ten minutes if no one is following behind you and bedding the tape.

Teamwork

INSIDE TRACK

On large finishing crews, one drywall mechanic operates the taper, another follows with a broad knife to wipe down excess compound, another spots fastener heads, another comes with the corner roller, and so on. If you are operating the automatic taping tool alone, however, you will need to put down the taper after taping a joint in order to wipe down excess compound.

Step 7 If you are taping alone, stop using the taper after about 10 minutes, close the gate valve, and place the head into the pail of water. Then proceed to wipe down all the tape you just installed to embed it and remove the excess compound.

5.3.6 Using the Corner Roller

After the tape and joint compound have been placed in ceiling and corner angles, use the corner roller tool (*Figure 37*). This device embeds the tape in the joint compound at inside corners. It forces out excess compound from the tape. At the corners, you will need to remove excess compound with a broad knife. Then, after the angles are rolled, you need to go back over the full length of the angle with a corner finisher called a plow. The general sequence at all corners is as follows:

- Taping
- Rolling
- Plowing
- Corner applicator finishing

Four metal rollers in the head of the corner roller will embed and smooth the tape while forming a sharp corner crease. The tool is easy to use. Work it from the middle of the taped joint out toward the ends of the joint. This will force any overlap of tape due to stretching out to the end of the joint, where it can be trimmed off. This method stretches the tape in place. It also helps to prevent bunching up, which can easily happen if you start rolling at one end and go toward the middle instead of the other way around.

5.3.7 Corner Plow Operation

The corner finisher, or plow, is used to take out excess compound from angles after the corner roller has embedded the tape. This tool is available in widths of 2" and 3". You may use either size to wipe down the excess compound after the corner roller is used. Simply snap one of these tools onto the associated ball on the end of the corner applicator handle and use it like a plow (*Figure 38*).

With the arrow end leading, work from top to bottom for vertical angles, and from one end to the other for ceiling intersections. Wipe down these angles further on both sides with a 6" taping knife.

The plow is also used for topping, together with the corner applicator (*Figure 39*). It smooths and finishes both taped and topped corners. The corner finisher feathers the joint compound out from the corner and onto the drywall. It finishes both sides of the corner at once.

105F37.EPS

Figure 37 Using a corner roller.

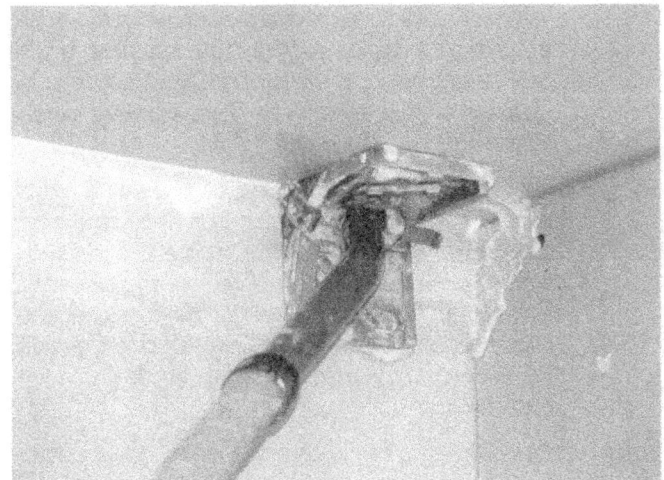

105F38.EPS

Figure 38 Using a corner plow.

105F39.EPS

Figure 39 Using a corner plow with a corner applicator.

A Different Type of Corner Finisher

With the MudRunner™ corner applicator, it is not necessary to apply pressure to the tool to feed compound to the plow. The operator merely turns the handle of the tool to dispense the compound stored in the reservoir just below the plow connection.

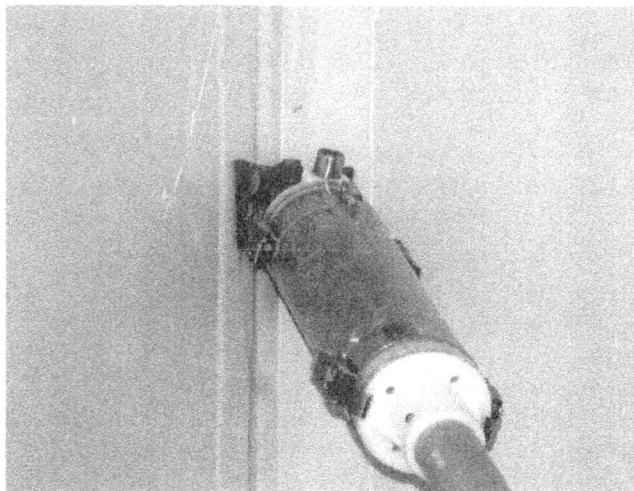

105SA03.EPS

Another nickname for this tool is a butterfly, probably due to its shape. It has four skimming blades and is designed to wipe down and feather both sides of inside corners and other angles at once.

The plow has a spring action that compensates for corners slightly over or under 90 degrees. The blade design produces a smoothly feathered joint. When you use this tool, be sure the three tips, or arrows, are pointing in the forward direction of travel.

5.3.8 Flat Finisher Procedures

The flat finisher (box) adjusts the amount of topping compound applied to the joint (*Figure 40*). The compound is automatically feathered out from crown to edges. The crown runs down the center of the taped joint. This raised area is balanced out by the shrinkage that normally occurs when the compound dries in the joint. You can use a box to apply topping to both tapered and butt joints.

Under the box by the handle connection, you will find a small dial. This dial controls the size of the crown left by the box on the surface. It also controls the amount of topping applied to the surface. As you turn this dial, you will see the blade

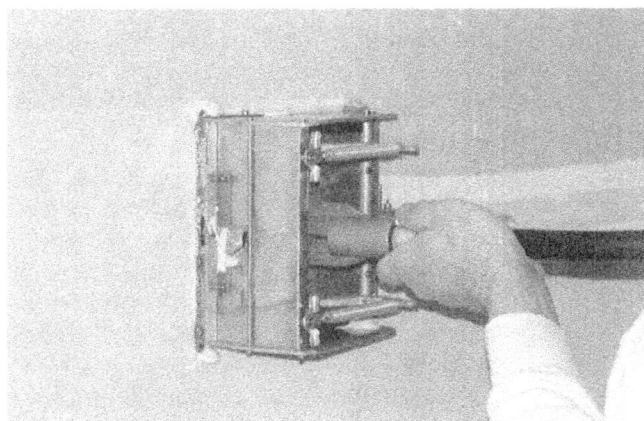

105F40.EPS

Figure 40   Using a flat finisher.

that controls the crown raise or lower. Establish the setting you will need for your particular application. In general, more crown is needed for earlier coats and less crown for later coats.

The flat finisher comes with handles of various sizes, usually from around 3' to 6'. Longer handles allow you to work at most higher wall and ceiling levels right from the floor without needing stilts or scaffolds.

Always run the box with the wheels leading and the blade trailing. Adjust your grip in relation to your body so that you lead the box with the handle, except at the joint ends. Also note that the end of the handle has a special gripping lever. This grip locks the box in whatever position you desire (in relation to the handle) as you move across the surface.

To finish flat joints, proceed as follows:

Step 1   Load the box through its opening behind the blade using the adapter spout, or nozzle, on the loading pump. The box loads in the same way as the automatic nail spotter. In general, the topping compound should be a little thicker than the taping compound.

Step 2   Apply topping compound on wall or ceiling taped joints by drawing the box steadily along the joint while applying pressure to the back of the box with the handle. This forces out the compound evenly through the opening, depending on the crown you set on the dial. The blade also serves to feather the compound thinly out to the edges, leaving the crown in the center. Always start from one end of the seam and go straight across to the other end without stopping. It should always be a smooth process, and you should always make sure you have enough compound in the box before you begin the run.

Step 3   Ceiling joints should be topped first. The first ceiling joints to receive the topping are the butt joints. Butt joints receive a first topping coat, one coat on each side of the butt, using the 12" box on each side. *Do not* put this coating across the center of the butt; leave the center alone for the first coat. Set the dial on the 12" box to #1 (fullest) crown. Think of this as giving each butt joint a double-wide treatment. The reason for this is that you deliberately finish butt joints much wider than regular tapered (flat) seams to hide the very slight crown already caused by the butted wallboard panels. The finished crown is so gradual and so slight that it will hardly be noticeable.

Step 4   The flat joints receive topping after the butt joints. Again, start with the ceiling. Use the 7" flat finisher set at #3 (medium) crown. Start at one end of the joint and

apply even pressure to the middle of the joint. Lock onto the grip as soon as possible after starting the run. Lead with the handle.

Step 5   When you approach the middle of the run, keep the lock on the handle and gradually release the pressure. Then remove the box in a sweeping motion from the surface.

Step 6   Next, reverse hand positions and start at the other end of the run. Repeat the process described above for beginning the run. Lead with the handle toward the middle of the joint where you stopped before. Again, keeping the lock on, slightly overlap the stopping point and remove the box with a sweeping motion.

5.3.9 Flat Finishing for Other Joints

For wall vertical flat joints, start at the bottom and lock onto the handle grip right away; then remove the box in a sweeping motion about 2' to 3' above the floor line. Start at the top of the joint and apply pressure down to your previous stopping point. Again, with the handle locked, slightly overlap that point and then sweep off the surface, which should neatly finish the two topped sections.

When applying topping to joints near doors, windows, and other openings, always work from the corner and move towards the opening. Just before the wheels reach the opening, keep the handle locked and lift the wheels. Then sweep away out into the opening so that topping compound is applied all the way to the edge.

5.3.10 Finishing Coats with the Flat Finisher

Second and third topping coats are applied in the same way as the first topping coat, but in thinner and wider layers. These finishing coats are applied to fill out minor shrinkage and unevenness that could produce shadows and other imperfections after painting. Always allow each coat to dry overnight. In cold or humid conditions, each coat might take even longer to dry.

Before starting each finishing coat, scuff the surfaces. This means lightly sanding the dried compound areas to remove any crumbs, burrs, globs, and so forth. This also prevents any of these things from producing scratches in the surface when you make another pass over them with a box. Be sure to wipe down all sanded areas.

Fill the 10" box, set the dial to #3 (medium), and then apply a straddle coat on all butt joints, starting with the ceilings. Apply this coat of topping compound right down the center (on top of the taped seam) between each of the previous double-wide coats. A correctly finished butt joint should always have a final width of at least 25".

Do the flat seams next. Use the 12" box if only one coat is required, or use the 10" box for the second coat, and the 12" box for the third. Reset the dial to #5 (least) crown and cover the ceiling and wall flat joints in the same way as for the first topping coat.

5.3.11 Corner Applicator Operation

In topping operations, the corner applicator provides a final finish for ceiling and inside corner angles. Use it after the bedding coat has dried. Attach the corner finisher to the corner applicator by snapping it in place. The plow becomes the skimming or troweling blade for the tool. This runs in the topping coat at the corner angles. The corner applicator smooths and feathers the final coat. The corner applicator does for angles and corners what the flat finisher does for flat and butt joints.

To use a corner applicator, proceed as follows:

Step 1 To load the corner applicator, first remove the nozzle end from the chain-slung filler adapter that goes on the loading pump. Inside the nozzle housing is a rubber O-ring that prevents leakage during filling. The filler valve on the corner applicator is inserted into the housing against the O-ring, which seals it. Pressure from the pumped compound pushes the tool's filler valve open.

Step 2 Once the corner applicator is full of topping compound, attach the 3" plow over the round opening. Place the tool at one end of the corner angle. Then with the nose of the plow leading, draw the tool along the angle, applying steady pressure with the handle on the back of the box. This forces the compound out where the plow distributes and feathers it along the run.

Step 3 As you near the end of the run, sweep the tool away from the surface in the same way as with a flat finisher. Reverse hand and body positions and start again at the unfinished end. Draw the tool back to the previous stopping point. Overlap just a little and then sweep away from the surface. This should neatly join both sections of the run. Make sure you apply the plow to neatly bisect the angle. Also, keep the tool at as close to a right angle to the corner as possible.

Step 4 To apply compound to vertical angles, start at the top of the wall and draw the tool downward, sweeping away from the surface at about knee height. Then place the head in the bottom of the angle seam and draw the tool back upward to barely overlap the previous stopping point. Gradually sweep away from the surface, neatly joining both sections of the run.

Step 5 Detail out the corner intersections with a broad knife. Feather any excess compound away from the angle on both sides, also with a broad knife.

When the plow is removed from the corner applicator and a ribbon dispenser attachment is fixed on, it becomes a tool known as a flat applicator. The flat applicator is an alternative tool to the automatic taper. It applies a bedding coat of taping compound and allows you to then attach tape to the surface by hand.

This semi-automatic taping method is useful in places where you physically cannot work with an automatic taper. There is also a mini-taper that might work just as well in confined areas. The flat applicator is convenient for emergencies, for hand operations, or for use by one mechanic when another mechanic is using the only taper on the job.

All these automatic taping and topping tools must be kept clean. Keep the tool heads submerged in a bucket of clean water whenever they are not being used.

5.4.0 Hand Finishing Procedures

Drywall finishing procedures start with gathering all the proper tools, equipment, and materials necessary to do the job. Decisions are then made about sequencing the various tasks: what is done first, second, and so on. These tasks include:

- Inspecting, repairing, and prefilling
- Taping flat joints, corners, and other angles
- Installing bead and trim pieces
- Spotting fastener heads
- Topping, scuffing, and sanding

To finish drywall, proceed as follows:

Step 1 Dimple the nail and screw heads, and cover them with joint compound. You can sometimes create a dimple by banging in the nail with the butt end of your knife handle. Damaged drywall must be patched. Do only a minimum of scuffing or sanding on the paper so as not to roughen it or raise the nap.

Step 2 Once the drywall has been properly installed, carefully inspected, and fixed where necessary, the next step is to prepare for spotting and taping. The usual finishing sequence is to spot the fastener heads; pre-fill gaps, damaged areas, and butt joints; tape the ceiling joints; and then tape the corners, other angles, and finally the flat joints.

Step 3 Prepare the joint compound according to directions and the job site conditions. Put down a suitable sheet of scrap wallboard first. Do all your mixing at this one place on top of the protective scrap board.

Step 4 For hand taping, load the mixed compound into the bladed mud pan or onto the hawk. Obtain the proper compound consistency. It should generally be the consistency of soft putty. Properly mixed compound does not fall off the hawk when it is tilted for a short time.

Step 5 Apply the compound to the bare joint with a broad knife. (If self-adhesive mesh tape is used, it is applied without a bedding coat.) While the joint compound is still wet, apply the joint reinforcing tape. Press the tape into the compound (*Figure 41*). Smooth the tape with a broad knife as it is applied. Force the excess compound out from under the tape and remove it with the knife. This bonds the tape to the compound. For perforated tape, force the excess compound up and out through the holes. Again, wipe away the excess with the knife. Make sure there is enough mud under the tape or bubbles will result.

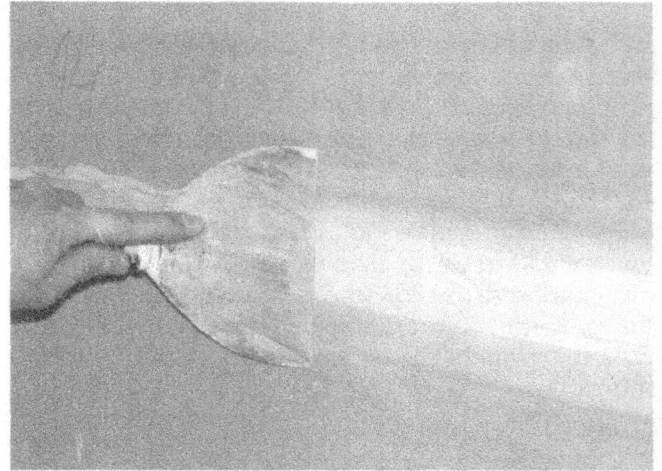

105F41.EPS

Figure 41 ✤ Applying tape.

Step 6 Spread a thin coat of joint compound over the top of the tape. Allow these bedding coats of joint compound to dry overnight. Special precautions need to be taken when finishing butt joints. For hand finishing, it is critically important to look at the taped butt before you finish it. Follow these guidelines when finishing butt joints:

- If the butt joint has a high side and a low side, coat the low side.
- If the wallboard butt forms a hollow, fill it back to the normal plane by adding much more compound than usual.
- If the butt joint is regular, it still needs to be finished with the double-wide method (*Figure 42*). For the first topping coat, do not cover the tape. On day one, make a crown of compound on each side of the tape. Then, on day two, apply the straddle coat to cover the tape between the crowns, thereby making one slightly larger crown in the middle of the seam.

Step 7 Apply a thin coat of joint topping compound. Feather the compound's outer edges. Feathering spreads the compound thinly from the center of the taped joint outwards beyond each edge of the tape, causing a slight crown over the center. However, the smaller the crown and the finer the feather, the less sanding will be required.

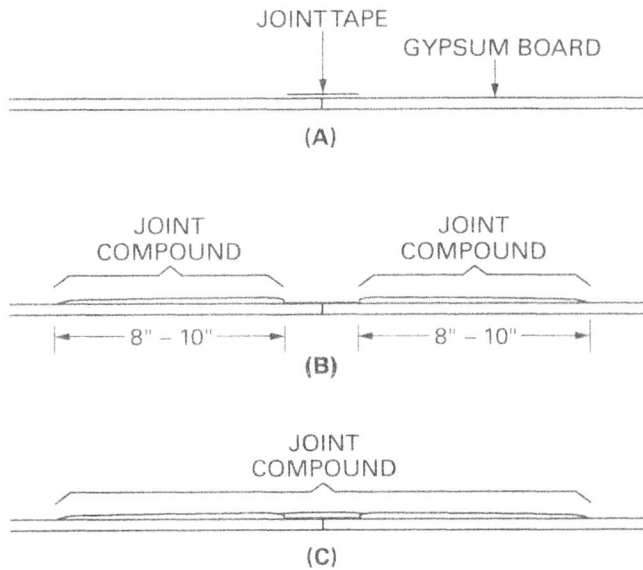

JOINT TAPE
GYPSUM BOARD

(A)

JOINT COMPOUND   JOINT COMPOUND

|← 8" – 10" →| |← 8" – 10" →|

(B)

JOINT COMPOUND

(C)

105F42.EPS

Figure 42   Finishing a butt joint using the double-wide method.

Step 8   Allow this coat to dry. Some compounds require 24 hours to dry; some take even longer. Drying times also depend upon atmospheric conditions within the structure. If the building is only partially closed off, the finishing work will be affected by the outside weather. If it is too cold or wet, the compound might not dry out at all until the weather changes. Finishing work depends upon a controlled interior.

Step 9   Once the coat has dried thoroughly, sand it smooth. Remove the sanding dust. Depending on the situation, you may be applying another topping coat, or one or more skim coats. These determinations are almost always made by your supervisor.

5.4.1 Sanding

Always wipe down after any light or heavy sanding to remove sanding dust and tiny particles of compound or other debris. Always check the coated surface to see if it is straight and smooth. Use a straightedge or level to do this.

Always have a hawk or mud pan available filled with topping compound, no matter what finishing procedure you are doing. This way, whenever you see something that needs a little touch-up, like a pit or scratch, you can take care of it immediately to avoid poor-quality finishing.

Correcting Oversanding

Excessive sanding or use of coarse sandpaper can cause the paper fibers of the drywall to stand up. If the problem is not too severe, light sanding with a very fine sandpaper or wiping the panel down with a damp sponge or cloth can correct it. Otherwise, use a light skim coat of topping or all-purpose compound to correct the problem.

For general sanding and scuffing procedures, 100-grit, 120-grit, or 150-grit sandpaper is recommended. Sand screen is also used for scuffing as well as for final sanding, because it does not raise the nap on drywall face paper. Sandpaper of less than 100 grit should be avoided, and any sandpaper coarser than 80 grit is unacceptable.

Remember, you are sanding a dried, porous wall joint covering material in order to smooth out tiny bumps and spaces so it will hold a final decoration as well as wallboard face paper. Guard against oversanding, which tends to grind out hollows.

WARNING!

Be sure to wear eye protection and appropriate respiratory equipment when sanding. Check the MSDS for the applicable drywall to learn about the safety hazards associated with that material.

5.4.2 Spotting Fastener Heads

Check the drywall nails. Be sure they have been dimpled (set below the substrate with a hammer). Apply the first coat of joint compound on top of the nail heads. Allow it to dry. Sand the dried coat with an abrasive cloth or sand screen. Apply a second coat of joint compound. Allow this coat to dry. Sand it smooth and apply a third coat. The covered area should be smooth and level with the substrate.

Many contractors prefer hand spotting and will not allow the use of a nail spotter tool for this task. Their reasoning is based on several factors:

• Fasteners may not always be below the substrate level, so the nail spotter's blade is frequently damaged. This causes downtime to change blades or tools. When there are no more replacement blades or tools, hand spotting will be the only remaining option.

- Spotting by hand is a reasonably fast method when done by an experienced mechanic.
- Hand spotting is thorough. It forces the compound more effectively onto the head and completely fills the dimple. It also packs compound down into the crisscross of Phillips-head screws better than the automatic tool usually does.
- Hand spotting is something even the newest apprentice can master almost at once. It helps them appreciate the nature of finishing work faster than any other process. It can also build confidence, speed, and an eye for detail.

All fastener heads need to receive three coats of topping so they are undetectable when the surface is finished. Use a 5" knife for the first two coats and a 6" or 8" knife for the third coat. Make sure the compound fills in the crossed indentations in Phillips-head drywall screws.

5.4.3 Outside Corners

To finish outside corners, proceed as follows:

Step 1 Attach the metal corner beads to the outside corner angles. Fasten them with drywall nails or a clinch-on tool or by applying tape with compound. Staples are sometimes used, although many contractors try to avoid them. Dimple the nail heads. Apply joint compound to each flange with a broad knife right after fastening.

Step 2 Spread the compound 7" from either side of the nose (center outside corner). Cover the metal edges with compound. Allow the compound to dry. Sand it lightly, then apply the next coat. Feather the coat out two inches from the first coat. Let this dry and sand it smooth. Also apply and smooth a third coat of topping at outside corners. Use an 8" knife for all three coats, a 6" knife for the first coat and an 8" knife for the others, or an 8" knife for the first two coats and a 10" knife for the third coat.

The goal with each coat is to fill the corner so it is flat, not concave. Too much compound will make the corner concave. Finish each outside corner so that the corner bead is completely invisible.

5.4.4 Inside Corners

To finish inside corners, proceed as follows:

Step 1 Cut the joint tape to the length of the corner angle. Apply joint compound to each side of the tape angles. Apply small amounts of compound to both sides of the corner angle. This prevents thin cracks from occurring at the angle. Crease the tape length along the center.

Step 2 Use a 4" broad knife to press and embed the tape in the compound. Apply enough pressure to wipe the compound from under the tape. Feather the compound 2" beyond the tape edges. Let it dry and then sand. A corner tool is available as an alternative (*Figure 43*).

Step 3 Apply and feather the next coat about 2" beyond the first coat. Dry and sand. Then apply a third coat where needed.

Step 4 For inside corners, it is better when hand finishing to apply compound first to one side, wait overnight, and then apply compound to the other side. With the automatic corner applicator/finisher tool, you only need apply one coat of topping to inside corners. This is called plowing or glazing the angles.

105F43.EPS

Figure 43 ♦ Using a corner taping tool.

You might find using trowels more comfortable than taping knives. There are many veteran drywall mechanics in the trade who were trained on trowels. They find taping knives too stiff and awkward. Other mechanics, however, cannot imagine finishing a taped joint with a trowel. Either tool can be used to produce equally attractive results. Try both types of tools, if you like, until you decide which is best for you.

5.4.5 Safety and Good Housekeeping

Follow all recommended safety practices for drywall finishing. Maintain your tools and equipment. If you use stilts instead of a ladder, practice safety; always put stilts on and take them off when you are able to lean against a wall, preferably in a corner. Never climb stairs or try to pick up

INSIDE TRACK

Using Stilts

Do not try to use stilts unless you have been properly trained to use them: If you plan to use stilts, make sure they are not prohibited by local codes.

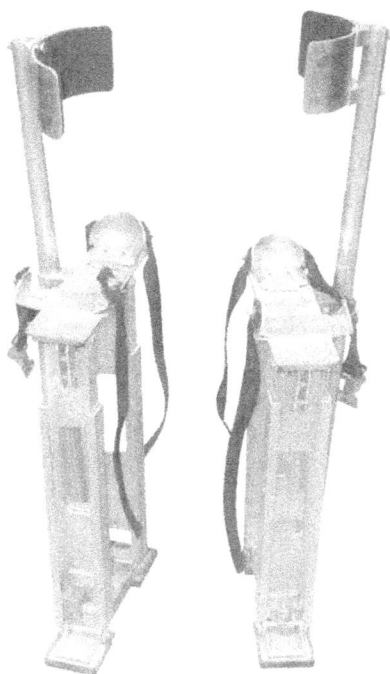

105SA04.EPS

something off the floor while wearing stilts. Ask for help from someone who is not wearing stilts.

Be aware that taping and topping tools, knives, and trowels carry with them a certain degree of hazard. For example, even a dull blade can cut or injure an eye or face. Any tool, if mishandled, can cause an accident or injury. If tools are allowed to clutter up a work area, they can also cause an accident or injury. Keep unused tools in a safe place out of everyone's way.

One of the most common hazards when finishing is slippery conditions, often caused by wet compound carelessly spilled on the floor. This is especially hazardous for someone on stilts. The best rule is if you spill something, clean it up—no matter what it is. If wearing stilts, either remove them and clean up the spill or ask someone who is not on stilts to clean it up for you; do not attempt to bend over on stilts. The same is true if you drop a tool.

Store finishing materials in a cool, dry, protected location. Provide adequate ventilation during dry sanding. Always wear a face mask or respirator to prevent inhalation of sanding dust. Wash hands after applying joint compound as well as after sanding. Proper safety and housekeeping procedures minimize illness and injury.

Final cleanup is always your responsibility. This means a complete sanding and wipedown of all ceilings and floors, and a thorough scraping and sweeping of all floors.

Keep in mind that sanding dust travels. If you are doing remodeling or repair in a finished building, secure the area in which you are working by covering doors and other wall openings with plastic. If you are working in a room that contains furniture, equipment, or other items, cover them.

6.0.0 ◆ PROBLEMS AND REMEDIES

The true test of your finished work will come not from how well you avoid making mistakes, but in how well you repair your mistakes. You may also have to repair those problems left for you by someone else. This section explains the areas where the most common problems are likely to occur and gives you information on their causes and how you can fix them. The four main problem areas are:

- Joints
- Compounds
- Fasteners
- Gypsum board panels

6.1.0 Finished Joint Problems

The common joint problems found in drywall work include:

- Ridging
- Tape photographing
- Joint depressions
- High joints
- Discoloration
- Tape blisters
- Cracks in the joint

These joint problems are described in detail in the paragraphs that follow.

6.1.1 Ridging

When a ridge occurs along a joint between two boards, it is often because there has been movement at the joint. Ridging is also sometimes known as beading or picture framing because a visible ridge that surrounds a board resembles a frame surrounding a picture. There are three probable causes of ridging:

- *High humidity, poor heat distribution, or not enough ventilation* – This results in expansion and contraction of the framing and boards.
- *Drywall that is not properly installed* – Improper installation includes misaligned framing and butt joints that do not fit well together. If you force two boards together, joint compound may be squeezed out, forming a ridge. Make sure the joint is not overstressed by too tight a fit.

  You may have to cut some of the drywall away at the joint to make a little gap between the boards. Do this by cutting with a knife, making several passes, or use a hand saw or chisel to remove a sliver of one of the boards. If the space left between the boards is too wide, the joint will be weaker than a properly-fitted joint. Ideally, the entire width of the joint tape is bonded to a drywall surface. The tape itself only needs to span a small space between the boards. If the space between the boards is too wide, less of the tape is bonded to the drywall. This can weaken the joint and promote ridging.

  You may have to overlap two pieces of tape to strengthen the joint. If the joint is very wide, it may be because one board, or both, is not securely fastened to the framing at the joint. Perhaps neither board is directly over the framing member. You may have to install fasteners at an angle through the boards into the framing in order to make the joint tight against the framing.

This presents a problem; one of the most basic rules is always to drive the fasteners straight so that the edges of the heads will not protrude. In this case, however, they will. Repair the problem by using a hammer to straighten the head (by bending the shank) after the fastener is installed. You may also need to cut and install a thin strip of gypsum board to bridge a space that is too wide.

- *Too much joint compound* – To correct ridging caused by too much mud, first sand the ridge smooth, then apply a finishing coat of joint compound. Hold a light at an angle to the area to make sure you have eliminated the ridge and left a smooth surface.

6.1.2 Photographing

If the joint is still visible even after the wall is finished and painted, this condition is known as photographing. The joint may show through as a slightly different color than the finished wall. Or it may be the same color as the finished wall but have a higher or lower gloss (shine) to it. Photographing can also occur over fasteners if there was insufficient joint or topping compound spotted on the heads.

The usual causes of joint photographing are:

- The installer failed to force excess joint compound out from under the joint.
- High humidity conditions delayed drying of the second and third topping coats of compound.
- The tape absorbed too much moisture from the compound, causing the mud to shrink and conform to the shape of the joint tape. Avoid this by wetting the tape before installation.

To correct photographing, sand the tape edges to feather them into the surface of the wall or ceiling. Then cover the tape with thin coats of joint or topping compound. Use thin coats so as not to rewet the tape too much, or seal the tape with a primer after sanding and before applying the final coats of mud. This keeps the tape from drawing too much moisture out of the mud.

6.1.3 Joint Depressions

A joint depression is a valley that occurs at a seam or joint. It will be most obvious when a light strikes the drywall at an angle. (Hint: Use a 10" or 12" knife to help find high and low spots.) There are two common causes of joint depressions:

- There may not be enough joint compound over the joint. This can happen when the joint compound mixture is too thin or when not enough joint mud is applied to the joint.
- The joint may be sanded too deeply.

The cure for joint depressions is to add more joint or topping compound to the joint. Then smooth it and sand again to get a flat surface at the joint. Make sure the joint is flush with the surface of the wall or ceiling.

6.1.4 High Joints

A high joint is the opposite of a joint depression. It occurs when a wide section of joint is raised above the rest of the drywall surface. Like the joint depression, a high joint is most noticeable when a light strikes it from an angle. High joints are the result of too much mud built up underneath or on top of the tape and/or improper feathering of each coat. The edge of the coat must be feathered into the wallboard surface. When this is not done, high joints will result.

To repair a high joint, sand the area as flush as possible without sanding into the tape. Then apply one or two final skim coats of topping compound. Feather each coat into the board surface. Make each coat wider than the previous coat to conceal the area of the joint (*Figure 44*).

6.1.5 Discoloration

Joints may discolor or turn lighter or darker than the rest of the finished surface. There are several common reasons for joint discoloration:

- Moisture may be trapped inside the joint. Until a joint is sealed, it can absorb water. If a joint is sealed before it is dry, water is sealed inside the joint. Trapped water will degrade the finish and discolor the surface. Be sure the joint is dry before sealing it.
- The joint was painted in conditions of excess humidity. To prevent this, reduce the room humidity before any painting is done.
- A poor-quality paint was used. Always be sure to use a good-quality paint. Cheap paint often gives uneven coverage and sealing. This increases discoloration.

6.1.6 Tape Blisters

A tape blister is really an air bubble in the surface of a joint. It can be several inches long or as small as a dime. Tape blisters occur when the bond fails

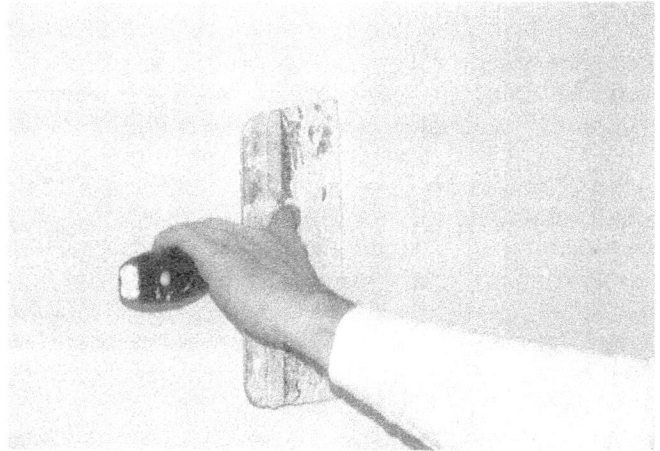

105F44.EPS

Figure 44   Coating with topping compound.

between the tape and the first bedding coat of joint compound. One way a bond can fail is when there is no bond to start with. This may happen if care is not taken when using an automatic taper and sections of tape come out without a mud coat underneath.

Tape blisters can also happen if the joint is too wide, because the tape was not properly embedded in the joint compound, or because the tape draws moisture too quickly from the mud. Another cause of a blister occurs when topping compound is used instead of joint compound to embed the tape.

To repair a tape blister, proceed as follows:

Step 1   Slit the blister with a knife. If the blister is large, cut and remove the entire section of tape that came unbonded.

Step 2   Sand or scrape out enough of the dried mud so you can embed a new section of tape.

Step 3   Work joint compound underneath the tape, smoothing the slit in the old or new section of tape into the joint compound as you go. This embeds the blistered section. This is a hand procedure only. Do not attempt to do this with another run of the automatic taper.

Step 4   Apply a skim coat of mud over the tape. When this coat is dry, apply the required number of topping coats, always allowing enough drying time in between coats. Sand enough to produce a smooth finish that is flush with the surrounding surface.

6.1.7 Cracking

The two common types of joint cracks are those that run along the edges of a joint and those that run along the center of a joint. Each has its own causes.

Edge cracks are cracks along joint edges that occur when the air temperature is high and the humidity is low when the joints are finished. This causes the joint compound to dry too quickly and unevenly, resulting in uneven shrinkage. To slow down the drying rate, run a wet roller over the joint or spray it with a fine water mist from an atomizer.

Edge cracks can also be caused by tape that has a thick edge or by joint compounds applied in coats that are too thick.

The procedure for repairing edge cracks depends on whether the crack is thin or wide. If the crack is small and thin, coat it with a latex emulsion or a thin coat of joint or topping compound. Then sand it as needed.

If the crack is wide, you may have to gouge out some of the joint compound to prepare the surface. Paint the gouged-out crack with a primer, then fill with joint compound and sand smooth.

Center cracks are cracks running along the center of a joint that occur for several reasons:

- If the tape is still intact, the crack is probably the result of applying joint compound too thickly. Also, low humidity may have caused the mud to dry too quickly.
- If the tape under the crack has been torn, it is possible that the structure is settling or another type of movement caused the crack.

To repair cracks along the center of the joint, follow these procedures:

- If the tape is still intact and the crack is narrow, apply spackling compound to the crack. If the crack is wide, use joint or topping compound to bridge the space.
- If the tape is torn, you may need to remove a section of tape and old joint compound before making repairs. Then retape the joint following the usual finishing procedures.

6.2.0 Problems with Compound

Compound has its own special set of problems. It can debond, grow mold, become pitted, sag, and shrink.

6.2.1 Debonding, Flaking, or Chipping

When the joint or topping compound will not bond to the tape or the board or becomes unbonded from either one, the condition is known as compound debonding, flaking, or chipping. Common causes of compound debonding include:

- A foreign substance was on the gypsum drywall surface or on the surface of the tape when the mud was applied. Examples of foreign substances include dirt, oil, sanding dust, and incompatible paint.
- The mud was mixed improperly, or the wrong ratio of water to dry powder was used to mix the compound.
- Too much water was added during mixing, or incompatible compounds were mixed with each other in order to add to the working supply, combine containers for storage, etc.
- Dirty water was used to mix the compound, or dirty tools were used to mix or apply it.
- Hot or heated water was used to mix the compound. One reason for avoiding hot water is because of possible sediment problems associated with hot water heaters.
- The installer used old or expired compound.

You can avoid most compound debonding problems by following the manufacturer's mixing and usage instructions exactly. Some manufacturers request that you let the compound sit for a while after mixing it. There is a good reason for this, so do not take any shortcuts. They will end up costing excessive time and waste in the long run.

Be sure to always use only clean, cold water to mix any compound. Also use clean mixing and application tools and equipment. Remember that automatic taping tools need to be kept in pails of clean water between uses. Always make sure the gypsum drywall surface, tape, and all mixing pails are clean, too.

Repairing compound debonding is very much like repairing tape blisters, only on a larger scale. First, separate the debonded section of tape from the dried mud. Then remove enough of the old mud to allow you to apply a new layer in which to embed the tape. If the old compound crumbles easily, remove all of it. You will also have to remove whatever mud was used to feather the joint. Apply new compound and tape as you would for a new joint.

6.2.2 Moldy or Contaminated Compound

Moldy compound smells terrible, so you will have no trouble recognizing it. Using contaminated water, dirty containers or tools, or letting the compound stand too long can result in mold, bacteria, and bad odors in the compound. Hot and humid weather also contributes to the growth of mold and bacteria.

Always be careful to examine every pail before mixing anything in it. You may be surprised at what you might find in a supposedly empty bucket. The best remedy here is simply to look before you mix. If you discover that your batch of compound has become moldy or contaminated, throw it out. Then be sure to soak your tools and containers in a solution of chlorine bleach and clean water at least overnight.

Be sure to clean and wash all tools and equipment components at the end of every working day. The mud intended for use the next day must be stored in covered containers and kept at room temperature overnight. That means warm room temperatures, not freezing cold or scorching hot.

6.2.3 Pitting

Small pits may appear in the finish of the mud after it dries. Pitting has several common causes:

*   Air escaped that was trapped in the mud mixture. This can happen if you mix the compound too vigorously or for too long.
*   The mud mixture was too thin.
*   Not enough pressure was used to apply the mud to the joint; that is, it was not embedded or wiped down properly.
*   Mud was not adequately mixed prior to application.

To prevent pitting, mix the compound thoroughly using a slow, steady motion. Set power mixers, if used, on slow speed. You are trying for a smooth mixture that is free of lumps. When you apply the mud, use enough force to establish a good bond to the surface, smooth it out, and feather the edges.

Repair a section of pitted compound by simply skim-coating with another topping layer in order to fill the pits. You may need to sand a little to form a smooth base for applying the new mud. Then apply the new compound as you would apply a topping coat to the joint. Feather it out to conceal the joint area. You may have to feather it wider than the original topping coat in order to completely hide the joint.

6.2.4 Sagging

When compound sags or shows evidence of runs, these conditions are usually present:

*   The mud was too thin. When mixed properly, compound is thick and smooth. Be sure to follow the mixing instructions exactly.
*   Water added to the mud or to the dry powder compound was too cold to mix completely. Again, cold water is essential, but do not use ice cold water. Remember that finishing is a room-temperature process. Anything colder than what normally comes out of a faucet in a warm room is just too cold.

To repair sags and runs, sand them very smooth after they dry. Then, recoat with layers of joint or topping compound as needed.

6.2.5 Excessive Shrinkage

If the mud shrinks too much when it dries, it is probably the result of one of the following:

*   Mud mixed too thin
*   Insufficient drying time between coats
*   Too much mud applied at any one time

To prevent this problem, use lightweight mud, which tends to shrink less. This problem is similar to the joint depression problem discussed earlier. As in that case, remedy excessive compound shrinkage by applying more mud. However, ensure that each previous coat is thoroughly dry before you begin any repair by adding more compound.

6.2.6 Delayed Shrinkage

Delayed shrinkage is caused when too much time elapses before the correct amount of shrinkage occurs. The mud is not shrinking enough and tends to resist drying out. Delayed shrinkage has several common causes:

*   Atmospheric conditions (slow drying capabilities and very high humidity)
*   Insufficient drying time between coats of compound (trying to rush the job before it is actually ready for each finishing procedure)
*   Excess water added to the mud mixture
*   Heavy fills (adding too much mud as a prefill or trying to fill large spaces in the gypsum drywall with compound instead of slivers or strips of wallboard)

One way to prevent delayed shrinkage is to use a faster-drying compound, perhaps a quickset compound, which sets up chemically and does not depend on water evaporation. Quickset compounds were discussed earlier in this module.

A remedy for this condition is to allow extra drying time and then to reapply a full cover coat of a heavy-mixed mud over the tape. Most shrinkage will generally take place on this heavy topping coat. With the right mud, the coat will dry faster and allow you to continue finishing procedures in the usual way.

The best defense against delayed shrinkage is to use a faster-drying compound in the first place. There is very little you can do to mud that needs more drying time, except to give it more time to dry.

6.3.0 Fastener Problems

Two common fastener problems that may be encountered are nail pops and fastener depressions. These problems are described in more detail in the paragraphs that follow.

6.3.1 Nail Pops

When drywall nail heads work up from under the finished surface after the installation is complete, the job is said to have nail pops. Nail pops are unsightly, protruding fastener heads.

If enough nails pop out, the drywall will loosen and sag. The nails can be driven in again and the hole refinished, but the best remedy is preventing nail pops before they happen. Here are the primary reasons for nail pops:

• Wood framing with relatively high moisture content will shrink as the lumber dries out. As the wood shrinks, the nails lose their tight holding power (*Figure 45*). When the wallboard is no longer securely attached, a space develops between the board and the stud; the nail shank is exposed at that point. Then almost anything that puts pressure against the wallboard will push it against the stud and the nail—which does not move—will actually pop right out of the panel along with the compound covering it.

• When drywall is fastened to framing that is out of alignment, stress on the drywall causes fasteners to work up above the surface (*Figure 46*).

• Gravity acting on ceilings and vibrations acting on walls will tend to work the nails loose (*Figure 47*).

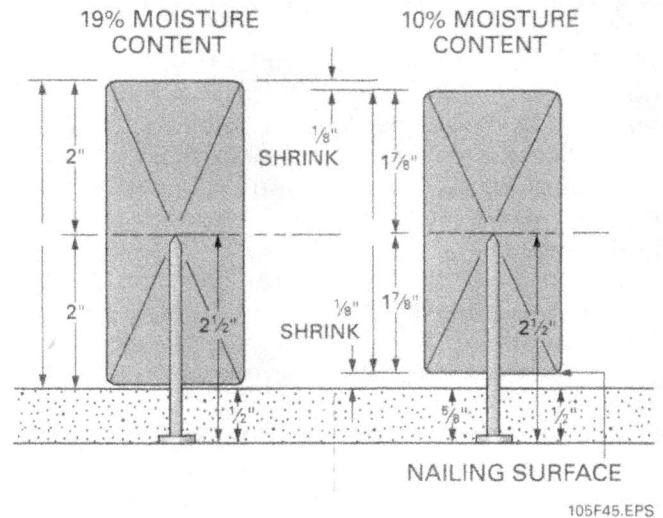

Figure 45 ⟶ Shrinkage contributes to nail pops.

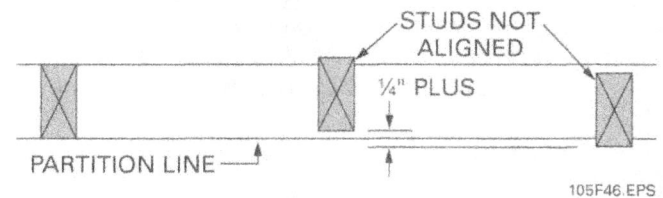

Figure 46 ⟶ Non-aligned framing.

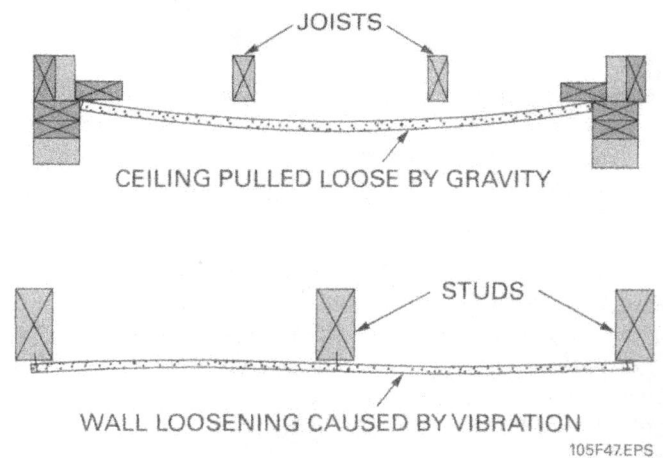

Figure 47 ⟶ Force of gravity can cause nail pops.

• The drywall may not have been installed properly (*Figure 48*).

• If a building has poor ventilation or an inadequate heating system, large temperature fluctuations will cause expansion and contraction of the framing and drywall. If there is too much of that, the fasteners will begin to loosen.

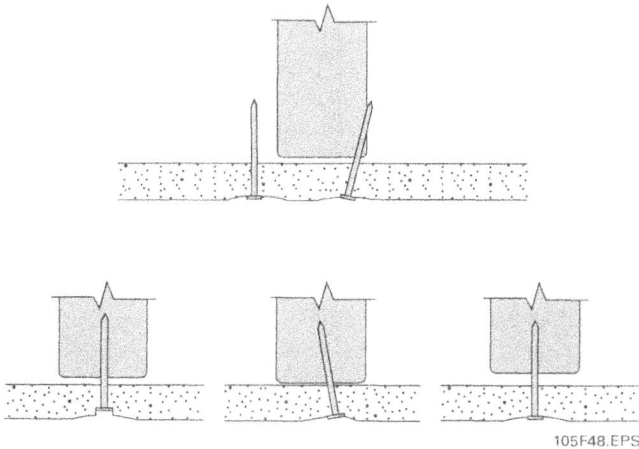

Figure 48   Improper drywall installation.

6.3.2 Preventing Nail Pops

Nail pops may show up days or weeks after installation is complete or gradually over a period of many months. How soon they appear depends on the degree of misalignment, the type of fasteners used, the moisture content of the framing at the time of installation, the amount of vibration present, and the temperature cycles to which the drywall and framing are exposed. Once you know the causes of nail pops, prevention is easier. To prevent nail pops, follow these key rules:

- Make sure your framing lumber is dry before you fasten any drywall to it. The builder should provide enough ventilation to speed up the drying process. When working in cold or humid weather, use portable heaters or blowers to warm or circulate the air. It may take only a few days to reduce the moisture content of lumber to an acceptable level, depending on temperature and humidity. Lumber is too green for hanging drywall if any wet spot appears when the lumber is hit sharply with the head of a hammer. Test several lengths before forming an opinion. The amount of moisture can vary from one piece to the next.

- Make sure the framing members are aligned in the same plane. Sight along the edges of the studs and joists to see that they are in a straight line. You can also check alignment by holding a long straightedge up against the studs and joists.

    If framing is out of alignment, repair it. Applying boards to framing that is out of alignment will prove to be a mistake. The framing will eventually spring back to its original position and the nails will pull out from the framing.

- Use ring-shank drywall nails or screws to fasten the drywall boards to the framing. They have more gripping power than plain-shank nails and offer more protection against nail pops. Also, consider using double rather than single fasteners. Double nailing gives more holding power than single nailing. Use floating corner angles to reduce stress on the drywall. Finally, you can use adhesive in addition to nails or screws to fasten the drywall boards.

- Always nail from the center of the drywall board toward the edges. If you nail from one edge of the board to the other, the board may not rest firmly against the framing for the entire length of the board.

    For example, assume that you are installing the final board in a wall. The other boards along the wall are already in, and so are the boards that cover the wall that forms the other half of the corner. On this final board, if you start installing the fasteners at one edge instead of at the center, the drywall may move slightly toward the opposite edge. This particular board is trapped by a corner and other drywall boards. Any movement after you start nailing will stress the board, eventually causing it to bow and pop the fasteners.

    A similar problem can occur if you work from the other edge toward the center of the board. The center may actually spring away from the framing. If you fasten the center of the board first and work toward the edges, the board will not be able to move once the first fasteners are in place. From the moment the first nail is driven home, the board is forced flat against the framing in the center and along the edges. Be sure to hold the drywall board tightly against the framing as you install the fasteners. This ensures that the board stays flat. Do not worry about installing a board with a slight bow. With proper installation, any stress put on the board by flattening its bow will relax in a short time.

Before you cover the nail heads on any drywall finishing job, check to be sure the nails are tight. Re-drive any loose nails. It is also a good idea to drive another nail on each side of a nail that has worked loose. Drive them about 1½" away from the old nail. After you re-drive any loose nails and add extra nails, go back and check all the nail heads again. The vibration caused by hammering may have loosened more nails. The few seconds you spend now can save you a few hours later. Also, if you nailed boards to both sides of a wall, driving the nails on one side may have loosened the nails on the other side. Be sure to check the first side again.

Attention to detail should prevent most nail pops. Take the time to check your framing lumber carefully and install your panels properly.

6.3.3 Repairing Nail Pops

When nail pops occur after you have finished a job, perhaps even after the texturing and painting are done, fixing them takes a little more time. When the nail head works out of the framing, it will show above the surface of the drywall. It may even lift the compound from the depression around the nail head. Repair it as follows:

Step 1 If the nail head has worked loose and become visible, just driving it back may not solve the problem. It is better to pull out the nail and replace it. The best and most permanent replacement is a drywall screw; otherwise, use a longer nail and/or another nail within 1½" of the first.

Step 2 Once the fasteners are in, fill the depression with mud and let it dry. If necessary, use a second coat of compound.

Step 3 You may have to repair the texture, depending on the surface's original texture. Be sure the texture of the patch matches all the surrounding texture.

INSIDE TRACK

Repairing Nail Pops

Another way to repair a nail pop is to drive a Type W screw about 1½" from the popped nail. Then, place a broad knife over the popped nail and hit the knife with a hammer to drive the nail back down.

Step 4 When the new texture has dried, paint it with an oil-based primer. If you skip this step, the paint may soak into the repaired spot when it is later repainted and will make the repaired area very evident.

6.3.4 Fastener Depressions

A fastener depression is a depressed area over the fastener head. This is the opposite of a nail pop. The joint compound over a nail or screw has sunk lower than the surface of the surrounding drywall.

Fastener depressions can be caused by several problems:

- Nails were dimpled too deeply or screw heads driven in too far.
- Not enough mud was applied to the fastener heads to cover them properly.
- The framing lumber was extremely dry. Dry lumber will absorb moisture, squeezing the board between the nail head and the edge of the stud or joist and pulling the fastener head deeper into the drywall.
- The installer used too few fasteners to hold the drywall firmly against the framing, allowing the drywall to flex independently of the framing and forcing the fastener heads deeper into the surface.

To prevent fastener depressions, avoid driving fasteners through the facing paper. Install the correct number of fasteners and space them properly. Spot the fastener heads with two coats of compound, sanding lightly between coats, if necessary.

Repairing fastener depressions is a simple matter. First, make sure you have installed enough fasteners. If you need more nails or screws to hold the drywall firmly against the framing, add them. Second, spot the fastener heads with joint compound to bring the surface flush with the surrounding drywall.

6.4.0 Problems with Wallboard

Common problems with gypsum drywall sheets include blisters, damaged edges, water damage, board bowing, board cracks, fractures, and brittleness.

6.4.1 Board Blisters

When the facing paper becomes unbonded from the surface of a piece of gypsum board, it is known as a board blister. It may be caused by a manufacturing defect, or it may be the result of careless handling or improper storage. The gypsum filler tends to break apart inside the wrapped board, causing the facing paper to loosen.

There are two common ways to repair board blisters:

- Inject an aliphatic resin glue, such as yellow or white carpenter's or wood glue, into the blister, and then press the paper flat. This is the best remedy where the blister is small or where the blister is not discovered until after the wall has been textured and/or painted.

((◉)) **WARNING!**

Before using any adhesive, check the manufacturer's instructions and applicable MSDS to identify any hazards. Wear protective equipment and apparel as specified by the manufacturer.

- Cut out the entire blistered area and finish it with tape and joint compound. Follow the usual procedure for embedding tape and finishing joints. If one width of tape is not going to be enough to cover the blistered area, add as many other strips as necessary.

6.4.2 Damaged Edges

Improper handling of gypsum drywall sheets is what generally causes damaged surfaces and edges. Such carelessness may cause the facing paper to tear or the gypsum core to crumble.

The only way to repair such damage is to cut off the damaged area back to sound gypsum board prior to installation.

If a board has already been installed and you detect damage along an edge or joint, cut out the damaged area back to sound board and prefill with mud. If this produces too large an area, install a filler strip of good gypsum drywall either laminated to a board layer beneath or attached with screws to the framing. Prefill around the strip and finish the joints as usual.

6.4.3 Water Damage

When gypsum drywall becomes wet, the core becomes soft and is easily deformed. Also, the facing paper may come unbonded (blistered) from the core.

If a board has been exposed to water, let it dry thoroughly before using it. Be very sure it is completely dry before installing it on the framing. If it

is so badly warped that even screw attaching will not straighten it, then, after it is dry, put it under a stack of new boards lying flat on the floor.

If a board is already installed and then becomes so wet that it warps away from the framing, drive in some additional screws to hold it. If additional screws do not help, take that board off the framing and replace it.

6.4.4 Board Bowing

Board bowing is similar to the warping problem discussed above. In this case, a board may have been forced into too small a space on the framing, causing the board to bow or warp.

Whenever you discover this problem, the best remedy is to trim the board edges in order to relieve the stress that caused the bowing. You may have to remove the board to do a proper trim job on the edges. Reattach the board when it has been trimmed to fit properly, so that you do not have to force or pry it into place.

6.4.5 Board Cracks and Fractures

A gypsum board can crack along its face, or it may fracture all the way through to the other side. There are various causes and cures for cracks and fractures.

Board cracks may occur along the face of any drywall board, but they are most likely to show up over a doorway, where there is a smaller and weaker section of board. If a crack is over ⅛" wide, treat it just as you would a regular joint. Repair it by taping and feathering the joint compound and topping compound until the crack does not show.

This type of cracking is often caused by movement or settling of the building. Many larger buildings, such as skyscrapers, have a built-in flexibility that may contribute to the cracking of interior drywall. In a building with a flexible frame, the best choice is non-bearing interior partitions that have a clearance at the top of every wall. The tracks are fastened to the ceiling to hold the tops of metal studs, which may or may not be actually fastened to the track.

There can be up to ½" clearance between the wallboards and the ceiling. Fill this space with caulk or a specialty gasket or trim such as the type shown in *Figure 49*. A control joint or expansion joint (*Figure 50*) might also be used. Place such metal or plastic trim around the appropriate board edges to give a finished appearance to the room and add protection to the walls.

Any one of three possible reasons can contribute to gypsum board fractures:

- The board was attached across the wide face of the structural framing members, such as the headers. If the framing is wood and the lumber shrinks, the board is compressed and it will crack. If the framing is steel and not very adequate, loads put on it may stress and crack some of the boards attached in this way.
- The wallboard was improperly handled or stored.
- The face paper was scored past the edge of a cutout.

To repair broken or fractured boards, completely cut out the damaged sections and replace them. If the damage was produced by scoring the facing paper beyond the cutout edges, simply repair this score with tape as you would any other joint.

6.4.6 Loose Boards

Loose gypsum drywall boards might be caused by any of the following:

- The boards were improperly fastened.
- The framing members were misaligned, uneven, or warped (in the case of wood).
- The screws or nails were not driven in all the way or else (with lumber) some shrinkage has occurred, pulling the framing away from the board, and thereby making it loose.

Improper fastening may be due to using incorrect types of screws or an improperly adjusted screwgun. This can result in screws being stripped or not seated properly, contributing to board looseness.

The remedy for fastener problems is generally to remove all faulty fasteners. Replace them with correct fasteners and properly drive them all the way in, so that they are well fixed into the framing and produce a good dimple in the surface.

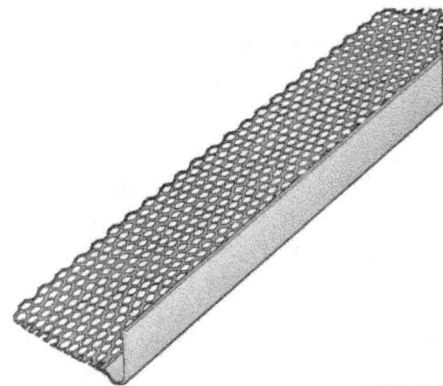

105F49.EPS

Figure 49 ◆ Veneer L-trim casing bead.

(A)

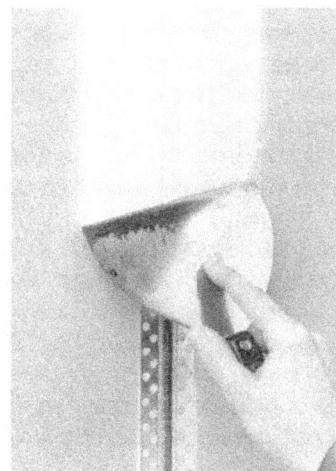

(B)

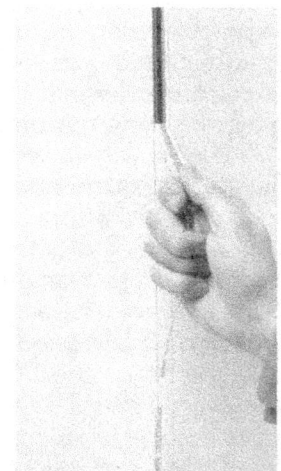

(C)

105F50.EPS

Figure 50 ◆ Applying an expansion joint.

When re-driving fasteners, make sure your free hand is pushing solidly against the board near the fastening point. It is important that the board be perfectly flat against the framing member while you are driving the fastener.

Double check to make sure you are using the correct type of drywall screw. Also, readjust the clutch on your screwgun to give you the proper depth into the board. You do not want to tear the face paper, but you need a dimple of about 1/16" to allow proper spotting and finishing. If nails are used, use the double nailing method.

If the cause of loose boards is poor framing, which may be out of alignment, twisted, or warped, your re-driven fasteners alone may not pull the board flush to where it should be. The only way to fix the problem may be to remove all the boards and correct the framing.

Another way to make a better board attachment is to use adhesive as well as additional screws to hold the board to the framing. However, if the framing is badly warped and you succeed in firmly fixing the board, your wall or ceiling might be just as warped as the framing. It is better to fix the framing.

One other possibility is to laminate an entire new layer of gypsum drywall over the warped layer using adhesive or other material. Not only will this provide new drywall laminate, but it will also fill in any spaces caused by the first layer's warping.

Finally and most easily, if loose boards are caused by loose nails or screws, drive them in farther. Check this before finishing any wall or ceiling. Pushing with your hands against the board (even while you are spotting with mud) will indicate if any board is loose. If it is, stop and re-drive the fasteners. You can often drive them by simply using the butt end of your broad knife. Add other fasteners if necessary, then continue spotting and finishing. The time you take to interrupt your finishing and fix the board hanging problem will prevent you or anyone else from having to do the job over again.

As in all repairs, you want the fix to stay fixed. Do not settle for shortcut methods. If you have to rip off the boards and reset the framing, it is better to do it now than to have the problem reported later. If the general contractor or customer discovers the poor framing, it could cost your employer their business and could cost you your job. Fix these problems right from the start.

6.4.7 Patching Drywall

Drywall defects such as holes and dents require patching. For holes two inches or less in diameter, apply joint compound and reinforcing tape over the hole. An additional tape layer may also be needed. Once the bedding coat and tape have dried, apply a topping coat, feathering the edges. Apply a finish coat, if necessary.

Large holes require a different method of repair. One method involves using a piece of drywall. Using this method, you would square off and cut out the defective area. Bevel the edges of the squared opening so that the bevels face you. Measure and cut out a patch of new wallboard to fit this opening. Bevel the patch edges to mate with the opening's edges. Use joint compound to cement the patch in place, then tape and finish the edges as you would normal butt joints.

There are also commercially available patching systems that use fiberglass or aluminum mesh. These patches can generally be used to repair holes up to 4".

For holes 12" or larger, square off and cut out a whole wallboard section back to the framing members (*Figure 51*). Cut a fresh patch to fit, cement the patch in place, and use fasteners through the patch into the framing members. Only one drywall screw in each corner of the patch should be necessary. Tape and finish the patch edges like butt joints.

To repair a wallboard dent, first sand the dented section. This raises the nap, but it also permits the joint compound to grip the drywall face paper. Fill the dent with one or more layers of compound. Allow each layer to dry before lightly sanding and then applying the next layer. Finally, sand the filled dent smooth and level with the surrounding wallboard.

Another patching technique is the hot patch or blowout patch, as shown in *Figure 52*.

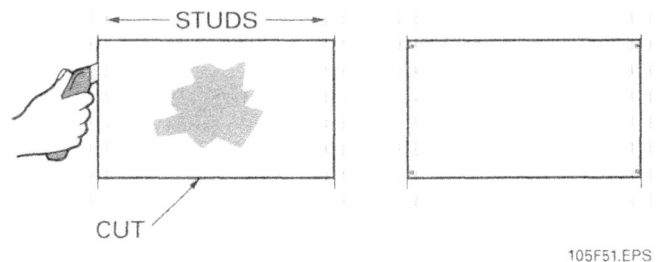

105F51.EPS

Figure 51   Patching a large hole.

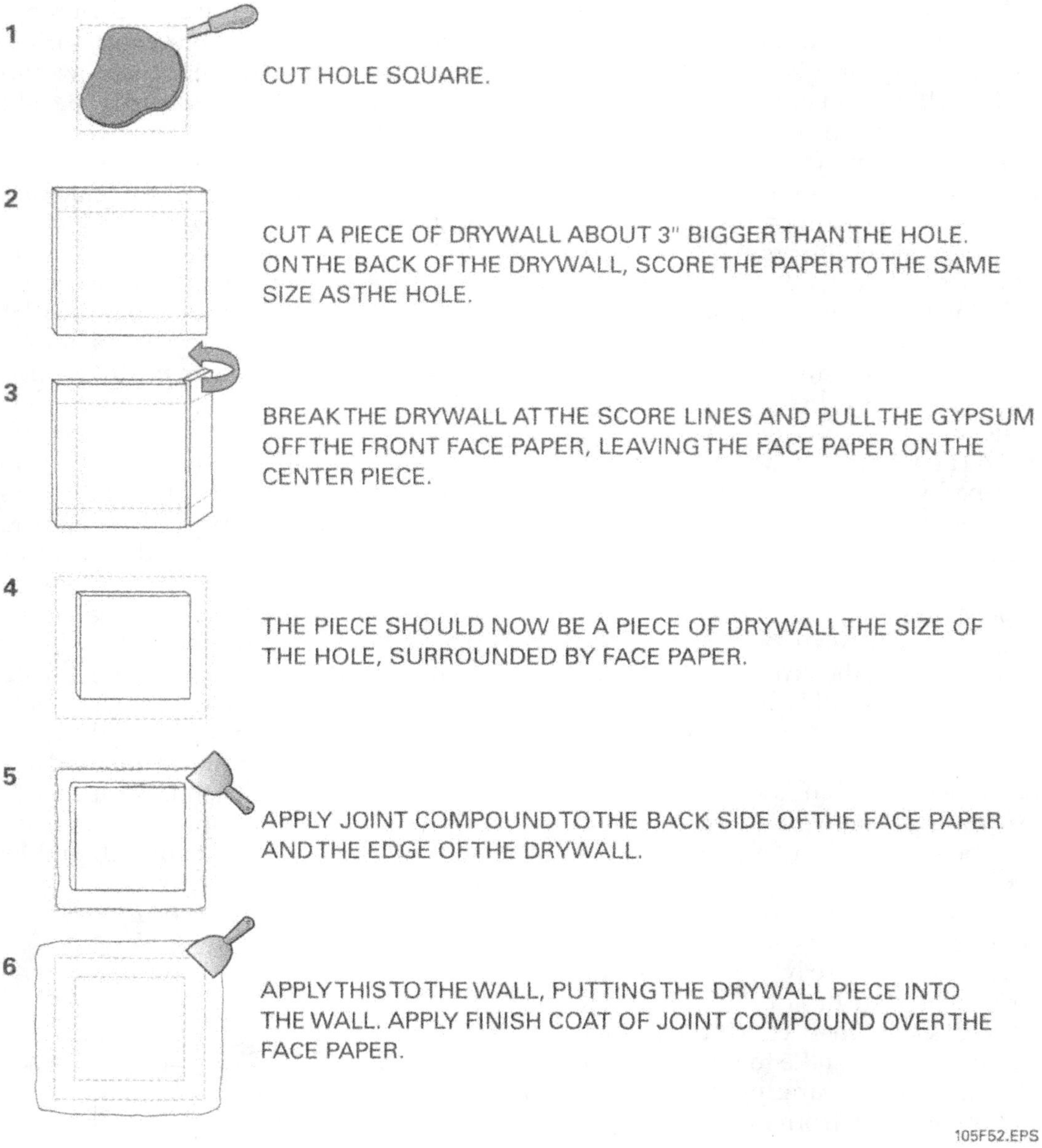

1

CUT HOLE SQUARE.

2

CUT A PIECE OF DRYWALL ABOUT 3" BIGGER THAN THE HOLE. ON THE BACK OF THE DRYWALL, SCORE THE PAPER TO THE SAME SIZE AS THE HOLE.

3

BREAK THE DRYWALL AT THE SCORE LINES AND PULL THE GYPSUM OFF THE FRONT FACE PAPER, LEAVING THE FACE PAPER ON THE CENTER PIECE.

4

THE PIECE SHOULD NOW BE A PIECE OF DRYWALL THE SIZE OF THE HOLE, SURROUNDED BY FACE PAPER.

5

APPLY JOINT COMPOUND TO THE BACK SIDE OF THE FACE PAPER AND THE EDGE OF THE DRYWALL.

6

APPLY THIS TO THE WALL, PUTTING THE DRYWALL PIECE INTO THE WALL. APPLY FINISH COAT OF JOINT COMPOUND OVER THE FACE PAPER.

105F52.EPS

Figure 52 ❖ Patching a hole in drywall using the hot patch (blowout) patching technique.

1. A drywall finishing job with tape embedded in joint compound, one separate coat of compound on joints and interior angles, and two separate coats of compound over fastener heads and accessories meets the _____ requirements.
   a. Level 1
   b. Level 2
   c. Level 3
   d. Level 4

2. A _____ is commonly used to cut drywall.
   a. hacksaw
   b. bandsaw
   c. circular saw
   d. utility knife

3. Which of these tools is used by striking it with a rubber mallet?
   a. Finishing trowel
   b. Corner clinching tool
   c. Pole sander
   d. Mud masher

4. The automatic tool that applies a smooth finish with feathered edges and a center crown to a taped seam is the _____.
   a. flat applicator
   b. banjo
   c. flat finisher
   d. nail spotter

5. Fiberglass mesh tape may be preferred over paper tape in _____ applications.
   a. high-moisture
   b. low-humidity
   c. high-temperature
   d. low-temperature

6. You would be likely to use metal edge tape in each of these locations *except* _____.
   a. outside angled corners
   b. a 90-degree inside corner
   c. the intersection of a ceiling and a radius wall
   d. an arch

7. Which of the following is an *incorrect* statement regarding topping compound?
   a. Topping compound is used for the second and third finishing coats.
   b. Topping compound is easier to sand because it dries softer than joint compound.
   c. Topping compound and joint compound are the same thing; you just add more water to joint compound to make topping compound.
   d. Topping compound shrinks less than joint compound.

8. When mixing dry-mix compounds, you are required to wear _____.
   a. gloves
   b. a respirator
   c. a hair net
   d. protective coveralls

9. Which type of texture material has good solution time, minimum-to-moderate fallout, and good bonding power?
   a. Premixed textures
   b. Powder joint compound textures
   c. Aggregated powder textures
   d. Unaggregated powder textures

10. Drywall fasteners should be installed so that _____.
    a. the head penetrates the paper
    b. there is a slight depression, or dimple, in the drywall
    c. the head protrudes ¼ of an inch from the drywall surface
    d. the head is exactly flush with the drywall surface

11. Which of these is the correct reason for applying multiple coats of compound?
    a. The compound shrinks as it dries, leaving depressions.
    b. A lot of the compound will flake and fall off as it dries.
    c. Extra buildup is needed to compensate for sanding.
    d. Walls look better when the seams are slightly higher than the wall surface.

12. If you find bubbles in tape joints after using the automatic taper, it probably means that _____.

    a. only one wheel was pressing on the drywall surface
    b. neither wheel was pressing on the drywall surface
    c. both wheels were pressing on the drywall surface
    d. the taper was upside down

13. Fasteners normally receive _____ coat(s) of topping.

    a. one
    b. two
    c. three
    d. four

14. When you are able to see a taped joint after the wall has been painted, it is known as _____.

    a. ridging
    b. photographing
    c. high joints
    d. discoloration

15. When mixing joint compound, you should use _____ water in order to avoid bonding problems.

    a. ice
    b. distilled
    c. cold, clean
    d. hot

Summary

Drywall finishing is the difference between craft and mechanics. Once it is finished and painted, the properly finished seam cannot be distinguished from the wall surface itself. Modern finishing equipment, if correctly used, can allow a crew to complete rooms much more quickly than a crew with hand equipment. However, you must use and maintain the equipment correctly in order to get good results. You also need to be able to finish by hand as well as with automatic tapers as there will be places where manual work is the only way to get good results.

The textured surface, more and more popular, eliminates finish sanding, but a poorly taped or fastened joint will still show through. Different kinds of corner beads and tape require different mud mixes and techniques. Drywall that is properly hung is much easier to finish well, preventing nail pops, alignment problems, and gaps. Remember, the walls are out in plain sight; any craftsman or customer will see and recognize shoddy work.

Notes

Robert T. Consroe

Vice President
C. J. Coakley Co., Inc.
Falls Church, Va.

The summer jobs Bob Consroe held while going to college led him to a highly successful career in the interior finish industry. His experience is a good example of the fact that you never know where life is going to lead you. His message is clear: find something you like and take advantage of any learning opportunities that come your way.

How did you choose a career in the construction field?
I interviewed with a number of firms in the spring of my final year in college. The only two offers that appealed to me were from a steel fabricator and an interior finish subcontractor. I had done some sheetrock and acoustical grid work during summer jobs and felt some familiarity with that work. I was attracted to the atmosphere at the office and loved the warehouse full of leftover material, scaffold parts and equipment.

What types of training have you had?
I spent six years pursuing a Bachelor of Science degree. When I graduated, I had taken math classes, physics classes, engineering classes, business class and an assortment of technology classes at two universities. The assortment of classes was a great start of a career. I loved going to school. Since then I have taken several classes on work related topics ranging from accounting to computers to safety. I still love learning.

What types of work have you done?
I spent my first few years as a junior estimator recording dimensions for a veteran quantity surveyor where I learned how to read blueprints. I then worked as an assistant to a project manager who priced drywall and acoustical work and assembled bids.

In addition to learning how to bid work, I learned how to set up jobs so that the carpenters could begin construction in the field. This involved making submittals, calculating stocking lists for materials, reviewing drawing revisions, attending meetings and communicating with the field crews working on the jobs. I have stayed in the interior finish business for my entire career, advancing to higher positions as I progressed and gained experience.

What do you like about your present job?
I currently oversee six project managers and two superintendents working on twelve field active jobs. There are another eight jobs being set up to start soon and four or five that are being closed down. I spend time helping the project managers with their problems and direction. I attend meetings at jobsites to plan and integrate our work with the other trades on the jobs and to visit with our employees doing the work. I occasionally have to deal with a problem or two that develop even though we try to plan against them. I also work with the local ABC office helping to manage the drywall apprentice program to train future workers for our industry.

What factors have contributed most to your success?
In my early years, I had a mentor who helped me learn about the drywall business, about business in general, about listening and working with people,

about being accountable. I also like to build things. I like working with my hands. I still get excited to go onto jobs and see the buildings being built. Every day the job changes as a result of the work of the builders at the site. Spending my career in the Washington D.C. area has afforded me an opportunity to work in many high profile DC landmarks.

What advice would you give to those who are new to the field?
It sounds trite but work hard every day and never stop learning. Listen to the old timers and learn. Ask questions. Take courses. Maintain relationships in industry groups. If you are proud of what you do and who you work with, going to work is easy.

All-purpose compound: Combines the features of taping and topping compounds. It does not bond as well as taping compound, but finishes better.

Bullnose: A metal corner bead with rounded edges.

Feathering: Tapering joint compound at the edges of a drywall joint to provide a uniform finish.

Joint compound: Patching compound used to finish drywall joints, conceal fasteners, and repair irregularities in the drywall. It dries hard and has a strong bond. Sometimes called mud or taping compound.

Lightweight compound: An all-purpose compound having less weight than standard compounds.

Mud: See *joint compound.*

Ridges: Slight protrusions in the center of a finished drywall joint that are usually caused by insufficient drying time. Also known as beads.

Skim coat: A thin coat of joint or topping compound that is applied over the entire drywall surface. Sometimes required under a high gloss finish.

Tape: A strong paper or fiberglass tape used to cover the joint between two sheets of drywall.

Tapered joint: A joint where tapered edges of drywall meet.

Taping compound: See *joint compound.*

Topping compound: A joint compound used for second and third coats. It dries soft and smooth and is easier to sand than taping compound.

Additional Resources and References

Additional Resources

This module is intended to be a thorough resource for task training. The following reference works are suggested for further study. These are optional materials for continued education rather than for task training.

Gypsum Construction Guide. Charlotte, NC: National Gypsum Company, 1994.

Gypsum Construction Handbook. Chicago, IL: United States Gypsum Company, 2000.

Painting and Decorating Craftsman's Manual and Textbook. Fairfax, VA: Painting and Decorating Contractors of America, 1995.

Figure Credits

The Stanley Works, 105F01, 105F03, 105F04 (bottom), 105F05 (top), 105F06, 105F09, 105F14, 105SA02, 105F43

Topaz Publications, Inc., 105F02, 105F04 (top), 105F05 (bottom), 105F07, 105F08, 105F10, 105F13, 105F16, 105F17, 105F23, 105F24, 105SA01, 105F25–105F28, 105F34–105F41, 105SA03, 105SA04, 105F44

Ames Taping Tool Systems, Inc., 105F15, 105F18–105F21

Porter-Cable, 105F22

USG Corporation, 105F29, 105F31, 105F50

Kraft Tool Company, 105F30

NCCER CURRICULA — USER UPDATE

NCCER makes every effort to keep its textbooks up-to-date and free of technical errors. We appreciate your help in this process. If you find an error, a typographical mistake, or an inaccuracy in NCCER's curricula, please fill out this form (or a photocopy), or complete the online form at **www.nccer.org/olf**. Be sure to include the exact module ID number, page number, a detailed description, and your recommended correction. Your input will be brought to the attention of the Authoring Team. Thank you for your assistance.

Instructors – If you have an idea for improving this textbook, or have found that additional materials were necessary to teach this module effectively, please let us know so that we may present your suggestions to the Authoring Team.

NCCER Product Development and Revision
13614 Progress Blvd., Alachua, FL 32615

Email: curriculum@nccer.org
Online: www.nccer.org/olf

❏ Trainee Guide   ❏ Lesson Plans   ❏ Exam   ❏ PowerPoints   Other _____

Craft / Level: _____   Copyright Date: _____

Module ID Number / Title: _____

Section Number(s): _____

Description: _____

_____

_____

_____

_____

Recommended Correction: _____

_____

_____

_____

_____

Your Name: _____

Address: _____

_____

Email: _____   Phone: _____

Admixture: Any material that is added to a concrete mixture to obtain additional properties.

All-purpose compound: Combines the features of taping and topping compounds. It does not bond as well as taping compound, but finishes better.

APA-rated: Building material that has been rated by the American Plywood Association for a specific use.

Blocking: A wood block used as a filler piece and support member between framing members.

Bridging: Wood or metal pieces placed diagonally between joists to provide added support.

Bullnose: A metal corner bead with rounded edges.

Cantilever: A beam, truss, or floor that extends beyond the last point of support.

Condensation: The process by which a vapor is converted to a liquid, such as the conversion of the moisture in air to water.

Convection: The movement of heat that either occurs naturally due to temperature differences or is forced by a fan or pump.

Corner bead: A metal or plastic angle used to protect outside corners where drywall panels meet.

Corrugated: Material formed with parallel ridges or grooves.

Cripple stud: In wall framing, a short framing stud that fills the space between the header and the top plate or between the sill and the soleplate.

Dew point: The temperature at which air becomes oversaturated with moisture and the moisture condenses.

Diffusion: The movement, often contrary to gravity, of molecules of gas in all directions, causing them to intermingle.

Dimension lumber: Any lumber within a range of 2" to 5" thick and up to 12" wide.

Dormer: A framed structure that projects out from a sloped roof.

Double top plate: A length of lumber laid horizontally over the top plate of a wall to add strength to the wall.

Exterior insulation finish system (EIFS): A protective and decorative coating applied directly to insulation board.

Feathering: Tapering joint compound at the edges of a drywall joint to provide a uniform finish.

Fire rating: A classification indicating in time (hours) the ability of a structure or component to withstand fire conditions.

Firestop: A piece of lumber or fire-resistant material installed in an opening to prevent the passage of fire.

Firestopping: A special putty, other material, or mechanical device used to block openings in fire-rated structures such as walls, ceilings, and floors to prevent the passage of fire and smoke.

Floating interior angle construction: A drywall installation technique in which no fasteners are used at the edge of the panel in order to allow for structural stresses.

Footing: The foundation for a column or the enlargement placed at the bottom of a foundation wall to distribute the weight of the structure.

Furring strips: Strips of wood or metal applied to a wall or other surface to make it level, form an air space, and/or provide a fastening surface for finish covering.

Gable: The triangular wall enclosed by the sloping ends of a ridged roof.

Girder: The main steel or wood supporting beam for a structure.

Green concrete: Concrete that has hardened, but has not yet gained its full structural strength.

Gypsum board: A generic term for paper-covered gypsum core panels; also know as gypsum drywall.

Gypsum wallboard: A generic term for gypsum core panels covered with paper on both sides. It is commonly used to finish walls.

Gypsum: A chalky type of rock that serves as the basic ingredient of plaster and gypsum wallboard.

Header: A horizontal member that supports the load over an opening such as a door or window. Also known as a lintel.

Glossary of Trade Terms

Joint: A place where two pieces of material meet. For example, the space between two drywall panels.

Joint compound: Patching compound used to finish drywall joints, conceal fasteners, and repair irregularities in the drywall. It dries hard and has a strong bond. Sometimes called mud or taping compound.

Joists: Equally-spaced framing members that support floors and ceilings.

Kerf: A groove or notch made by a saw.

Lath: Thin, narrow strips of wood used as a base for plaster.

Lightweight compound: An all-purpose compound having less weight than standard compounds.

Millwork: Various types of manufactured wood products such as doors, windows, and moldings.

Mud: See *joint compound.*

Nail pop: The protrusion of a nail above the wallboard surface that is usually caused by shrinkage of the framing or by incorrect installation. Also applies to screws.

Oriented strand board (OSB): Panels made from layers of wood strands bonded together.

Perm: The measure of water vapor permeability. It equals the number of grains squared of water vapor passing through a one square foot (sq ft) piece of material per hour, per inch of mercury difference in vapor pressure.

Permeability: The measure of a material's capacity to allow the passage of liquids or gases.

Permeable: Porous; having small openings that permit liquids or gases to seep through.

Permeance: The ratio of water vapor flow to the vapor pressure difference between two surfaces.

Plaster: A compound consisting of lime, sand, and water used to cover walls and ceilings.

Plastic concrete: Concrete in a liquid or semi-liquid workable state.

Plenum: A sealed chamber for moving air under slight pressure at the inlet or outlet of an air conditioning system. In some commercial buildings, the space above a suspended ceiling often acts as a return air plenum.

Post-tensioned concrete: Concrete placed around steel reinforcement such as rods or cables that are isolated from the concrete. After the concrete has cured, tension is applied to the rods or cables to provide greater structural strength.

Pre-stressed concrete: Concrete that is placed around pre-stressed reinforcing steel in a casting bed. This type of concrete cannot be cut without first consulting a structural engineer.

Rabbeted: A board or panel with a groove cut into one or more of its edges.

Rafter: A sloping structural member of a roof frame to which sheathing is attached.

Reinforced concrete: Concrete that has been placed around some type of reinforcing material, usually steel.

Ribband: A 1 × 4 nailed to ceiling joists to prevent twisting and bowing of the joists.

Ridges: Slight protrusions in the center of a finished drywall joint that are usually caused by insufficient drying time. Also known as beads.

Shakes: A handsplit wood shingle.

Sheathing: The sheet material or boards used to close in walls and roofs.

Shiplap: Lumber with edges that are shaped to overlap adjoining pieces.

Sill plate: A horizontal timber that supports the framework of a building. It forms the transition between the foundation and the frame.

Skim coat: A thin coat of joint or topping compound that is applied over the entire drywall surface. Sometimes required under a high gloss finish.

Slurry: A thin mixture of water or other liquid with any of several substances such as cement, plaster, or clay.

Soleplate: The lowest horizontal member of a wall or partition. It rests directly on the rough floor.

Striated: A surface design that has the appearance of fine parallel grooves.

Stringer: The support member at the sides of a staircase; also, a timber used to support formwork for a concrete floor.

Strongback: An L-shaped arrangement of lumber used to support ceiling joists and keep them in alignment. In concrete work, it represents the upright support for a form.

Stucco: A type of plaster used to coat exterior walls.

Studs: The vertical support members for walls.

Sub floor: Panels or boards fastened to the tops of floor joists.

Substrate: The underlying material to which a finish is applied.

Tape: A strong paper or fiberglass tape used to cover the joint between two sheets of drywall.

Tapered joint: A joint where tapered edges of drywall meet.

Taping compound: See *joint compound*.

Top plate: The upper horizontal member of a wall or partition frame.

Topping compound: A joint compound used for second and third coats. It dries soft and smooth and is easier to sand than taping compound.

Trimmer joist: A full-length horizontal member that forms the sides of a rough opening in a floor. It provides stiffening for the frame.

Trimmer stud: The vertical framing member that forms the sides of a rough opening for a door or window. It provides stiffening for the frame and supports the weight of the header.

Truss: An engineered assembly made of wood or metal that is used in place of individual structural members such as the joists and rafters used to support floors and roofs.

Underlayment: A material such as plywood or particleboard that is installed on top of a subfloor to provide a smooth surface for the finish flooring.

Vapor barrier: A material used to retard the flow of vapor and moisture into walls and prevent condensation within them. The vapor barrier must be located on the warm side of the wall.

Vaulted ceiling: A high, open ceiling that generally follows the roof pitch.

Veneer: The covering layer of material for a wall or the facing materials applied to a substrate.

Water stop: Thin sheets of rubber, plastic, or other material inserted in a construction joint to obstruct the seepage of water through the joint.

Water vapor: Water in a vapor (gas) form, especially when below the boiling point and diffused in the atmosphere.

Index

Index

V

Vacuums, 5.11–5.12
Valley (rafter), 2.30, 2.31, 2.34
Vapor barriers
  aluminum foil backing on drywall, 2.8, 2.9, 3.2
  for cold storage and low-temperature facilities, 3.25
  under concrete slab, 3.2, 3.18, 3.20, 3.23
  crawl space, 3.22–3.23
  installation, 3.13, 3.15, 3.16, 3.21–3.24
  overview, 3.2, 3.19, 3.33
  between sill and foundation, 2.15
  and subfloor, 3.17, 3.19
  in walls, 2.13, 3.2, 3.16, 3.23–3.24
Vapor diffusion retarders (VDR), 3.21–3.24
VDR. See Vapor diffusion retarders
Veneer (architectural), 2.10, 2.12, 2.13, 2.65
Veneer (plywood), 2.3
Ventilation
  attic and under the roof, 3.20
  basement, 3.19–3.20
  with cathedral ceiling, 3.14
  problems from overinsulation, 3.4
  and skylights, 3.14
  work area, 5.21, 5.35
Vents, 3.19, 3.20
Vermiculite, 3.6, 3.11, 5.20
Vibration, 5.40, 5.43

W

Waferboard, 2.32
Wallboard, gypsum, 2.64. See also Drywall
Wall coverings, 2.10, 5.2
Walls
  concrete, 2.11, 2.14, 2.15, 2.55
  curved, 2.10, 4.4
  exterior
    components, 3.4, 3.19, 3.21, 3.24
    curtain, 2.36–2.38, 2.39
    R-value, 3.4, 3.34
    shear, 2.45
  fire rating, 2.40, 2.55, 2.56, 4.23–4.24
  firestopping material in penetrations, 1.3, 2.25, 2.56, 4.25
  framing, 2.21–2.27, 2.42, 5.22
  heat loss or gain through, 3.8, 3.9
  interior
    in commercial construction, 2.39–2.45, 2.46
    load bearing, 2.19
    thickness, 2.40
  multi-dwelling (party), 2.35–2.36, 2.55, 4.23, 4.24
  residential construction, 2.21–2.28, 2.35–2.36
  R-value, 3.4, 3.5, 3.34
  stone, 2.13
  vapor barriers, 2.13, 3.2, 3.16, 3.23–3.24
Warehouses, 3.25, 5.2, 5.15
Warping, of drywall, 5.43
Waste disposal, pressure-treated lumber, 2.3
Water. See also Moisture; Moisture control
  for cleanup, 5.24
  in concrete preparation, 2.11
  damage to drywall, 5.43

  ground water, 3.24
  in joint compound preparation, 5.38
  vapor, 3.19, 3.33
Waterproofing, 3.24–3.25, 4.33
Water stop, 3.25, 3.33
Weather
  climate zones in the U.S., 3.5, 3.34
  cold. See Cold conditions
  hot conditions, 3.5, 3.9, 5.21
  humidity, 4.11, 4.32, 5.17, 5.21, 5.36, 5.39
  wind, 2.6, 2.10
Web, truss, 2.33
Windows
  building wrap around, 3.28
  control joints around, 4.31
  drywall around, 4.22
  flashing, 3.28, 3.29
  framing, 2.22, 2.24–2.25, 2.34
  heat loss or gain through, 3.8, 3.9
  largest tilt-up panel, 2.37
  LSL, 2.6
  skylights, 2.32, 3.14
  sound isolation around, 4.28
Wiring and cabling, 2.18, 2.19, 2.38–2.39, 2.53, 4.26
Wood
  bark, sawdust, or shavings as insulation, 3.6
  effects of using green, 5.40
  major types in lumber, 2.2, 2.3
  R-value, 3.3
Work area, 4.10–4.11, 5.21–5.22, 5.24, 5.35
Work ethic, 1.10–1.11, 1.12, 1.19
Wrap, building, 3.25–3.28

Y

Youth Apprenticeship Program, 1.9

www.ingramcontent.com/pod-product-compliance
Lightning Source LLC
Chambersburg PA
CBHW061400210326
41598CB00035B/6046